LAND USE AND ENVIRONMENT IN INDONESIA

WOLF DONNER

Land Use and Environment in Indonesia

PHOTOGRAPHY BY ERIKA DONNER

Published in association with the
Institute of Asian Affairs, Hamburg, by

C. HURST & COMPANY, LONDON

First published in the United Kingdom by
C. Hurst & Co. (Publishers) Ltd.,
38 King Street, London WC2E 8JT
© Wolf Donner, 1987
ISBN 1-85065-011-X
Printed in England on long-life paper

CONTENTS

FIGURES

TABLES

ABBREVIATIONS

ADB	Asian Development Bank
ASEAN	Association of South East Asian Nations
BAPPENDA	*Badan Perancangan Pembanguan Daerah* (Regional Development Planning)
BAPPENAS	*Badan Perencanaan Pembangunan* Nasional (Authority for National Development Planning)
BIES	Bulletin of Indonesian Economic Studies
BIOTROP	University of Bogor Programme for the Study of Biological Subjects in the Tropics
BKKBN	*Badan Koordinasi Keluarga Berancana Nasional* (National Family Planning Coordination Board)
BPS	*Biro Pusat Statistik* (Central Office of Statistics)
CIDA	Canadian International Development Agency
CITES	Convention on International Trade in Endangered Species of Wild Flora and Fauna
DI	*Daerah Istemewa* (Special Territory — Yogjakarta)
DKI	*Daerah Kusus Ibukota* (Capital Special District — Jakarta)
EEZ	Exclusive Economic Zone (maritime)
EPTA	Expanded Programme for Technical Assistance (United Nations)
ESCAP	Economic and Social Commission for Asia and the Pacific (United Nations)
FAO	Food and Agriculture Organisation of the United Nations
HIPPA	Water Users' Organisation
HPH	Forest Exploitation Licences
HYV	High-Yielding Varieties (plant)
IBRD	International Bank for Reconstruction and Development (World Bank)
ICID	International Commission on Irrigation and Drainage
ILO	International Labour Organisation (United Nations)
INMAS	*Intensifikasi Masal* (Intensification of Production Programme — without a credit component)
ITB	*Institut Teknologi Bandung* (Institute of Technology, Bandung)

ITC	Intertropical Convergence Zone
IUD	Intra-Uterine Device
JABOTABEK	Jakarta-Bogor-Tangerang-Bekasi Town Planning Project
KIK	*Kredit Invesasi Kecil* (Small-Scale Investment Programme)
LBN	*Lembaga Biologi Nasional* (National Biological Institute)
LIPI	*Lembaga Ilmu Pengetahuan Indonesia* (Indonesian Institute of Science)
LKBN	*Lembaga Keluarga Berencana Nasional* (National Family Planning Office)
LNG	Liquified Natural Gas
NEDECO	Netherlands Engineering Consultants
NUFFIC	Netherlands Universities Foundation for International Cooperation
OSR	Organisation for Scientific Research in Indonesia
PKBI	*Perkumpulan Keluarga Berencana Indonesia* (Indonesian Planned Parenthood Association)
PPA	National Environment Protection and Conservation Service
PUSKEMA	*Pusat Kesehatan Masyarakat* (Public Health Centres)
REPELITA	*Rencana Pembangunan Lima Tahun* (Five Year Development Plan)
SMEC	Snowy Mountains Engineering Company
TAD	Transmigration Area Development, Kalimantan
UNDP	United Nations Development Programme
UNEP	United Nations Environment Programme
WTO	World Tourism Organisation
VOC	*Vereenigde Oost-Indische Compagnie* (Dutch East-India Company)

GLOSSARY OF INDONESIAN, JAVANESE
AND DUTCH TERMS

adat	traditional prescriptive law in Indonesia
afspoelen	being washed away
alang-alang	cover of *Imperata* or a similar grass
ani-ani	rice-cutting knife
bahasa Indonesia	official Indonesian language
banjar	village assembly (on Bali)
banjir	flood
belukar	secondary forest, often degraded bush
boerenbossen	peasants' forest
bupati	local representative of the landlord
cultuurstelsel	forced cultivation system under Dutch colonial administration
daerah istimewa	special territory (of Yogjakarta)
daerah kusus ibu kota	special capital district (of Jakarta)
desa	village, local community
embung	artificial 'hill lake'
erfpacht	long-term lease of land or hereditary tenure during Dutch colonial period
gelandangen	loafers/idler
gogo	dry field for subsistence
gotong royong	mutual aid and cooperation
herbebossing	reforestation
herbeplanting	regreening
huma	dry field for subsistence
jati	teak tree
kabisu	clan, kinship group
kabupaten	district
kampung, kampong	living quarter, settlement (Jawa)
kaya putih	oil of cajaput tree (*Melaleuca leucodendra*)
kebun, kebunan	fruit gardens
kebun campuran	mixed garden
kecamantan	sub-district
koffiecultuur op hoog gezag	Dutch government coffee cultivations
kolonisatie	term for early Dutch *transmigrasi* efforts
ladang	shifting cultivation, dry-field cultivation

lahar	mudslide
lamtoro	lead tree (*Leucanea leucocephala*)
lamtoronisasi	the planting of *lamtoro* hedges along contour lines
niet-blijvers	temporary residents
padas	hard-pan
pasang surut	tidal rice cultivation
pekarangan	mixed garden (Jawa), with multistoreyed cropping
penghijuan	regreening
polowijo system	farming on dry *sawah* fields
propinsi	province
raja	local prince, king
reboisasi	reforestation
sawah	irrigated rice field
stervend land	waste or 'dying' land
surjan	technique of cultivating swampy land using the camberbed system
swankarya	the village in transition
swidden	cleared plots of land during shifting cultivation
tahu, tempé	soya bean cake
talun	perennial crops grown around the village
taungya	agro-forestry (Burmese)
tebasan system	cultivation or field work with hired labour
tegal, tegalan	non-irrigated field (Javanese)
telaga	small natural lake often enlarged by man-made earth wall or dam
tipar	semi-permanent cultivation of dry rice with dibble
transmigrasi	resettlement of Javanese on the Outer Islands
tuan taneh	head of kinship group, big landlord
tumpangsari	agro-forestry
uitboeren	depriving the peasants of land
wonosobo	an indigenous form of crop rotation

PREFACE AND ACKNOWLEDGMENTS

Second only to the danger of a nuclear holocaust, it is the deterioration of our natural environment that threatens the future of mankind. But whereas a global nuclear catastrophe is not very likely to happen — except unintentionally — critical observers get new evidence every day that the human race is rapidly losing the basis of its physical existence, the very ground on which it stands and which provides the basis of life. Optimists, on the other hand, maintain that until now human intelligence has always found a way out of danger and even catastrophe and we should be able to identify the problems ahead of us, to analyse their origin and to remedy their consequences. In other words, if man can destroy his environment, he can also save it; if technology can destroy nature, it can also reconstruct it.

However, destruction of the environment is an uncontrolled process which no-one actively intends but which occurs as an indirect result or side-effect of actions aimed at quite different goals. We do not produce carbon dioxide because we want to destroy our atmosphere, or sulphur dioxide in order to cause acid rain: they are both by-products occurring when fossil fuels are converted into energy — energy needed to run our machines, light and warm our homes, and move our cars. We do not cut down trees in order to cause soil erosion, a sinking ground-water table or desertification; we do it because we need arable land to grow food or because we need wood to build a shelter, make furniture or cook meals for the family. But in many cases such goals could be achieved without causing so much damage. Of course, measures to avoid or reduce environmental destruction require firm intentions and specific action — action for example, to avoid or reduce air pollution, or halt soil erosion or desertification. Such action is generally not regarded as economic or profitable, whereas the production of food, energy, consumer goods and the like not only meets an immediate demand but also brings income to the producer. Measures to *prevent* air and water pollution, noise pollution and the destruction of soil and vegetation add to the costs of production and reduce profit margins; or they increase the price of the final product and thus weaken the position of the producer in the market. Action to *remedy* actual damage usually requires more expenditure, drawn mainly from public funds and thus burdening the taxpayer.

Throughout a long period of economic history, man regarded many resources as a free gift of nature, particularly air and water, and although he often had to pay for others like land and minerals, he used them as though they would always be available and could never be exhausted. Nor did waste disposal present problems; dumps were filled with solid waste and liquids simply discharged into rivers, lakes or the sea. This consumption of the environment did not cause much harm as long as there were comparatively few people living in it. A hundred years ago no more than a thousand million people polluted our planet and relatively few of these used more of the environment than was needed for simple survival. When industrialisation spread and became the dominant method of production, a rather small part of the world's population started to consume more and more of the environment. It not only used what was available locally but conquered other parts of the world in search of the raw materials and energy-bearing fuels needed to satisfy their rapidly growing demand for consumer goods.

In the period that followed, the northern hemisphere divided into what we now call the 'First' and the 'Second' World: intensely industrialised countries, consumers of raw materials and exporters of industrial goods. Ports full of vessels disgorging their steam into the air; industrial zones shrouded in clouds of smoke and dust, pumping waste into rivers and interconnected by railways and roads which polluted the environment with exhaust fumes and noise — such was for a long time the proud image of enterprising man. He presumed to follow the Biblical command 'Be fruitful and multiply, and replenish the earth, and subdue it; and have domination over the fish of the sea, and over the fowls of the air, and over every living thing that moveth upon the earth.'[1]

However when man became fruitful and multiplied and replenished the earth at an ever-increasing rate, the relationship between him and his environment began to deteriorate. The urge to subdue the earth has led to a situation where the demand for air, water, soil and all sorts of raw materials and fuels has increased to such an extent that they have become more and more difficult to obtain. And yet the bulk of the earth's population gains very little from this race for more raw materials, but merely acts as supplier to the industrialised North. Suppose the countries that we now designate as the 'Third World' were to begin consuming the same amount of raw materials and energy as the First and Second World — those resources would be exhausted in no time.

The process of loading the environment with pollutants has been accepted as a danger only during the twentieth century, but the destruction of the earth's soil has been going on for several millennia. Archaeology has discovered that in many regions where deserts or semi-deserts prevail today, millions of people once lived and fed on a soil that is now saline, waterlogged or completely dried out. The whole Mediterranean basin is an example of how man and his animals destroyed a stable eco- logical equilibrium through deforestation and overgrazing, and regions like Mesopotamia and the Indus valley show the results of inappropriate irrigation. Today the process of deforestation and soil erosion continues on a global scale but it is already centuries old, and is partly the work of ancient civilisations such as those of China and the Andes.[2] More recent events on the North American continent helped to bring about the new awareness of soil degradation, as when, in May 1934, large dust clouds swept out of the mid-western plains eastward to the Atlantic seaboard.[3]

Soil degradation can now be seen in varying degrees around the globe. Usually it is a consequence of population pressure on the last land reserves, together with the destruction of cultivated soils, but another factor is the avarice of the First and Second World's industry in collusion with the ruling class of the Third World. It can be assumed that for a long time the majority of people in the Third World lived a hard but decent life in harmony with their natural environment. This is no longer so. The question of whether colonialism should be blamed for what has happened in these countries does not concern us here. The question for us is rather: what were the effects of the North's aid in environmental terms after these countries became politically independent? The fact is that the Third World no longer lives in harmony with nature, that the population is growing rapidly despite hunger and misery, and that the demand for food, soil, wood etc. grows proportionately. In addition, land and water are required for the production of export goods to pay for necessary imports. Thus a process can be observed in most tropical and subtropical countries which starts with the cutting back of forest cover to gain more cultivable soil. This soil is thus prone to erosion, salinity or waterlogging, and eventually its fertility declines to the extent that cultivators abandon it and cut new forest, leaving behind an exhausted piece of land that may never recover.

Under temperate climatic conditions forests recover comparatively quickly after being cleared, but tropical forests, once the area cleared and cultivated exceeds a certain size, are in danger of extinction. The biological reasons for this are only partly known, but the results are visible everywhere.[4] At present the tropical forests of the world are being destroyed at an unprecedented rate. Although hard data are lacking, the most conservative estimates put the destruction at 11 ha. per minute, and if it continues at this pace, almost all undisturbed tropical forest will

be eliminated early in the next century.[5] If the present trend continues
— and there is no basis for hoping otherwise — the forests in the
temperate zone will shrink by 5% by the year 2000 and in the tropical
zone by 40%. In 2020 the total forest area of our globe will stabilise at
1,800 million ha. with 80% in the temperate and a mere 20% in the
tropical zone. And since these 370 million ha. of tropical forests are
hardly accessible or of no economic value it can be predicted that the
entire valuable tropical forest will be extinct by 2020.[6] Unfortunately,
although the loss of the forest is a great evil in itself, it is only the
beginning, because deforestation initiates a whole chain of environmental
disasters such as soil erosion, lowering of the water-table, floods and
landslides, sedimentation, deterioration of the micro-climate — in
brief, the deterioration of people's living conditions.

In former times when men were few and land was plentiful, the
deterioration of soils caused little concern: migration into new lands,
actually or allegedly underpopulated, was the way out of misery, but
this has changed radically: 'The lands of the earth are occupied; frontiers
to new lands have disappeared. The only new frontier that appears is
underfoot, in the maintenance of productivity of lands now occupied.'[7]
Unfortunately, no one who actually has the power to influence this
course of events seems to care very much. Scientists speak, write and
warn but their influence on policy-makers is marginal.

A visitor to tropical countries today cannot fail to observe the loss of
topsoil through erosion, the loss of organic matter, the porous structure
of the soil and the accumulation of toxic salts and chemicals. Remote
sensing is now an excellent means of discovering and measuring changes
on the earth's surface, and no one can ignore any longer the expansion of
deserts and semi-desert areas which are growing by approximately 6
million ha. every year.[8] Estimates of the loss of irrigated land by waterlogging, salinity and alkalisation range between 125,000 ha. and
200,000–300,000 ha. annually. This may seem relatively little, but the
figures refer to the most fertile soils, and this annual loss means that food
for some 15 million people can no longer be produced.[9]

The fate of the earth's soils depends entirely on *us*, and the question is
whether our society is ready to bear the costs necessary to conserve the
land which is the basis of our life and of those who come after us.
Whether or not these soils deteriorate further or are saved depends
largely on whether or not governments are ready to act. But this often
requires unpopular political decisions, so that only a stable society with
sufficiently developed institutions can be expected to take the necessary

steps. A society wracked by war, hunger, internal unrest and corruption, or one infatuated by 'Modernisation' and no longer interested in rural development, will quickly face inextricable difficulties.

Anyone who works in agricultural development in the Third World will be confronted with deforestation and soil erosion and all their consequences. As an international civil servant I had gained some experience of these problems in many developing countries, particularly in Nepal,[10] Haiti[11] and to some extent Thailand,[12] and I felt that for several reasons Indonesia would be an ideal country to demonstrate the problem of soils in the tropics. First, the population density on the island of Jawa is one of the highest in the world and is still growing, thus placing increasing pressure on the cultivable soils. Secondly, although vast parts of Indonesia, the so-called 'Outer Islands', are sparsely populated, they are now the target of substantial immigration followed by deforestation and the cultivation of difficult soils. Thirdly, a tremendous amount of research has been done over the last century on the problems of forest and soil in Indonesia and the population's impact on them, and fourthly the Indonesian Government is extremely conscious of environmental problems and has a minister of state specially responsible for them. Dr Emil Salim, the minister in charge, is keen to do everything possible. In addition, the Indonesian press devotes a comparatively large amount of space to environmental problems and projects. After travelling in Indonesia and studying the press for some years, I must say that I have never visited a country where so much attention is given to environmental problems either by official bodies or by educators. Writing, painting and photographic contests have been held and prizes awarded to people who have been involved in environmental protection. Similarly, on visiting some of the provincial universities I was surprised by the eagerness with which environmental problems were dealt with, particularly in the faculties of geography and agriculture. Students spend part of their time in villages studying the soil question and, as far as is possible, teaching farmers protective techniques. Certainly compared with advanced Western standards many practices may appear superficial or inadequate, but compared to other developing countries Indonesia does much to spread environmental consciousness among politicians, scientists and — last but not least — ordinary people.

This study is meant to contribute to these efforts — not so much by offering new scientific findings, but by presenting a survey of research done so far, of experience gained through fieldwork, and of certain political strategies all dealing with the natural environment in connection with land use. It is intended to help those, Indonesians as well as foreigners, who work in the field of environmental protection and acquaint them with work already done.

I would like to thank various institutions and individuals who have supported this study: the Institute of Asian Affairs, Hamburg, for their institutional backing; the Stiftung Volkswagenwerk, Hanover, for financially supporting my study tour to Indonesia in 1981; and the Indonesian Institute of Sciences (LIPI), Jakarta, for unbureaucratically helping me to collect the maximum information. Furthermore I am indebted to many Indonesian, Dutch and German specialists who helped me in the preparation and execution of the tour and study, and especially to those attached to the Agricultural University, Soil Research Institute, Bogor; Gadjah Mada University, Yogyakarta; Mulawarman University, Samarinda (East Kalimantan); East Kalimantan Transmigration Area Development Project (TAD), Samarinda; Area Development Project West Pasaman, Padang (West Sumatera); the Food and Agriculture Organisation of the United Nations, Jakarta; Vrije Universiteit, Instituut voor Aardwetenschappen, Amsterdam, and Landbouwhogeschool, Vakgroep Bosteelt, Wageningen. I am particularly indebted to my wife Erika who supported my work in Indonesia, contributed the pictures and patiently did the first editing of the manuscript, and finally to the British publishing house Hurst & Co. which did the later editorial work.

As usual when writing about developing countries, one has to stress that the available statistics are not always reliable; this is especially understandable in a country as large as Indonesia. The few data included in this study should therefore be regarded as indicators of magnitudes and trends rather than the result of exact measurements. As for the spelling of proper names, I have tried to follow the official Indonesian version as far as possible, úsing Sandy's *Atlas Indonesia. Buku Pertama Umum*[13] as a guide for geographical names. Otherwise the English spelling is used. This explains why the names *Jawa* and *Sumatera* are used for those islands and *Javanese* and *Sumatrans* for their inhabitants.

From my own experience, scientific studies and project working papers (generally crammed with the most valuable information), more often than not disappear into the desks of bureaucrats and are soon regarded as out of date, never getting a chance to influence political or technical decisions. Thus, specialists who arrive later often have no idea what work has already been done, and start from scratch. This study, with its extensive but by no means exhaustive bibliography, seeks to rescue from oblivion some such studies, to the advantage of all who take on these tasks today. At the same time it is my hope that the study may encourage the Indonesian authorities to continue expanding their work of protecting the environment, and that it will support the individuals and strengthen the institutions involved in this vital effort.

Köln-Porz (Fed. Rep. of Germany),   WOLF DONNER
January 1987

NOTES

1. Genesis i.28.
2. Lowdermilk, 1935, pp. 409–19; Stoddart in Chorley, 1969, pp. 43–65; Carter and Dale, 1976; Kinzelbach, 1981, pp. 754–8.
3. Lowdermilk, 1935, pp. 417–18.
4. Gómez-Pompa *et al.*, 1972, pp. 762–5, and readers' mail, loc. cit. 1973, pp. 893–5.
5. Rubinoff, 1982, p. 253.
6. *Global 2000*, German version, 1981, pp. 674–700.
7. Lowdermilk, 1935, p. 419.
8. Biswas, 1978.
9. *See Global 2000*, loc. cit., fn. p. 31; Kovda, 1972; Kovda in Polunin, 1972.
10. Donner, 1972.
11. Donner, 1970.
12. Donner, 1978.
13. Sandy, 1977. In addition, map sheets issued by P.T. Pembina were used. Thus minor differences in the Indonesian spelling may occur.

Part I. INTRODUCTION

1

THE GEOGRAPHICAL SETTING

Dimensions and Structure

The territory of the Republic of Indonesia and its geographical structure is in many ways unique. To begin with, the distances are vast. The westernmost (95°E) and easternmost points (141°E) are 5,200 km. apart. From north (6°N) to south (11°S) the distance is 1,900 km.[1] However the land area consists of no less than 13,677 islands ranging from parts of the world's largest islands such as Borneo (Kalimantan) or New Guinea (Irian) to mere coral atolls. 6,044 of them are believed to be inhabited, but only about 3,000 are substantially settled. According to the Topographical Agency of the Indonesian Army the land area of the country amounts to 1,919,443 sq. km., approximately equal to that of Mexico or eight times that of the United Kingdom or West Germany. The sea area claimed by Indonesia, including the 12-mile-zone, amounts to 3,272,160 sq. km. Thus land forms only 37% of its economic territory.[2]

The present shape of Indonesia is the result of plate tectonic movements, the submergence of very old land masses, ancient and recent volcanic activity and relatively recent erosion. Even today the area is still geologically mobile. According to the displacement theory, it is not the actual continents but six extended plates and a few smaller ones which are drifting away from the mid-oceanic ridges. Three of them, namely the Eurasian plate, the Indo-Australian plate and the Pacific plate meet in the Indonesian area. Since it is known that most geological activity occurs along the boundaries of the plates, Indonesia can be regarded as a rather disturbed landscape.[3]

Two major continental shelves meet in the vicinity of Indonesia. As an extension of the Eurasian continent the Sunda Shelf (some 1.8 million sq. km.) stretches towards the south-east, comprising the southern part of the South China Sea and the rather flat seas around the Malacca peninsula and between the islands of Sumatera, Jawa and Kalimantan. From Australia the Sahul Shelf projects into the Indonesian area encompassing New Guinea and the Aru islands and covers a mere 1.3 million sq. km. Both the shelves have average depths of no more than 40–80 m.,

1

but they break abruptly into deep seas, troughs and trenches. The southern and south-western coasts of Sumatera, Jawa and the Lesser Sunda islands situated on the margin of the Sunda Shelf fall, after a short distance, into the Java Trench, with depths down to 7,054 m.; east of Kalimantan runs the Makassar Strait where the shelf plunges to 2,000 and 2,500 m, and the Celebes Sea with a maximum depth of 6,220 m. The Sahul Shelf slopes down into the Timur and Aru trough (3,300 m.) and finally into the Banda Sea with a maximum depth of 7,440 m., the so-called Weber Deep.

As mentioned earlier, the edges of the plates and shelves are particularly prone to tectonic activity. During the Tertiary era, when the Alps and the Himalayas were being formed, a range of folded mountains were lifted from deep troughs and became part of the Sunda mass. The Sunda mountain range is one of the greatest coherent mountain belts in the world connecting the old land masses of the northern and southern hemispheres. Two main mountain arcs run parallel to each other of which the inner one is volcanic and the outer one sedimentary. The main chain or inner arc starts in the knot of mountains in the Himalayan and Tibetan region and stretches south and then east over no less than 7,000 km., passing through the Patkai and Arakan mountains in northern and western Burma, over the Andaman and Nicobar islands, through Sumatera, Jawa and the Lesser Sunda islands to the Southern Moluccas and, turning north-west, ending with the Banda islands south of Seram. The outer arc, separated from the inner by a geosyncline, consists of two separate parts. West of Sumatera there is a chain of islands composed of the Nias and the Mentawai islands. Beyond the Enggano islands the arc is interrupted but reappears south of the Lesser Sunda islands in a second chain comprising Sumba, Sawu, Timor and the Tanimbar islands. From here this arc also turns north and west through the Kai islands to Seram and Buru, where it ends. Two important branches have yet to be mentioned. There is a mountain arc which passes through the highest peaks of northern Kalimantan and of Sulawesi and, striking north, is at its most prominent on the Philippine islands before joining the East Asiatic arc system. And finally the Melanesian arc system coming from New Zealand enters the Indonesian area via the main chain of New Guinea and, passing through the North-Moluccan island of Halmahera, turns north to the Philippines.

On the outer edge of the archipelago, folded and partly volcanic mountain chains alternate with geosynclines. The first of these runs between Sumatera and the islands west of it and between the northern

and southern chains of the Lesser Sunda islands where it is called the
Suwa Sea (3,470 m. deep) and the Wetar Strait (3,430 m.). The southern
coast of Jawa falls immediately into the deep sea. The second geosyncline
lies mainly under the ocean and reaches depths of 7,450 m. in the Java
Trench and 3,300 m. in the Timor Trough and the Aru Trough north of
the Sahul Shelf.

In the interior of the archipelago the Java Sea, with its flat bed aver-
aging 60 m. deep, still belongs to the Sunda Shelf. Its 200 m. isobath runs
from the north-eastern edge of Jawa along the east coast of Kalimantan
and east of this are various deep troughs: the Bali Sea (1,500 m.), the
Flores Sea (5,150 m.) and the Banda Sea (5,400 m.) lying north of the
Lesser Sunda islands; and the Makassar Strait (2,500 m.) and the Celebes
Sea (5,500 m.) between Kalimantan and Sulawesi. The Moluccan Sea
and the Ceram Sea are shallower. Thus between the Sunda and Sahul
Shelves the eastern archipelago forms a welter of contorted mountain
ridges together with deep troughs and basins.[4]

Since the mountains were folded in the Tertiary period, there have
been other decisive changes. Following a general glacial retreat, the
ocean level rose to such an extent that the land bridge between Asia and
Australia was partly submerged, becoming an island arc while the whole
area which is now Indonesia became an archipelago. This cataclysm put a
stop to the easy exchange of animal and plant life between the two
ancient continents with consequences we can still see today. Furthermore a gigantic process of erosion transported enormous masses of sediment from the mountains on the neighbouring shelves into the trenches
and troughs, filling up depths of more than 1,000 m. At the same time
extended alluvial lowlands were formed on Sumatera, Kalimantan and,
to a lesser extent, Jawa. This process is still going on, with alluvial
islands off the east coast of Sumatera gradually becoming part of the main
island.

Thus a long and continuing geological process is forming the land
from which the Indonesians have to get their living. The country is a
good example of how land formation affects the potential as well as the
problems of an area, and the fact that 63% of Indonesia's territory is sea
gives the country a very peculiar structure. Furthermore the relief of
Indonesia is very pronounced. The highest point is the Puncak Jaya in
New Guinea (5,030 m. high) and the deepest known points are in the
Sunda or Java Trench (7,455 m.) and in the Weber Deep in the Banda
Sea (7,440 m.); a difference of roughly 12,500 m. On many islands there
is a difference of several thousand metres between the highest summit

and the adjacent sea-bed.

Since the collapse of the continental bridge between Eurasia and Australia, the exchange of animals and plants has been so much reduced that it is possible to distinguish two biological areas in Indonesia separated by a boundary now known as Wallace's Line. From Eurasia the elephant came to Sumatera, the tiger to Sumatera and Jawa, the tapir and orang-utan to Sumatera and Kalimantan and the rhinoceros to Sumatera, Jawa and Kalimantan, but none of the big mammals reached Sulawesi or the Lesser Sunda islands. In contrast five marsupial species came from the Australian continent to the Kai Islands and two to Sulawesi. Of course, certain animal or plant individuals succeeded in overcoming Wallace's Line by passive or active travelling. This happened for instance when driftwood was used as a raft or when seeds were transported by water, wind, or birds and nowadays the division is no longer seen as a line running along the Makassar Strait, but rather as a zone called the Wallacea, enclosing Sulawesi, the Moluccas and the Eastern Sunda islands where species from both the continents co-exist, except of course for the larger mammals.

This separation, the result of the submersion of the archipelago and the existence of the deep ocean trenches, had an important influence on later economic development in relation to the natural environment. For instance, after the people had domesticated the wild ox or *banteng* that came as far as Jawa and Kalimantan from the Eurasian continent, they were able to take up wet-rice cultivation on a large scale and this became typical of the islands on the Sunda Shelf. Not till much later did the technique cross Wallace's Line and spread east to Timor. On the other hand root crops such as taro and yam and the sago palm that were typical of the Melanesian culture approached Wallace's Line from the east and eventually passed over it.

The extent to which plants, animals and people established themselves on the different islands depended very much on the prevalent environmental conditions, particularly the terrain, climate, water and soil conditions. Their present distribution is partly the result of political decisions and historical events, but till recently the environmental factor played the most important part.

Geographers usually subdivide the Indonesian archipelago into the Greater Sunda islands, comprising Sumatera, Jawa, Kalimantan and Sulawesi (though the latter lies off the Sunda Shelf); the Lesser Sunda islands (Nusa Tenggara) comprising a chain of smaller islands between Bali and Timor; the Moluccas (Maluku), including the tiny islands south of the Banda Sea — although they are actually a continuation of the Lesser Sunda islands; and finally Irian Jaya, the western part of New Guinea. The Moluccas and Irian Jaya are sometimes referred to as the 'Great East'.

Table 1. LAND AREA AND POPULATION OF THE INDONESIAN ISLANDS, GROUPS OF ISLANDS AND PROVINCES, 1980

	Surface (sq. km.)[a]	Population (1980)	Density (inhab./sq. km.)
Sumatera	473,606	28,016,160	59
— Aceh	55,392	2,611,271	47
— North Sumatera	70,787	8,360,894	118
— West Sumatera	49,778	3,406,816	68
— Riau	94,562	2,168,535	23
— Jambi	44,924	1,445,994	32
— South Sumatera	103,688	4,629,801	45
— Bengkulu	21,168	768,064	36
— Lampung	33,307	4,624,785	139
Jawa	132,187	91,269,528	690
— Jakarta	590	6,503,449	11,023
— West Jawa	46,300	27,453,525	593
— Central Jawa	34,206	25,372,889	742
— Yogyakarta	3,169	2,750,813	868
— East Jawa and Madura	47,922	29,188,852	609
Nusa Tenggara	88,488	8,487,110	96
— Bali	5,561	2,469,930	444
— West Nusa Tenggara	20,177	2,724,664	135
— East Nusa Tenggara	47,876	2,737,166	57
— East Timor	14,874	555,350	37
Kalimantan	539,460	6,723,086	12
— West Kalimantan	146,760	2,486,068	17
— Central Kalimantan	152,600	954,353	6
— South Kalimantan	37,660	2,064,649	55
— East Kalimantan	202,440	1,218,016	6
Sulawesi	189,216	10,409,533	55
— North Sulawesi	19,023	2,115,384	111
— Central Sulawesi	69,726	1,289,635	18
— South Sulawesi	72,781	6,062,212	83
— South-east Sulawesi	27,686	942,302	34
Maluku	74,505	1,411,006	19
Irian Jaya	421,981	1,173,875	3
Indonesia, total	1,919,443	147,490,298	77

[a] Surface area according to Topographical Agency of the Indonesia Army.

Source: Business News (Indonesia), 8 July, 1981.

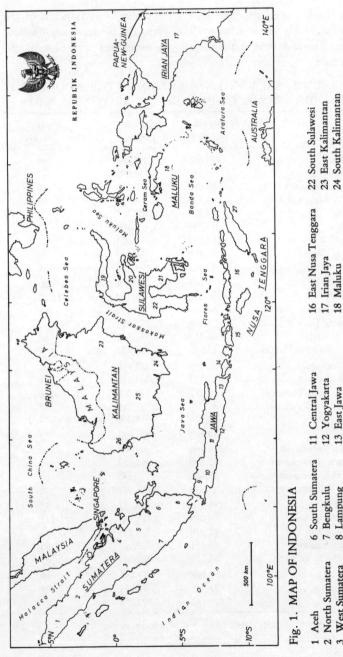

Fig. 1. MAP OF INDONESIA

1 Aceh	11 Central Jawa	22 South Sulawesi
2 North Sumatera	12 Yogyakarta	23 East Kalimantan
3 West Sumatera	13 East Jawa	24 South Kalimantan
4 Riau	14 Bali	25 Central Kalimantan
5 Jambi	15 West Nusa Tenggara	26 West Kalimantan
6 South Sumatera	16 East Nusa Tenggara	27 East Timor
7 Bengkulu	17 Irian Jaya	
8 Lampung	18 Maluku	
9 Jakarta	19 North Sulawesi	
10 West Jawa	20 Central Sulawesi	
	21 South-east Sulawesi	

The Indonesian administration divides the country into 29 provinces (*propinsi*),[5] each comprising 20–50 districts (*kabupaten*). Next come sub-districts (*kecamantan*) and communities (*desa*). Some have special status for political or historical reasons; for example the provinces of Aceh and Yogyakarta, are designated *daerah istimewa* (special territory) and Jakarta *daerah kusus ibu kota* (special capital district).

In the development of world trade and particularly during the Dutch colonial period, Jawa became the administrative and commercial focus of the archipelago. The other islands were called the 'Outer Islands', and their inhabitants felt neglected, oppressed and exploited by Jawa and the ethnically distinct Javanese. The absence of any focal point other than Jakarta, together with the centralised administrative system, has done little to alter this feeling up to the present time.

The Islands

The problems which concern us may be better understood if we begin by describing the main characteristics of the islands and groups of islands.

JAWA

Jawa is the main island, although with an area of 132,187 sq. km., it ranks only fifth in size compared with the other islands or part-islands of the republic. On the other hand, with 91.3 million inhabitants it ranks first in population. It is more than 1,000 km. long; at its broadest it is 81 km. wide. Situated on the margin of the old Sunda platform it is part of the inner volcanic arc. Most of the island consists of Tertiary and Quarternary material, but there are also folds, volcanic deposits and sediments, which are geologically very young. Geological activity finally led to Jawa being formed from a number of separate islands.

The northern part is a geosynclinal lowland consisting mainly of Tertiary developments modified by very young folds that stretch as far as the small island of Madura. In the south there is an area of Quaternary volcanoes on a Tertiary rock base and finally the southern mountain range. This chain of mountains extends along the southern half of the island, interrupted by valleys and basins which broaden towards the east. In places, mainly in the west and in the east, the mountains fall abruptly into the sea whereas in the centre, part of the mountain chain is under water and there are extended coastal plains.

In western Jawa the mountains make transport and communications difficult but in the centre and the east the terrain is easier. The distance separating the mountains is sufficient for flat and well-drained areas to be developed into intensively cultivated land; and roads and railways are able to span the mountains without much difficulty.

Volcanoes are a prominent feature on Jawa (see above) as elsewhere in Indonesia. Estimates of their number and of the extent to which they are active vary greatly because conditions change and interpretations often differ. Indonesia has about 300 volcanoes of which at least 200 have been active in historical times. Since AD 1600, 76 of them have been or are still active and another 49 have fumaroles or solfataras from which issue hot vapour and sulphurous gases respectively.[6] Jawa alone has 121 volcanoes, more than any other country on earth;[7] some twenty years ago, thirty-five of these were classified as active,[8] but in 1983 the Indonesian Department of Mines and Energy stated that of 127 volcanoes in Indonesia, 70 (17 on Jawa) were active.[9]

Only a little more than 1% of Jawa's land surface consists of older rocks; two-thirds comprise volcanic deposits such as lavas, tuffs and pumice, and the rest is young sandstone, conglomerates, clay, marl, limestone, coral limestone, gravel and alluvial deposits along the rivers and coasts. These latter formations are found in the southern parts of the island, in the north-east and on the island of Madura. On the central southern coast, particularly in the Gunung Sewu and the Gunung Kidul, impressive mountain areas of tropical karst can be seen, consisting of permeable limestone with extremely poor soils and a lack of water. The increasing population of these areas faces great difficulty in surviving on this land, and many emigrate to the cities and to other islands.

The small island of Madura, separated from the north-east coast of Jawa by a narrow sound, is the continuation of the Northern Chalk Range (Pegunungan Kapur Utara) of Jawa. The soils which developed on this limestone base are poor, but the island has the advantage of a flat terrain. Though there is no volcanic activity, ashes blown over from the Javanese volcanoes have improved soil fertility in certain places and the people are ingenious in building terraces and making the best of the soils and the long annual drought. This island has a population of more than 450 per sq. km. who live in extreme poverty and therefore represent a large proportion of the transmigrants settling on the Outer Islands.[10]

The history of vulcanicity has often meant catastrophe for the local inhabitants, but it also helps to explain why the extremely fertile soils of Jawa and some of the other islands are able to sustain huge populations. The matter thrown out by Javanese volcanoes is carried by rivers or by winds, and can thus renew soil fertility even in quite distant parts. On Jawa at any rate volcanoes have helped to create several fertile alluvial belts. Jawa's fertile volcanic soils have for centuries attracted wet-rice cultivators. Unlike islands such as Kalimantan or Sumatera, there were no dense jungles, and settlers could quite easily establish permanent clearings for cultivation. Shifting cultivation in the proper sense was hardly known, in contrast to densely forested tropical countries and some other Indonesian islands. Soil fertility seems to be inexhaustible and is renewed not only by occasional eruptions of volcanic ash but by the weathering of lava material. Leaching is moderate since heavy rains are rare and there is a clear dry season that grows longer towards the east. When rain and floods do occur, they transport precious material down to the plains and improve the fertility there.

Only the fertility of the soils can explain why hundreds of thousands remain in and return to areas that are so susceptible to volcanic eruptions. Volcanic catastrophes are by no means a thing of the past. When the Semeru volcano in East Jawa erupted in 1981, some 300 people were immediately buried under masses of mud, rubble and lava; and when in 1982 the Galunggung volcano in East Jawa became active, 57 violent and 400 light eruptions spread some 50 million cu. m. of lava over the neighbouring lowlands; 60,000 people were said to have lost their homes and fields.[11] It is worth remembering that the Krakatoa eruption in 1883, the most catastrophic in historical times, took place in Indonesia — in the Sunda Strait between Jawa and Sumatera.

Of the numerous Javanese rivers two-thirds flow north to the Java Sea, draining the plains and carrying excess silt into the ocean where pronounced tides limit the formation of swamps. The north coast advances annually a mere 9 m. into the sea, which is negligible compared with other islands. Land use has developed accordingly. The valleys and basins of the interior are carefully terraced and irrigated, as are the northern slopes and plains. The irrigation water carries valuable sediments, and contributes to a slow process of levelling. Thus the character of the original landscape is continuously changing. But, as already mentioned, there are also disadvantaged areas where people have to contend with poor soils and a shortage of water.

Forest regions have been reduced to a few remote areas and whereas the levelled and terraced land is mainly kept under wet-rice up to an altitude of 1,400 m., tea plantations and vegetables can be found higher up.

With the growing population pressure on the soils, the balance between man and the environment is deteriorating rapidly. Even the poorest soils are now being cultivated and the remaining forests cut back. Thus floods and soil erosion are on the increase and the limitations of Jawa's capacity to sustain its population are becoming obvious. In times of volcanic eruption in particular there is no longer spare land in which the affected population can settle.

SUMATERA

With a surface area of 473,606 sq. km. Sumatera is the second largest of the Sunda Islands after Kalimantan, and with 28 million inhabitants (1980) it ranks second after Jawa. Between its north-eastern extremity near Banda Aceh and the Bay of Lampung in the south is a distance of some 1,750 km. The average width is 300 km. In the west a chain of islands lies about 100 km. off the coast stretching from Simeulue in the north to Enggano in the south, and including Nias and the Mentawai; beyond these islands to the west the shelf falls abruptly into the deep sea of the Indian Ocean. To the east Sumatera is separated from the Malay peninsula by the Strait of Malacca, between 300 km. and 50 km. wide and between 30 m. and 100 m. deep. Off this coast there are a number of very flat and swampy islands including the tin-producing islands of

Bangka and Belitung. Here Sumatera touches the Java Sea.

An east-west cross-profile of Sumatera shows a pronounced mountain spine in the west sloping steeply into the Indian Ocean, while to the east it meets a vast flat plain stretching from the mountains to the coast. Therefore, physiographically Sumatera can be subdivided into coastal swamps in the east, the hilly zone to the west of it, the high mountains of the Bukit Barisan and the rather narrow west coast.

The eastern lowlands with their extended swampy coast have accumulated over the course of time on the shallow Sunda Shelf because heavy rains have washed enormous quantities of sediment down from the young mountains in the west by way of the numerous rivers which drain the eastern slope of the mountain spine, the foothills and the upper parts of the coastal plains. Some of these rivers are important: for example, the Musi on which 10,000-tonne ocean-going vessels can navigate 130 km. upstream; the Hari, navigable nearly 800 km. upstream; the Kampar and the Rokan — just to name only a few. All carry substantial quantities of sediment down from the mountains and deposit them either along their own banks, on the coast or in the sea. Thus Sumatera's eastern coastline is extended annually by up to 90 m.[12] so that former islands have been absorbed; flat swampy islands such as Rupat, Bengkalis, Padang and Rantau which are still separated from Sumatera by narrow sounds will be connected to it in the near future. This new plain hardly rises above mean sea-level; parts of it are swampy all the year round and densely overgrown with mangrove thickets.

Historically and economically this region has only been of interest to people who are closely involved in shipping and maritime trade. Between the fifth and thirteenth centuries, the Srivijaya Empire had its centre where Palembang, once Sumatera's largest town, is now situated. From there control was established over the sea lanes through the Malacca and the Sunda Straits, and ships were required to call at the Empire's market-ports and pay dues.[13] Nowadays this area, though still thinly populated, is partly given over to agriculture. The land between the foothills and the swampy lowlands which becomes dry during the months of low rainfall produces abundant wet-rice. The situation on the natural levees that stretch along both banks of the larger rivers is even more favourable. Due to fresh sediment being deposited every year, these strips of land are very fertile and have thus been prime locations for settlement, particularly near Palembang, over a long period of history. They are now densely populated (sometimes as densely as Jawa) and produce several crops which are then marketed along the natural waterways.

Finally, the so-called tidal rice (*pasang surut*) is grown near the estuaries of the larger rivers draining into the Java Sea. This technique exploits the tidal range of 1.0 to 3.5 m. when the sea water dams the river water, which then can be diverted into irrigation canals and to the fields. Protective dykes and proper drainage facilities are essential.[14] Tidal irrigation is regarded as having great potential for stimulating rice production in the lowlands of eastern Sumatera, which could be settled by transmigrant farmers. An area of 600,000 ha. of marshland in the province of Jambi is estimated to have good potential as tidal rice land,[15] and the Upang tidal farming project in the province of South Sumatera covers 235,000 ha.[16]

Obviously, the economic use of the swampy lowlands is confined to very limited areas where favourable conditions prevail. Between them are large areas unsuitable for human settlement, undeveloped and almost deserted. And although the soils, alluvia and organisols are rich in nutrients, permanent waterlogging prevents an exchange of water and thus reduces the supply of oxygen, so that even wet-rice yields are poor.[17] The mangrove thickets are cut for timber and firewood and to produce charcoal.

Water and temperature have almost as great an influence as soil on agricultural production. Because of the equator, which runs through the centre of Sumatera, the climate is moist and very hot; the hottest period is from March to May, with an average temperature of 26.7°C on the lowland plains. But with an annual range of 1.4°C even the coolest months, January and February, do not show a marked difference. Similarly, rainfall induced by the mountain barriers reaches an annual average of more than 3,800 mm. and falls throughout the year, being at its heaviest from October to January. Of course, the physical relief of the island brings some variations in temperature and precipitation, with annual rainfall varying between 2,438 mm. in the north and more than 3,000 mm. in the west. In contrast to Jawa and the Lesser Sunda islands there is no dry season, and the extensive cloud cover is a handicap in the production of plants, like wet-rice, that require a good deal of sunshine.[18]

The eastern plain rises gradually into the foothills in the west, a hilly belt reaching up to 500 m. This area is the most extensively cultivated, and the not very pronounced profile does not pose any serious problems. The soils, however, are latosols and red-yellow podzolic soils that are highly leached because of the heavy rainfall (between 2,000 mm. and 4,000 mm. annually). Therefore, plant nutrient levels are greatly reduced and the number of useful plants that can be cultivated is limited.

From the air, a lush vegetative cover gives the impression of a highly fertile region, and the jungle, where it remains untouched, is impenetrable except by river. But once it is cleared and the soil is exposed to sunlight and rain, soil fertility decreases rapidly. Therefore, shifting cultivation (*ladang*) has been the only method of exploiting land in this area. People abandoned the clearings where sweet potatoes, cassava, rice and maize were grown after one or two years, to permit the jungle to recover before the soil fertility was completely lost, and they did not return to clear the plot again for ten to fifteen years.

This was a viable technique which did not lead to irreversible ecological damage as long as the population remained small — but this is no longer the case. The period of cultivation is now being extended, and where the clearing is abandoned too late, a useless secondary bush formation (*belukar*) or *Imperata* grass (*alang-alang*) spreads.

Early in this century the Brazilian rubber tree (*Hevea brasiliensis*) was found to be useful in this hilly zone. *Hevea*, a jungle tree which does not require great soil fertility, still seems to be one of the most useful plants in replacing the original jungle. Therefore, many farming families now grow rubber trees alongside their subsistence crops: mainly wet-rice together with coconuts, tobacco, coffee, tea, pepper and kapok. This development indicates that the hilly region may well become a quite productive area provided the selection of plants is attuned to the ecology of the hills.[19]

In North Sumatera, in the hinterland of Medan, the hilly zone bears extremely fertile soils due to the fact that volcanic material has been carried there from the volcanoes of the nearby Barisan mountains, and because there are no swamps in this area. The area, still thinly populated at the end of the nineteenth century, was developed by the Dutch into one of the world's most productive plantation zones. Tobacco, tea, rubber and other export goods such as coffee and sisal were produced and shipped from here; recently the oil palm (*Elaeis guineensis*), introduced at the beginning of the twentieth century, has become a favourite crop.[20]

The main characteristic of Sumatera's landscape is Bukit Barisan, the mountainous spine along the west coast. This belongs to the young Tertiary and Quarternary folded mountains that have developed along the margin of the old Sunda mass, and is characterised by a number of volcanoes, some of which are still active. It contains the island's highest point, Mt Kerinci (3,805 m.). Vulcanicity has contributed to relief formation with explosions and tectonic depressions, whereby a number of lakes came into existence, e.g. Lake Toba with a surface area of 1,131 sq. km. and a depth of 433 m.

In contrast to the mountain ranges on many of the other islands, Bukit Barisan consists mainly of volcanic material which has turned into quite fertile andosols; thus, again in contrast to other mountain ranges, we find old and extended settled areas in the Barisan mountains with intensive agricultural production. Furthermore, tectonic basins and trenches were filled in the course of time with alluvial material and thus offer additional areas with high soil fertility. A fault-block depression stretching through the mountains from north-west to south-east forms the home of quite a number of old ethnic groups — from the Gajo people in southern Aceh to the Orang Lampung in the Ranau depression. This testifies to early settlement and colonisation, and although its importance cannot be compared with that of Jawa, the cultivation of wet-rice was no doubt its economic basis also.

However, it would be incorrect to classify volcanic soils *a priori* as fertile; the case of Sumatera shows that this is not so. The irregularity of population density in the highlands indicates also that the volcanic soils are of varying quality. In the humid tropics, as soon as acidic lavas and ashes weather to soils, they cease to show the fertile properties developed from basic material. Areas of the highlands of Sumatera where such less favourable soils prevail are numerous.[21] According to Hardjono, 'the major problem from the farmer's point of view has been the infertility of soils throughout the greater part of Sumatera', and she refers to the foothill region east of the Bukit Barisan ranges where undulating land and the absence of swamps would in normal circumstances permit agriculture, but where the soils have developed from acidic volcanic material. In addition, leaching has occurred widely since most of Sumatera south of Medan receives heavy rainfall.[22] Likewise, the predominance of volcanic soils should not cause us to forget that in some regions Tertiary limestone and sandstone which are low in minerals prevail, and are often denuded and exposed to erosion, and that permeable sandstone and marl soils, covering parts of northern and central Sumatera, offer only a moderate ability to bear crops.[23]

Therefore, in addition to the *sawahs*, the irrigated rice fields, most people practised a supplementary *ladang* economy on the mountain slopes to harvest additional rain-fed hill rice. But now that the area has access to the market economy, dry rice has disappeared, giving way to marketable bush and tree products. In Sumatera coffee has become the most important permanent upland crop. There are other crops, but most are confined to specific areas. For instance, on the slopes of the volcanoes Mt Karinci and Mt Marapi, at heights above 1,000 m., the cinnamon tree is grown; similarly, cultivation of pepper (*Piper nigrum*) has developed on the eastern slopes of the Barisan mountains in the south of the island. Finally, with the growth of the urban and petroleum-extracting centres that offer expanding markets, and with the chance to export to Singapore, the production of vegetables and potatoes at higher altitudes has spread substantially. This branch is particularly concentrated in the highlands of the Batak and Minangkabau peoples and on the slopes of the volcano known as Mt Dempo. In places, vegetables alternate with wet-rice in crop rotation and in some cases have even replaced it.[24]

The Barisan mountains slope steeply towards the west coast, leaving only a narrow coastal plain rarely more than 20 km. across; the area is reminiscent of the coastal plains of Sulawesi and the smaller islands. On these alluvial lowlands the basic food of the inhabitants is produced on the wet-rice *sawahs*, except in the permanently swampy parts where it is provided by the sago palm. Animal protein is supplied by coastal fisheries and the oil gained from coconut growing is also marketed in the highlands. For more than a decade, the cash crop grown on the often very steep foothills has been the clove tree (*Eugenia caryophyllata; Syzygium aromaticum*).

KALIMANTAN

The largest of the Great Sunda islands, Borneo, is of compact shape measuring 1,500 km. (north-south) by 1,300 km. (east-west). With an area of 737,000 sq. km. it is the third largest island in the world after Greenland and New Guinea. Whereas the north consists of part of the Federation of Malaysia and the Sultanate of Brunei, the southern 73% (539,460 sq. km.) belongs to Indonesia, with the name of Kalimantan. With 6.7 million inhabitants (1980) Kalimantan is the second least densely populated island of the Republic, after Irian Jaya. The island forms part of the Sunda Shelf, and the surrounding Java and South China Seas are comparatively shallow. To the east, however, the Makassar Strait and the Celebes Sea indicate the end of the shelf with substantial depths. The land is structured by a number of mountain ranges and lowlands containing Indonesia's largest river systems.

The larger western part of the island, already subject to folding in Mesozoic times and fractured into individual blocks during the Tertiary, shows the continuation of older chains striking from the Malacca Peninsula and younger folds from Sumatera. The eastern part belongs to a younger geosyncline.[25] The mountains bordering on East Malaysia, Pegunungan Kapuas Hulu and Pegunungan Iban, reach 1,767 m. (Mt Lawit) and 2,160 m. (Mt Harun) respectively. From a kind of central mountainous knot around Mt Cemaru (1,681 m.) the Muller mountains (Mt Liangpran, 2,240 m.) and the Schwaner mountains (Mt Raya, 2,278 m.) run south-west, and the Meratus mountains (Mt Besar, 1,892 m.) run south. Thus the mountains of Kalimantan barely exceed 2,000 m.; the highest peak, Mt Kinabalu (4,101 m.), is at the north-eastern tip of the island mass, in Sabah. The river systems with their alluvial deposits are of greater economic importance than the mountains. Since most of Kalimantan is still covered by dense tropical rain forest, hardly ever penetrated by roads let alone by railways, the rivers have been the only means of transport and communication.

The (western) Kapuas, 1,143 km. in length and thus Indonesia's longest river, flows west from Mt Cemaru and forms a huge delta of 5,400 sq. km. before it merges into the Sunda Sea near Pontianak. Together with a few medium-sized rivers, the Kapuas drains the greatest part of western Kalimantan, and small steamers can navigate as far as 500 to 600 km. upstream. The south, between the Schwaner and Meratus mountains, is drained by about half a dozen larger rivers, among them the Barito originating in the Muller Mountains and at 900 km. the second largest river of the republic; its mouth is near Banjarmasin. The east is less flat and various smaller river systems drain into the Makassar Strait. Indonesia's third largest river, the Mahakam, like the Kapuas, has its source in the centrally situated Cemaru and cuts through an extended mountain system before flowing into the sea near Samarinda after a course of 775 km.

The importance of the river systems is manifold. First, they have built up the vast forest-covered plains of the island in the course of geological time and continue to do so. Well over half the total area of the island is below 150 m. and water can be tidal up to 300 km. inland.[26] Until recently they were the only routes of penetration into the interior, although rapids in the upper courses and sand bars blocking their mouths sometimes make shipping difficult or impossible. Finally, settlers could to a certain extent use the river to penetrate and use the land once they reached the more elevated and inhabitable banks. The coastal areas, particularly the zones around the estuaries, are widely covered by perma- nent and seasonal swamps, and mangrove thickets form an additional hindrance for those who want to go ashore and settle. The swamp areas will be exploited economically for the production of tidal rice, and some projects involving several thousand hectares, partly managed by resettled farmers, are being implemented in West and South Kalimantan.[27]

Kalimantan, astride the equator like Sumatera, also has a hot and moist climate. The average temperature is 25–30°C, occasionally rising to 35°C. Usually rain falls between late afternoon and early morning throughout the year with an average of 3,810 mm; as a minor variation, July to September experience somewhat less rainfall.

There has been no recent vulcanicity on the island, and the soils are generally poor in nutrients. As a result, the development of the agricultural sector is moderate. The predominant parent materials are shale and sandstone which are naturally poor, and in the humid tropics high temperatures and moisture result in intense weathering and leaching, thus depleting alkaline earths and silica while iron and aluminium compounds remain. Wherever there is a hilly or mountainous relief, there is a high chance of erosion as soon as the vegetative cover is broken and the less weathered parent material is exposed.[28]

The swampy coast, the jungle-covered terrain, the murderous climate, the malarial infestation and the poor soils, all hostile to settlers, have till recently prevented economic development in Kalimantan comparable to that in Sumatera, let alone Jawa. Even today, many of the small number of inhabitants are aborigines, the Dayaks, who live in the jungle and are well adapted to hunting and gathering but who occasionally also engage in shifting cultivation.[29]

Land use is thus restricted to a number of crops and, of course, forestry. In the recent past, logging and the export of logs has made Kalimantan an important foreign exchange earner. Rice is the staple food for the people of Kalimantan, although growing conditions are not good and the production does not meet local demand. Sago is the second most basic food prevailing along the swampy coast.

Nearly all settlements are found along the coast and the main rivers; only a few have more than 25,000 inhabitants.[30] For a long time, the population was concentrated only in West Kalimantan in the delta of the Kapuas river with Pontianak as its focal point, and in South Kalimantan in the catchment area of the Barito river — at the centre of which is Banjarmasin.

In the more densely populated areas, the soil is used more intensely and the development of the petroleum industry has increased purchasing power and thus demand on the markets. Next to the export crops copra and pepper, the farmers (and particularly new settlers from the transmigration projects) grow fruit, vegetables and coffee, and along the coast fishing is of some importance. Rubber plantations cover an increasing area of the deforested ground, thus offering a certain protection against soil erosion as in Sumatera.

SULAWESI

Sulawesi is the third largest of the Great Sunda Islands; with 189,216 sq. km. it ranks third after Borneo and Sumatera, and with a population of 10.4 million (1980) it is third to Jawa and Sumatera. The population density (55/sq. km.) is close to that of Sumatera (59/sq. km.).

Whereas Sumatera and Borneo are situated on the Sunda Shelf, Sulawesi is separated from it by the deep Makassar Strait to the west, and a number of deep troughs and basins to the north, east and south. Thus it emerges from the deep sea. In the north it is structurally and geologically connected with the Philippines via the Sangihe islands, and although the extreme south reveals elements of the Sunda mountain system, the island may well be regarded as the southern end of the Samar arc of the Philippine archipelago.[31]

Part of the island consists of fragments of the old Sunda mass consisting mainly of granite and belonging to the East Asiatic island system. Furthermore there are areas of limestone mountains with impressive karst forms, in the cavities of which loamy weathered soils have accumulated. Finally, there are two regions where vulcanicity prevails, one in the Minahasa region of the northern peninsula (Mt Soputan — 1,661 m.) and the other on the southern peninsula around Ujungpandang (Mt Lompobatang — 2,871 m.). Although the latter has shown no recent activity, the island is never safe from sudden eruptions. Mt Colo on Una-Una island in the Bay of Tomini, for instance, was dormant for 90 years until it erupted in the summer of 1983, destroying five villages, 700,000 coconut palms and driving away the 7,000 inhabitants.[32]

Sulawesi is an island of uplands cut by deep rift valleys; this is explained by the tectonically labile zone where the Eurasian and the Pacific plates meet. Except for some narrow stretches of coastal and intermontane plains, the terrain is rugged with steep slopes, and little of the island lies below 450 m. above sea level, while quite a large area is higher than 2,500 m. The south-eastern peninsula achieves a height of 2,790 m. (Mt Mengkoka), the north-eastern 2,835 m. (Mt Katoposa). Mt Lompobatang at 2,871 m. is the higest peak on the southern peninsula, and in the north Mt Sonjol reaches 3,225 m. In the centre of the island, several summits rise above 3,000 m., Mt Rantemario (3,440 m.) being the highest point on the island. The high mountains, together with the deep surrounding seas, result in a difference of relief in adjacent areas of between 7,000 m. and 8,000 m.[33]

In the central part of the island a number of mountain lakes have accumulated in depressions, some of which are extremely deep (e.g. Lake Poso — 1,500 m.). The rivers are short, rarely longer than 100 km. and mostly with a steep gradient.

Because of the unusual shape of the island with four peninsulas emerging from the centre, thus forming a coastline of no less than 4,800 km., narrow coastal plains, together with the deep sea around the peninsulas, have prevented the formation of both alluvial plains of any significance and coastal swamps. Occasionally, the coast has been altered by marine erosion as a result of changing sea currents; only recently, five villages had to be moved from the north coast to the nearby foothills for this reason.[34]

There is hardly any point on the island more than 40 km. from the sea, and because of the difficulty of travel by land, one would expect most transport to be by coasters. However, because of widespread and treacherous coral reefs offshore, the coastal waters are dangerous for shipping and there are few good anchorages, obviously not enough to encourage sea transport. The peninsulas form three huge and deep bays: Teluk Tomini to the north, Teluk Tolo to the east and Teluk Beno to the south; the west coast of Sulawesi has only a few small bays. Thus, with its extremely rugged interior which does not favour communication and an inhospitable coast, Sulawesi is extremely compartmentalised and the parts are isolated from each other.[35]

Climatically, most of Sulawesi belongs to the permanently humid belt, and only the southern and south-eastern tip have a pronounced dry season. Most of the island is hot and humid all the year round. Temperatures are in the range 22–30°C, showing extremes of 19°C and 34.4°C; in the higher mountains, temperatures of less than 10°C are found. There are certain seasonal variations in rainfall, and whereas December, January and February are the rainiest months in the north as well as in the south, the differences between the wettest and the driest months (August and September) are far greater in the south. Similarly, the pronounced relief in the interior results in extreme differences in rainfall from one place to another. Between 1931 and 1960, the annual average rainfall amounted to 3,352 mm. in the north (Manado), 3,188 mm. in the south (Ujungpandang),[36] and only

533 mm.[37] in the west (Palu).

Like the other Greater Sunda islands, Sulawesi was probably once covered with tropical rain forest, except perhaps for the drier parts of depressions or basins where grassland prevails. So far, the lack of transportation facilities and the absence of rivers suitable for rafting logs has prevented the forest from being exploited on a large scale for the export of wood. Instead, the main forest products are gum, resin and rattan, as well as ebony, which is however of minor importance. The destruction of the primary forest stems from another cause. While it is true that the population density in the interior is low, the less fertile soils derived from sandstone and marl offer only a moderate crop-bearing capacity, and thus the local people practising shifting cultivation have caused a substantial degradation of the original vegetation. Over vast areas the primary forest has been replaced by secondary bushland on lateritic and podzolic soils, and *alang-alang* grassland is spreading rapidly.[38]

Only two minor zones, one in the north and one in the south, profit from fertile volcanic soils, in contrast to most of the island where since they have been formed from non-basic parent material, poor soils prevail. Wherever the soil is exposed due to clean cutting, heavy rain causes leaching even where there is a fairly long dry season. The short rivers have a large volume as soon as they are swelled by rain. Along the steep slopes heavy erosion takes place and the more the denudation of the interior proceeds, the less balanced is the downflow of the rivers and their load of eroded material increases.[39]

A major handicap for agricultural development is the scarcity of level land. Most of the coastal plains are narrow and discontinuous, but there are some larger areas at the head of the Gulf of Bone, on the southern peninsula around Lake Tempe and on the south-eastern peninsula in the valley of the Sampara river. However, a lack of water in the dry season makes intensive land use difficult even in those areas.

In the interior a comparatively small population depends on shifting cultivation for its main crop, mountain rice. More recently, maize was introduced, and cassava is cultivated as a secondary crop. This limitation to subsistence crops is explained by the lack of transport which is needed for the marketing of surplus staple crops. The boom in spices, however, has changed the picture and more recently the upland farmers have specialised in growing clove and nutmeg trees, almost entirely in smallholdings. Since spices are a low bulk commodity, marketing is relatively easy even under prevailing conditions.[40]

The most intensively used land is on the northern peninsula in

Minahasa, and in the south around Ujungpandang, where fertile volcanic soils make possible the production not only of subsistence but also
of cash and export crops, which are encouraged by the greater possibilities
for maritime transport. Here, where the island's population is concentrated, wet-rice and sugar-cane, coconuts and maize, coffee and some
rubber are produced.[41] The main commodity, particularly in the north,
is copra, produced by a large number of smallholders living along the
coast, whereas southern Sulawesi, mainly around Lake Tempe, is the
main producer of wet-rice with even a surplus for the market. The
coffee-growing area is in the Toraja uplands of the south.

The fact that the long-established agricultural productivity on the
volcanic soils of Sulawesi seemed to hold great future promise, fostered
the idea of making the island another Jawa. Even during the colonial
period this prospect was discussed, but its supporters did not realise that
by far the largest proportion of the land had experienced no volcanic
activity since the Tertiary period and thus no natural rejuvenation of the
soils. Most soils are extremely poor and are subject to serious leaching
due to high temperatures combined with heavy rainfall (more than
2,500 mm. in most of the valleys). The lack of water during part of the
growing season is another handicap, particularly since the greater part
of the island with its rugged relief makes irrigation impossible.

However, the island's potential should not be underestimated. Next to
mining and forest economy, the absence of unhealthy and useless swamps
leads people to settle on the small coastal plains and other flat land. The fact
that two of the most successful transmigration projects are on Sulawesi
shows that the island offers a potential when risks are taken and possibilities exploited. Although it may not be advisable to expand the area
under cultivation, it may well be possible to use it more intensively.

MALUKU

The province of Maluku, commonly known as the Moluccas, consists of
a large number of small islands between the Philippines in the north, the
Lesser Sunda Islands in the south, New Guinea in the east and Sulawesi
in the west. The largest island is Halmahera (16,800 sq. km.). The islands
stretch from about 3°N to about 6°S straddling the equator. The surface
area is given as 74,505 sq. km., or barely 4% of Indonesia's territory, and
is inhabited by 1.4 million people (1980), giving a population density of
19 per sq. km.

The geological structure and relief are extremely confused, since it is here that the East Asiatic island system and the Sunda mountain system meet and interpenetrate. Some of the island arcs dealt with earlier end here in a spiral: the outer arc runs from Timor over the Taimbar and Kai islands into the Moluccan islands of Seram and Buru; the next emerges from Wetar island and runs via a few small islands to the Banda islands (the remains of what was once a gigantic volcano) south of Seram where it ends.[42] These together form the southern Moluccas with the island of Ambon as the administrative seat. The northern Moluccas are separated by an east-west striking mountain ridge emerging from the eastern peninsula of Sulawesi and following the Banggai, Sula, Gomumu and Misool islands to the Vogelkop peninsula of New Guinea. The island group of Halmahera where there are still active volcanoes is a bizarre shape similar to that of Sulawesi. Here on Halmahera the Melanesian mountain system stretching over the ridges of New Guinea meets the East Asiatic island system coming from north Borneo via north Sulawesi and continuing on to the Philippines.

The islands of Maluku are separated by very deep seas. The Molucca Sea reaches a depth of 4,971 m. in the Bacan basin south-east of Halmahera; however, the Halmahera basin east of the island is only 2,072 m. deep. The Sulu basin between Sulawesi and Buru is 5,580 m. in depth, the Buru basin itself 5,319 m. The Banda Sea reaches depths between 5,400 m. and 5,800 m. but drops to 7,440 m. in the Weber Deep some 200 km. west of the Kai islands. In contrast, most of the elevations on the islands are not so impressive, though the highest point of the province, Mt Binaiya on Seram, rises to more than 3,000 m., Buru island rises to 2,736 m., Bacan to 2,111 m. and Ternate to 1,726 m. The large island of Halmahera has various summits of more than 1,000 m. but its highest, Mt Gamkonora, reaches only 1,615 m. Most of the smaller islands are rarely more than 500 m. in height; for instance, the Banda islands do not surpass a maximum elevation of 656 m.

Thus, we find in the province of Maluku, as in Sulawesi, a very pronounced relief from deep sea to high mountains. The highest point in Seram, for instance, and the lowest point of the sea north of the island, only some 150 km. away, are 7,710 m. apart. Consequently, the shelf of these islands is very small since they emerge abruptly from the sea and there is little chance for the accumulation of alluvial coastal plains worthy of the name. Apart from a few bays and broader valleys, the islands are covered with mountain ridges which form the spine of the arcs; though they are sometimes twisted and interrupted for 100 km. and more.

Similarly, their composition is varied due to its orogenesis. Seram and Buru, the mountains of which rise to 2,500–3,000 m., consist of older crystalline rocks, mesozoic and eocene limestone, but neogenic folding can also be found. Other islands such as Misool, Obi and the Sulas also consist of old slate and jurassic limestone. Halmahera, where there is still some volcanic activity, shows relatively little recent folding. Nearly everywhere, however, recently emerged coral limestone can be found.[43] Because the province covers 1,000 km. from north to south, across the equator, the climatic conditions vary from the equatorial to the sub-equatorial. In Ambon, at its centre, the annual average rainfall is 3,449.3 mm. with the wettest month being July (589.3 mm.) and the driest November (104.1 mm.). The temperature is more balanced with an annual mean of 26.7°C and the hottest months being December and January (27.2°C) and the coolest July (25.0°C).[44] Towards the south, conditions come closer to those of Nusa Tenggara.[45] The increasing rainfall towards the north results, on most of the islands, in dense tropical rain forest that is rarely accessible by road. Towards the south the rainfall decreases and thus Seram, for instance, is covered with extended savannahs.

The Moluccas were among the earliest of colonial acquisitions and played an important part in the East Indian trade since they offered spices which were in great demand in Europe. This caused them to be known as the 'spice islands' — cloves, nutmeg and mace were shipped from there — but cultivation of these plants also spread to the other islands of Indonesia and to other tropical regions, and thus spices lost their importance for Maluku.

The obstacles in the way of economic development are numerous. First, the islands are small, spread over a large area of sea and far from all sizeable potential markets. Sea transport is inadequate since the simple boats are ill-adapted to rough seas. On the rugged islands themselves, the construction of motor roads began only recently.

The population is sparsely distributed over the islands, the relief of which is not such as to encourage the agglomeration of families. Therefore, the population density reaches reasonable figures on only a few favoured islands, such as Ambon with 82 per sq. km., Saparua with 81, Banda with 70 and Ternate with 32. Elsewhere the density is below 5 per sq. km.[46] Thus land use in Maluku is nowadays largely confined to subsistence production, and in the most remote places such as the Kai islands the population is generally malnourished.[47] Lava rocks and coral limestone permit the growth of a luxuriant rain forest, but forest clearings result in little arable soil. Hence wet-rice cultivation is of hardly any importance although its production was being promoted recently. But this is within the Melanesian cultural area, and therefore rice as a staple food is replaced by tuber crops such as yams, complemented by the starch of the sago palm which grows in the coastal swamps and riverine lowlands.[48] Coconuts, occasionally marketed, grow all over the coasts and add to the people's basic food, which is completed by maize, sugar-cane and tropical fruit. Goats, sheep and, in Christian villages, pigs are raised. Fishing is, of course, widespread throughout the islands. Forest products, apart from sandalwood, play no significant part in the islands' economy. Spice production still occurs on the Banda islands, on Ternate and Tindore west of Halmahera, and of course on Ambon, from where nutmeg, pepper and cloves are shipped.

It therefore seems that the potential of Maluku is limited. Nevertheless, efforts to settle transmigrants on the islands date back to 1954. Before 1972, nearly 2,000 settlers were established in two projects on Seram, but great difficulties arose over land ownership and the protection of forests. The greatest handicap, however, was marketing. The settlers' plots were well selected, prepared, irrigated and cultivated, but when the rice surplus had to be sold, the normal boats in use could not be relied on to cover the short distance to neighbouring Ambon where there was a great demand for food. Plans to settle 700 families in northern Halmahera were therefore abandoned.[49]

NUSA TENGGARA

Two island arcs continue from Jawa eastward, running partly under water, and form the Lesser Sunda islands, known locally as Nusa Tenggara. They are subdivided administratively into three provinces: Bali, with the island of that name (5,561 sq. km. and [1980] 2.5 million inhabitants); West Nusa Tenggara, comprising the islands from Lombok to Sumbawa (20,177 sq. km. and 2.7 million inhabitants); and East Nusa Tenggara, reaching from Komodo to Timor (47,876 sq. km. and 2.7 million people). Geologically, the island arcs continue east to the province of Maluku.

The two island arcs are easily distinguished. The inner arc to the north is a continuation of the chain of mountains and volcanoes from Sumatera and Jawa right through Bali, Flores and on through Wetar and ending in the Banda islands. The outer arc is a continuation of the Mentawai islands west of Sumatera, disappearing south of Enggano and reappearing with Sumba whence it continues through Timor and the Kai islands to Seram. These islands are built up of limestone and show no trace of vulcanicity. North of the chains the Flores Sea reaches a depth of 5,060 m.; to the south the Java Trench reaches 4,123 m. south of Bali; the Suva Sea between the two island arcs reaches 3,382 m. and the Timor Sea between Timor and Australia 3,312 m.

As with the Moluccas, the Lesser Sunda islands emerge from deep sea and are mountainous; peaks of 3,000 m. and more are found on Bali, Lombok and Timor, and 2,000–3,000 m. on Flores and Sumbawa. Sumba and the smaller islands all rise to less than 1,000 m. All the islands are rugged and have a pronounced relief, so that there are very few alluvial plains. Consequently, the potential for agricultural production, particularly of field crops, is limited, and the development of an infrastructure is costly. In addition, rainfall diminishes eastwards; Bali has a mean annual rainfall of 2,149 mm. whereas Timor receives only 1,456 mm. The dry season is pronounced, August and September being the driest months. They bring no more than 58 mm. each in Bali, but only 2 mm. in western Timor where there are seven to eight dry months a year. This again is a severe handicap for agricultural production.

The features of Bali, and to some extent also of Lombok, show great similarities to Jawa. The further east we go, however, the more conditions change and land use changes accordingly, even though soil quality is good on all the islands, volcanic as well as non-volcanic.

Passing from Bali to Lombok we cross Wallace's Line and enter the westernmost point of the ancient Australian-Pacific continent; thus there are changes in flora and fauna, although the type of forest does not change abruptly. We saw in East Jawa the advance of the tropical dry and monsoon forest, which thus confines the tropical mountainous rainforest to the higher altitudes; this feature becomes even more pronounced on the Lesser Sunda islands. In the interior of some of the larger ones like Sumba and Timor, extended grass savannahs with *alang-alang* have spread. The forests of Flores, Sumba and particularly Timor once supplied much sought-after sandalwood for centuries, others produce teak, dye-wood and cinnamon, but these activities are no longer important. Nowadays, the dry forests and pastures are used to graze cattle, the only export of the islands with which the people pay for their food imports. But although the rearing of animals is carried out as extensive farming, overgrazing has already severely damaged the natural potential, particularly on Sumba and Timor where capacity has been outstripped.

Traditionally, people use the soil for three to five years consecutively before leaving it fallow for a period of ten years. But the increasing pressure of population on the soil has reduced the fallow period to a mere three years. The consequences are visible everywhere: reduction in soil fertility, hence a decline in yields, and hence the fact that about one-third of the basic food supply for Flora, Sumba and Timor has to be imported.

Taking the islands individually, it can be said that the well-being of their people depends on specific factors: the soils and the possibility of natural rejuvenation from volcanic deposits; the gradient of the terrain which often barely permits the growing of field crops in reasonable quantities; the damming of rivers for irrigation; rainfall and length of the dry season, and, last but not least, the infrastructure (particularly for the marketing of agricultural products).

Bali is an outstanding example of an island favoured by nature, which in 1980 maintained 444 people per sq. km. By contrast, West Nusa Tenggara supports 135, East Nusa Tenggara 57 and East Timor only 37 per sq. km., and even some of these islands cannot feed their population unaided. Bali has extended alluvial plains stretching down from 600 m. above sea level to the coastal area. Comparatively large rivers irrigate the highly volcanic soil, the climate is equable, the natural vegetation luxuriant, wet-rice is produced wherever possible and coconuts, sugar-cane, coffee, cocoa, tobacco and peanuts grow well.

Conditions often change from one place to another, even on the small islands, and the rugged relief creates microclimatic pockets with very different living conditions. This occurs on most of the islands apart from Bali, particularly those of the outer arc where there are zones of extreme drought, and even subsistence crops often fail.[50]

With a few exceptions, subsistence land use tends to predominate on these islands with maize as the staple crop. Cassava is often planted as a reserve crop in case the maize fails, and beans are also sown in the maize fields; upland rice is cultivated only under favourable conditions. Therefore, the characteristic picture of the cultural landscape is that of dry farming. Among the tree crops, coconut palms dominate along the coasts; the nuts generally being locally consumed. Furthermore, there is the kapok tree furnishing upholstering material and the lontar palm (*Borassus flabellifer*) used in the production of palm wine. At higher altitudes with sufficient moisture some coffee, rubber and tobacco are grown, and the consumption of fish adds to the protein intake of the islanders.

Thus the potential of the islands is generally limited due to various factors, and because of growing population pressure some islands are beginning to resemble Jawa's zones of poverty.

IRIAN JAYA

Since 1962 *de facto*, and 1969 *de jure*, the western part of the island of New Guinea, till 1962 under Dutch colonial administration, has been part of Indonesia. It has recently been subdivided into three provinces so as to facilitate development.

The island of New Guinea is the largest in the world after Greenland (831,000 sq. km.). The western part, now called Irian Jaya or West Irian, accounts for roughly half of it. In its geological structure as well as the ethnic origin of its population it belongs to Melanesia, and is in fact the most important western link of the inner Melanesian insular arc.[51] Irian Jaya, with a surface of 421,981 sq. km., accounts for 22% of Indonesia's national territory but, with less than 1.2 million inhabitants in 1980, for only 0.79% of the national population. With an average of three people per sq. km. it is very sparsely populated, in parts hardly explored, and with hardly any infrastructure over vast areas.

Tectonically New Guinea is redolent of the Sunda islands, and it seems that they were once part of a continental bridge between Southeast Asia and Australia. Since the island is on the Sahul shelf it is not immediately surrounded by deep seas. The main physical feature is a folded central mountain spine running through the whole island from west to east.[52] In Irian Jaya it is composed of the Sudirman and Jayawijaya mountains, of which the highest is Mt Jaya (5,030 m.) and ten others are over 4,800 m. The snowline runs along the 4,300 m. isohypse, but the glaciers and snowfields are contracting. In 1850 the Meren glacier covered 5.1 sq. km. and the Northwall firn 9.1 sq. km.; in 1972 the figures were 1.9 and 3.6 sq. km. respectively.[53] A substantial part of the mountain area extends beyond the tree-line along the 3,200 m. isohypse.

North of the central mountain spine run the Northern or Coastal mountains, also called the Van Rees mountains. Rising in places as high as 2,272 m., they form a fracture zone resulting from volcanic activity; in 1907 an earthquake lowered the northern coast by several metres in one night, and the existence of limestone at altitudes of 950 m. and even 1,700 m. indicates colossal geological movements.[54]

In the north, the central mountains slope rather gently into a depression that runs east-west and is bordered by the Van Rees mountains.

Here the north-bound streams converge to form the Memberamo river which then breaks through the coastal range and debouches into the Pacific Ocean. All these valleys are swampy and so largely is the narrow coastal strip. Towards the south, the central mountains descend steeply into a vast plain, actually a mesozoic part of the Australian land mass. Most of this plain is swampy right down to the coast and is drained by a number of rivers which emerge from the central mountain spine, the most important being the Digul and the Pulau-Pulau and their tributaries. Both coasts, apart from the Vogelkop (Bird's Head) Peninsula in the north-west, show but little structure.

The climate depends on the trade winds. In January they bring abundant rain from the north-east, in July the south coast is touched by the somewhat drier south-east winds, and the north coast comes under the influence of the doldrums. Rainfall is heavy and falls during most of the year with minor variations. Along the north coast there is precipitation of 2,540 mm. per year, and in the high mountains up to 8,000 mm. has been measured. Only a few places have less than 1,500 mm. in a year; Merauke, for instance, in the extreme south-east has as little as 20 mm. in a dry month. The temperature remains constant between 26°C and 27°C. throughout the year (in the high mountains, of course, it drops to freezing-point).

Vegetation shows great variations according to the elevation. The extended moist lowlands are covered with mangroves and casuarinas, swamp forest and swampy grassland. On slightly higher ground, the evergreen tropical rain forest begins, rises up the mountains and turns into mountain rain forest and cloud forest at about 900 m. Above the tree-line is grassland that consists partly of *Imperata cylindrica* or *alang-alang*; the varieties reveal a mixture of Asiatic-Malayan and Australian species.[55]

The soils of Irian Jaya, being without volcanic rejuvenation, are poor in nutrients and subject to leaching once the protective forest cover is destroyed. No mountain soil can resist the power of up to 8,000 mm. of rain per year. But although the natural conditions resemble those of Sumatera and Kalimantan, the population follows a different economic pattern. Outside the few larger settlements, where a more modern style of living is encroaching, the majority of the population live in isolated tribal groups scattered over vast areas. Parts of the mountain areas are supposed to have been settled for as long as 5,000 years, but these people have barely begun to touch the potential of the island. Confined to hunting, gathering, fishing and very little field cultivation, they have developed techniques which spare nature. They established mud-bank farming in swampy stretches and protected cleared slopes against soil erosion by planting shelter belts of casuarina or by a specific form of terracing. Next to bush-fallow farming, a very protective method of shifting cultivation, there is intensive home-gardening. The main cultivated plants are sweet potatoes (*Ipomoea batatas*), taro (*Colocasia esculenta*), sugar-cane and various local vegetables. Recent studies have shown that the mountain-dwellers still live in relative harmony with their natural environment and that customs and taboos largely protect the ecology of their living space.[56] For this reason, and because the population density is low, Irian Jaya is still covered with more than 70% forest, and areas under *alang-alang* are far fewer than on other islands.[57]

The development of Indonesia's sparsely populated easternmost provinces is, of course, a tremendous challenge to those responsible for transmigration policy in Jakarta. After some minor efforts were made — only 267 families were moved to a few settlements before 1969 — the lack of infrastructure hampered further development.[58] Recently, however, there has been news that Jakarta is planning a transfer of several million Javanese and Balinese to Irian Jaya. The main areas for new settlement would be the district of Merauke in the south-east, the north coast and the Van Rees mountains as well as the Vogelkop Peninsula. The authorities in neighbouring Papua New Guinea already view this prospect with alarm.[59]

No doubt such an influx would cause ecological problems, even if it could be organised and financed. Considering that half of the island must remain forested for protective purposes and a certain amount will remain as productive forest (3.1 million ha. have been allocated to a dozen large timber companies), only about 500,000 ha. can be regarded as suitable for settling people from Jawa and Bali. Observers are afraid that the lowlands might soon be overcrowded and that the new settlers would move up into the mountains and drive away the indigenous people, who in any case would no longer have enough land to continue the ecologically protective form of land use. Ethnically and culturally, the incoming Muslim rice-growers would meet animistic and partly Christianised forest-dwellers.[60] Thus Irian Jaya may well become an experimental area in the field of land use and inter-ethnic adjustment, presenting greater problems than those so far met on the other islands.

30   *Land Use and Environment in Indonesia*

NOTES

1. Khan, 1973, p. 220.
2. Rutz, 1976, p. 172, fn. 5.
3. Ulrich in Dietrich, 1970, pp. 129–40; Seibold, 1977, pp. 18–28.
4. Khan, loc.cit., p. 221.
5. After Irian Jaya had been split into three provinces in 1984.
6. Uhlig in Kötter *et al.*, 1979, p. 44.
7. Brüning, 1954, p. 474.
8. Speed, 1971, p. 9.
9. *Indonesia Times*, 27 July 1983.
10. Hardjono, 1977, p. 11.
11. Hamzah in *Indonesia Times*, 16 Jan. 1984.
12. Soekmono, quoted by Hardjono, 1977, p. 10.
13. Speed, 1971, p. 25.
14. Rieser, 1975 (ICID); Dequin, 1978, p. 135.
15. *Indonesia Times*, 15 Aug. 1981.
16. Loc.cit., 19 Aug. 1981.
17. Birowo *et al.* in Kötter *et al.*, 1979, p. 397.
18. Uhlig in Imber and Uhlig, 1973, pp. 19–42.
19. Kötter *et al.*, 1979; Speed, 1971.
20. Kötter *et al.*, 1979.
21. Uhlig in Imber and Uhlig, 1973, pp. 31–2.
22. Hardjono, 1977, p. 12.
23. Kündig-Steiner, 1964, p. 1288; Uhlig/Imber and Uhlig, 1973, p. 33.
24. Birowo *et al.* in Kötter *et al.*, 1979, pp. 400–01.
25. Machatschek, 1941/2, pp. 39–40; Brüning, 1954, p. 473.
26. Speed, 1971, pp. 12–13.
27. Rieser, 1975 (ICID).
28. TAD, 1980, p. 17.
29. Hardjono, 1977, pp. 12–13.
30. Speed, 1971, pp. 12–13.
31. Khan, 1973, p. 222.
32. *Indonesia Times*, 30 July 1983.
33. Khan, 1973, p. 221.
34. *Indonesia Times*, 6 Nov. 1981.
35. Fisher, 1971, p. 232.
36. Hardjono, 1977, p. 13.
37. Speed, 1971, p. 14.
38. Uhlig in Imber and Uhlig, 1973, pp. 33–4.
39. Hardjono, 1977, p. 13.
40. Birowo *et al.* in Kötter *et al.*, 1979, p. 402.
41. Neef, 1977, p. 288.
42. Loc.cit., p. 289.
43. Machatschek, 1941/2, p. 39.
44. Fisher, 1971, p. 214.
45. Speed, 1971, pp. 14–15.
46. Mohr, 1938, p. 491 (more recent figures not being available).

47. Everingham, 1983.
48. Uhlig (ed.), 1975, p. 321; Birowo et al. in Kötter et al., 1979, pp. 403–4.
49. Hardjono, 1977, pp. 87–90.
50. Fox, 1977.
51. Krug, 1963, II., pp. 1423–7.
52. Speed, 1971, pp. 15–18; Khan, 1973, p. 222; Neef, 1977, p. 496.
53. Hope, 1976, p. 34.
54. Krug, 1963, II., p. 1424.
55. Loc.cit.
56. Michel, 1983; Aditjondro in Indonesia Times, 29 Dec. 1983.
57. Hardjono, 1977, p. 13.
58. Loc.cit., p. 89.
59. Callick in The Times of Papua and New Guinea, 18 Nov. 1983.
60. Aditjondro, loc.cit.; Callick, loc.cit.

2

THE DEMOGRAPHIC SETTING

Dimensions and Structure

The Indonesian population (147.5 million) is at present the third largest group in Asia belonging to a single political entity after the People's Republic of China (976.7 million) and India (673.2 million).* But whereas in 1970–81 the annual population growth amounted to 1.5% in China and 2.1% in India, in Indonesia it was 2.3%. According to the latest estimates, a static population will be reached in China in 2040 at 1,435 million, in India in 2140 at 1,838 million and in Indonesia in 2140 at 400 million.[1] These hypothetical figures, based on the present situation, point to a grim outlook, but much can change in the course of time and it seems that there will be some changes.

It is extremely difficult to count a population occupying an area of 1.9 million sq. km. spread over more than 5,000 km., and consisting of more than 13,000 islands, of which 6,000 are inhabited. However, the censuses and surveys carried out by the Republic of Indonesia have a good reputation for reliability: cross-checks and microcensuses made in periods between nationwide censuses have largely confirmed the latter's findings.[2] Thus we may assume that the more recent population figures can be accepted as exact enough, at least for the purpose of this study.

In the early days of population statistics, the Dutch administration made little attempt to count the people living in its colony. The nineteenth-century statistics are wholly unreliable, and even the figures for 1905 (37.4 million) and 1920 (49.3 million) have to be treated with great caution.[3] The Dutch conducted their last census before the Second World War in 1930, when a total population of 60.7 million was shown. In earlier days the growth of the Indonesian population was irregular. Bad harvests followed by famines, cholera, influenza, small-pox and other epidemics resulted in annual growth rates that rarely exceeded 1.6%. In modern times the Japanese occupation during the Second World War and later the fight against the Dutch colonial system brought the growth rate below 1%. However, after independence and the resumption of normal conditions, the annual growth rate quickly rose above 2%: Indonesia was experiencing its population explosion.

*Figures valid in 1980.

32

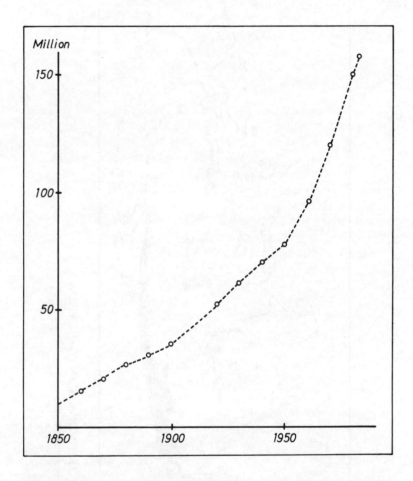

Fig. 2. GROWTH OF INDONESIAN POPULATION (1850–1983)
Source: Based on Indonesian government statistics

The census figure for 1961 was 96.4 million, for 1971 119.2 million and for 1980 almost 150 million.

The reason for this development was no doubt a rapidly decreasing death rate, increasing life expectancy and a more or less stable birth rate. Between 1960 and 1980 the crude death rate declined from 23 to 13 per thousand (or by 40.9%), while the crude birth rate only dropped from

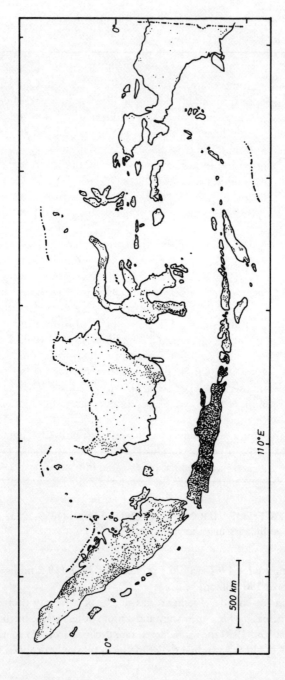

Fig. 3. DISTRIBUTION OF THE INDONESIAN POPULATION
Source: Based on Indonesian government statistics.

46 to 35 per thousand (or by 22.7%); during the same period the life expectancy of a new-born child increased from 41 to 53 years.

In relation to the total area of the country, even a population of 150 million corresponds to a density of 79 per sq. km., an alarming figure. However there are between 60 and 200 ethnic groups in Indonesia, ranging from highly developed Javanese to Stone-Age tribes, but for most of them, particularly the younger generation, the official Indonesian language, *bahasa Indonesia*, is a common base and 90% share a common faith, Islam. Other religious groups include Christians and animists who enjoy freedom of religion. Thus the numerous islands have developed demographically in quite distinct ways. Hence 61.9% of all Indonesians occupy only 6.9% of the national territory, namely the island of Jawa (including Madura), whereas only 38.1% inhabit the other 93.1%.

It was on the island of Jawa that the Dutch colonialists focused their interest. It was here that they built their capital Batavia (now Jakarta) and administration, trade, industry and thus population were concentrated. But it seems that long before the Europeans arrived the island was already relatively densely populated. An early figure, released by the Dutch in 1815, of 4.6 million Javanese is believed to be a great underestimate. By 1920 it had grown to 35.9 million, in 1971 to 82.1 million and in 1980 to 91.3 million. This gives an average population density on Jawa of 690 persons per sq. km., a figure that conceals the fact that the island is unevenly settled. The Special District of Jakarta (D.K.I. Jakarta Raya) has a density of 11,023 people per sq. km., including the metropolis of Jakarta with more than 7 million inhabitants; it is indeed a highly urbanised area. But even provinces with preponderantly rural populations have 593 (West Jawa), 609 (East Jawa), 742 (Central Jawa) and even 868 (Yogyakarta) per sq. km. Population densities in particular areas of East Jawa reach 1,500–2,000 per sq. km. whereas the Ruhr, the most industrialised zone of West Germany, has no more than 1,585 people per sq. km.[4]

The situation on the other islands is quite different: Bali, with 444 people per sq. km., is more densely populated; Sumatera (59) and Sulawesi (55) show much lower figures; and Kalimantan and Irian Jaya have only 12 and 3 people per sq. km. respectively. Again, these are global figures for the islands, and on the larger ones the population density varies from province to province, e.g. North Sumatera 118 and North Sulawesi 111 people per sq. km.

Similarly, the annual population growth-rate varies from island to

Table 2. ANNUAL POPULATION INCREASE IN INDONESIA,
1971–80

	Growth rate %		Growth rate %
Sumatera	3.32	*Kalimantan*	2.98
— Aceh	2.93	— West Kalimantan	2.31
— North Sumatera	2.60	— Central Kalimantan	3.43
— West Sumatera	2.21	— South Kalimantan	2.16
— Riau	3.11	— East Kalimantan	5.73
— Jambi	4.07		
— South Sumatera	3.32	*Sulawesi*	2.22
— Bengkulu	4.39	— North Sulawesi	2.31
— Lampung	5.77	— Central Sulawesi	3.86
		— South Sulawesi	1.74
Jawa	2.02	— Southeast Sulawesi	3.09
— Jakarta	3.93		
— West Jawa	2.66	*Maluku*	2.88
— Central Jawa	1.64		
— Yogyakarta	1.10	*Irian Jaya*	2.67
— East Jawa and Madura	1.49		
Nusa Tenggara	2.01	*Indonesia, total*	2.32
— Bali	1.69		
— West Nusa Tenggara	2.36		
— East Nusa Tenggara	1.95		
— East Timor	—		

Source: Biro Pusat Statistik, Jakarta.

island and from province to province. This is determined not only by
natural increase but also by inter- and intra-island migration.
Between 1971 and 1980, the Indonesian population increased by 2.32%
per year; the corresponding figures for the main islands were Sumatera
3.32, Kalimantan 2.98, Sulawesi 2.22 and Jawa only 2.02. The provincial figures are even more impressive: namely Lampung 5.77, East
Kalimantan 5.73, the capital area of Jakarta 3.93 and only 1.10 for
Yogyakarta (see Table 2).[5]

These figures are the result of two important political measures: the
transmigration and family planning policies. The rapid population increases in Lampung and East Kalimantan provinces, and indeed Jakarta,
indicate that they have been the target of migrant streams during the
years in question. Low growth rates indicate out-migration combined
with successful family planning.

Population Policy

Family planning is of crucial importance for most countries in the Third World, and because the Indonesian experience has been so successful, it should be examined briefly.

The official population policy of Indonesia after independence was strictly one of encouraging fertility. Losses during the war and in the independence struggle, the idea of a vast but empty country with huge potential, and finally the desire to become strong and politically important made the Sukarno administration prohibit even the distribution of contraceptives.[6] A high birth rate policy during those years throughout the Third World was only too readily accepted by the population. Children were a status symbol for men as well as for women, particularly in an agrarian society where they constituted part of a system of social security. High infant mortality required a large number of births so that a sufficiently large number of children i.e. sons would survive to care for the old and disabled members of the family.

When General Suharto took over power in Indonesia in 1966, things changed. Politicians became aware that the Javanese proverb 'Many children — much wealth' was no longer valid. Many children died young and those who survived did not find paid work and could barely support themselves, let alone their parents. Politicians and academics expressed their concern over the rapidly growing population that would soon negate all efforts to administer and develop the country: 'The success of our nation depends on the success of our family planning policy. With our abundance of children we produce our own problems. If we do not succeed in curbing the population growth rate we will not solve the other problems of our country.'[7] Fewer children would mean a higher per capita income, better food and better education.

The new population policy was not enforced by broad government activities, laws and the like; the subject was too delicate and feelings could easily be violated. On the other hand, an Indonesian Planned Parenthood Association (PKBI) had existed since 1957 and traditional contraceptive techniques were well-known, in particular the urban population used abstinence, the rhythm method, withdrawal and douche, and even jungle tribes such as the Dayak Bawos in central Kalimantan had ways of preventing over-population. Modern techniques like the pill, injections, IUDs or condoms, and sterilisation were less well-known. Doctors implemented PKBI activities with the aim of improving the health of mothers and children, treating sterile mothers and convincing people that longer-spaced births were necessary. But government support was lacking.

A new period started in early 1967 when the PKBI held its first congress and the government lent support to family planning. A pilot programme was launched in Jakarta, and in October 1968 the Ministry of Public Welfare set up a National Family Planning Office (LKBN), which however was closed down two years later. The year 1970 is regarded as the official start of family planning policy in Indonesia with the foundation of the National Family Planning Coordination Board (BKKBN), which is in charge of activities all over the country. Various amendments underline its importance and the government has adjusted its policy according to need and accumulated experience. In 1972, the President declared the BKKBN non-governmental and non-departmental but under his own authority, and in 1978 he reduced the activities of the Central Board (BKKBN Pusat) to mere co-ordination. The activities were transferred to the provincial and lower levels, with responsibility devolved to the governors and lower authorities; a policy of decentralisation unique in Indonesia and which observers believe to be one of the main reasons for the overall success of the family planning programme.[8] However, the programme's general administration and development policy still suffer from many shortcomings due to centralisation of decision-making.[9]

Geographically the programme was extended throughout the country in three stages. In 1970 it covered Jawa, Madura and Bali, and made contact with 78.2 million people or 80% of all Indonesians. In 1974, it was extended to the provinces of Aceh; North, West and South Sumatera; Lampung; Western Nusa Tenggara; West and South Kalimantan and North and South Sulawesi, encompassing another 30 million people. By 1979 the remaining provinces had been included.[10]

The decentralisation of family planning activities made it possible for the peculiarities of each region or ethnic group to be considered. As it turned out, there were tangible differences from island to island and from province to province regarding customs and religious beliefs, attitudes towards family and children, and indeed attitudes towards the authorities. By delegating responsibility to the lowest possible level it became easier not only to recruit volunteers at such levels, but also to consider the variations in enthusiasm, cultural values and other typical local characteristics in favour of or against family planning.

In Bali, for instance, where there is a high degree of cohesion in village social structure, the *banjar* or village assembly was the point at which the programme could be successfully propagated among the people. Once convinced of its necessity and advantages, the *banjars* channelled modern family planning practices into the homes of the villagers, and supervised its continuous implementation. Apparently the Balinese and their authorities are fully aware of the ever shrinking land-to-man ratio. Moreover, the ancient Hindu religion prevalent in Bali has values different to those of Islam. Taken together, this resulted in the 'miracle of Bali'. After the general introduction of the IUD, the annual population increase dwindled from 2.3% in 1970 to 1.6% in 1980.[11]

On other islands where the social structure of the villages is looser, different ways of approaching people had to be found. The allies of the programme were provincial governors, army, police, civil servants, religious leaders and so forth. It was important to choose an ally who was generally respected in the region; for instance, it seems that the army and police have the greatest authority in East Jawa, whereas in West Jawa uniforms are disliked and social leaders are preferred. Sometimes the programme is run by women, and sometimes — as in Bali — by men; mothers' and acceptors' clubs support and keep it running.[12] Frequently the press reports on those agencies which support the BKKBN officials. The army may participate in building a family planning clinic in a remote place, or a well-known person may offer himself as an example so that others will follow and also become acceptors.

In fact, the family planning programme uses every opportunity to spread its ideas among the people and reach its targets. Apart from ample funds, of which Indonesia itself contributes 65–70%, and efficient logistics including the national production of contraceptives, they use a good deal of 'salesmanship' to propagate the idea. Every year there is a new slogan or trick, and strong official support is forthcoming from the President down. The government encourages the programme's implementation among its employees, the civil servants, who are rather strictly controlled in matters of divorce and re-marriage.[13] From the fourth child on, no family allowance is paid,[14] and there are reports that paymasters hold back the salaries of civil servants who cannot produce their wives' Family Planning Card.[15]

It is remarkable that although explicit sexual literature is still banned in Indonesia, methods of family planning are freely discussed all over the country, at all levels and in most institutions. It is a topic of conversation at social gatherings, discussed on radio and television and even a theme in traditional shadow plays. Protestant and Catholic conferences deal with it and even the Muslim authorities propagate the three-child family; the wooden drum next to most mosques that calls the faithful is now often used to remind the village women to take the pill.[16]

Some family planners hope that the progress in women's education will popularise the small family to the advantage of all its members. But education and a professional career do not necessarily influence women's fertility. Because educated women generally have a higher social status than women in other Asian countries, particularly Muslim ones, the planners hoped that they would show a better understanding than they actually do for the idea behind family planning and would care more for professional progress than for a house full of children. In practice, however, not so many women are really interested in a professional career, and in addition many husbands do not like their wives to have careers. Those who are financially better off have enough servants for children not to disturb their careers, and since children are still a status symbol, women's education will not alter the situation unless society begins to accept other values.[17]

Family planning efforts among the urban population have not had the same success as in the countryside, even though in the towns a large number of Public Health Centres (PUSKEMAS) were put at the disposal of those wanting contraception. Certainly the centres are frequented by the lower income classes, but as soon as men or women achieve somewhat better social positions, they disdain queueing up for mass treatment. Middle-class people prefer to consult private physicians or clinics, and it is now planned to offer clinics, graded according to the cost of the treatment they provide. Townspeople are usually familiar with contraceptive methods and have a higher standard of education than village people, but this does not necessarily mean that their general attitude towards small families and children as a status symbol is more modern. Recently, the family planning programme admitted the necessity of concentrating on the urban population, since their response is generally lagging behind that of rural people.

Hitherto, target groups of the family planning programme have been the lower classes, especially in the countryside: farmers, daily-paid workers, craftsmen — groups which, surprisingly, accepted the new policy with particular enthusiasm. The experience of Indonesia shows that even in purely Muslim areas family planning can succeed without rapid industrialisation, and that it can happen before family incomes grow tangibly or education and health services change the social situation.[18]

The targets of the family planning policy (expressed in the percentage of couples of reproductive age [PUS] participating in any form of contraception) have been fixed repeatedly, then exceeded and revised upward; in fact, the positive response of the people approached has been stronger than anyone dared to expect. But the results vary greatly between the provinces. In the region of Jawa/Bali, for instance, less than 5% practised contraception before the programme started. By 1978, 28% of the women concerned had become acceptors.[19] By the end of 1982, the provincial differences became clearly visible: whereas in West Jawa no more than 30–35% of the couples concerned had become acceptors, around 65% had done so in the rest of the island and Bali.

The reason for West Jawa lagging behind is a shortage of funds, working facilities and personnel among a large population of difficult character in a disadvantageous geographical location. Moreover, a first attempt to introduce the pill was a failure. However, it seems that the visit of the Deputy Chief of BKKBN at the end of 1982 mobilised the province: more than 100,000 people were won over within one month, and the programme is now supported by all social groups including ulemas, students, people's leaders, housewives and members of various organisations.[20]

Family planning in Indonesia should not be seen merely as a reduction in the number of births but rather as a new style of living together — a transformation, particularly, of the village structure. Before the programme started, it was the rule that a married woman had to bear a child whenever she conceived, and this was often once a year. Thus early marriages resulted in women dying early from plain exhaustion as a reproductive unit, leaving motherless children and a husband who married again only to send his next wife to the same fate. Indonesian family planners maintain that women who give birth too early and too late may also get higher than normal incidences of child deformity, besides reducing their life expectancy. So the programme aims at 'spaced' births, or fewer births from each woman, giving mother and children a chance to live together for a longer time.[21] Fewer children means healthier children and mothers, with better educational and employment opportunities for the offspring and a higher per capita family income. This, of course, does away with the old standard of large numbers of children being a status symbol.

Reviewing the spectacular success of the Indonesian family planning policy, we still cannot hide the fact that — at least in Jawa — it will not solve the problem of over-population. Enthusiasts often overlook the size of the problem and the time-lag before even the best birth control programme can have results. Let us suppose that the programme had introduced the two-child family as an obligatory upper limit in 1971 (although such a measure would only be possible in a dictatorship), Jawa had about 76 million inhabitants and rather more than half, about 40 million, were in the under-20 age group. Thus, approximately 20 million couples would be entering the reproductive age in the coming twenty years, and this number would double during that time provided that the two-child family policy was strictly observed. This in turn would lead to an annual average increase of 2 million people in Jawa alone. Later, of course, the net increase would dwindle towards a balance between the birth and death rates; but this would not happen before 2020 (the World Bank, less optimistically, forecast 2115).

But even with the most improbable hypothesis of a two-child system, what would Jawa look like? A stable population would be arrived at with between 140 to 150 million people, or 1,000–1,100 persons per sq. km. — double the density of the Netherlands.[22] Disregarding the lack of space, how would these people be employed unless the whole island were transformed into a gigantic industrial landscape with labour-intensive factories? And this on the island with the best soils for agricultural use. Jawa would become the largest megalopolis on earth. This would have been the case had enforcement of the two-child family policy begun in 1971. But the policy was not enforced and Jawa's future development is now even more bleak. Even taking into account the success of family planning and the intention to strengthen it, this will not be enough to save the island's physical environment from destruction by excessive land use.

Transmigration

There is yet another possible means of saving Jawa and other over-populated islands from strangulation or suffocation by their populations. The existence of vast, sparsely inhabited areas in the Outer Islands suggests the need for a less unequal population distribution by organised — or at least supported — resettlement of farmers from overpopulated islands on 'empty' islands. In fact, such a policy has existed for a long time and is widely known by the Indonesian designation of *transmigrasi* (transmigration).

If we include officially organised intra-insular migration, and the resettlement of farmers to ease the burden on over-populated areas and to make use of hitherto neglected land reserves, we can trace back transmigration well into colonial times. When the Dutch conquered the island of Lombok in 1894, they quickly implemented such measures, and the word 'transmigration' was already in use then.[23] The organisation of inter–insular migration came somewhat later, around 1905.

The Dutch administrators had become aware of the growing population pressure on islands such as Jawa and the need for some action, but there was another reason for dealing with migration concepts. In those years, early in the twentieth century, foreign companies began to establish plantations on the Outer Islands, particularly Sumatera, for the production of exportable goods such as rubber, but also tea, tobacco, pepper and spices. Similarly, drilling for oil and tin mining were started. These enterprises soon suffered from a shortage of labour because the Outer Islands were only thinly populated and most of the inhabitants were farmers whose labour was largely absorbed by work in their own fields. The Dutch administration thus initiated a policy called *kolonisatie*, but occasionally also *transmigrasie*, meaning the resettlement of farmer or landless families from overcrowded Jawa to the Outer Islands. They envisaged that some of these people, after having completed the work on their own 1-ha. family holdings in the resettlement project, could easily be hired to work on nearby plantations.

The first area to be settled with a substantial number of migrant families, between 1905 and 1940, was Lampung province in southern Sumatera, opposite Jawa; new settlements were also established in other parts of the island and a few on Kalimantan and Sulawesi. By the end of 1940, before the Second World War had begun to have an impact on the colonial system, a total of more than 200,000 people had been resettled under the Dutch *kolonisatie* scheme.[24] After the war and the end of Dutch supremacy, the Indonesian government soon took up again the idea of redistributing the national population.

It is not our intention here to analyse and offer a detailed critique of the Indonesian transmigration policy, but rather to judge its impact on the environment and the natural potential of the country. The impact is felt in two directions: first in the reduction of population pressure on the densely populated islands such as Jawa, Madura, Bali and Lombok; and secondly, in the settling of a large number of families on land on the Outer Islands hitherto covered with natural vegetation, primarily forest.

By relieving the pressure on densely populated land, soils that are of marginal economic use could be returned under ideal conditions to afforestation. Thus, more protective forests could be used to fight erosion by acting as reservoirs of humidity, and so forth. By moving large numbers of people to neglected forest areas, arable soils could be taken

Indonesia is widely believed to be a hilly tropical paradise covered with dense rain forest. Some of the trees that grow in the rain forest are economically useful. There are irrigated rice fields in the valleys and the whole area is overlooked by volcanoes which occasionally boost soil fertility.

into agricultural production, providing food and income for the cultivators. Properly executed, such a policy could well lead to a better and more protective use of the nation's natural resources.

If the target is to reduce population pressure on Jawa, the transfer of a mere 200,000 people within 35 years is not particularly impressive.

Although the number of migrants has increased year by year, it has remained unsatisfactory in relation to population growth. For instance, the number of people who moved from Jawa in 1941 was less than one-twentieth of the island's population increase during the same year.

The Second World War put an end not only to the Dutch *kolonisatie* scheme but to the Dutch administration as well, and when the independent Republic of Indonesia was finally acknowledged in 1949, the national administration had already given consideration to a new resettlement policy. The scheme, that now became known exclusively under the Indonesian name of *transmigrasi*, foresaw the transfer of people from Jawa and Bali to the Outer Islands mainly to reduce the over-population. But, like most Third World countries at that time, Indonesia went through several years of trial and error, and thus the organisational framework of the transmigration scheme and to some extent the concept itself were frequently revised.[25]

Nevertheless a transmigration programme was implemented in 1950 — with results that compared badly with the unrealistic targets set by the administration. The first programme, conceived in 1947, aimed at moving no less than 31 million people within fifteen years, and when it was revised in 1951, there was a target set of more than 48 million to transmigrate within thirty-five years (1953-87). It was thought that 1 million would be moved during the first five years, 2 million during the second five years, 3 million during the next five years, and so on. The administration soon realised that this would never be practicable, and finally aimed at moving 100,000 in 1953; in fact, only 40,000 migrated. Between 1950 and 1969, less than 500,000 people left Jawa, Madura, Bali and Lombok under the auspices of government-supported transmigration programmes; 84% went to Sumatera, 10% to Kalimantan, 4% to Sulawesi, and the rest to other islands. In those years, however, the flow of individual migrants from the Outer Islands to Jawa and particularly to Jakarta was so strong that between 1950 and 1972 the net outflow from Jawa was little more than 46,000 migrants a year, a figure which grew to 120,000 in the period 1975-80.[26]

With the consolidation of the Indonesian administration and with growing experience, the number of migrants increased, but it was not till the early 1980s that it reached a substantial annual volume of more than 250,000 people. By 1980 the total number of transmigrants in one of the government schemes reached 1.1 million — according to official figures. Those who went on their own, the so-called 'spontaneous' transmigrants, bring the figure for those who have left their over-populated homeland so far during this century to about 2 million.[27] This

On the overcrowded island of Jawa, hoardings are erected to encourage landless farmers or farmers with small farms to register for transmigration to the Outer Islands.

is an insignificant figure when we remember that the aim is to ease the population pressure on the soils of Jawa, Bali and Lombok. The annual population increase in Jawa alone in 1980 amounted to no less than 1.8 million people. Even if we accept the 1980/1 figure of 278,263 official transmigrants, that represents no more than one-sixth of Jawa's population increase.

One of the aims of the transmigration scheme is to avoid further population growth on the overpopulated islands. In practice, this means transferring the surplus population to other islands. With an annual surplus of about 2 million, 5,500 people would have to be settled every day on one of the Outer Islands in order to balance the two figures. This is manifestly impossible.[28] The reason why none of the ambitious targets can ever be reached is not so much the lack of readiness of people to go but rather the difficulty of financing their transfer, settling them suitably, and offering them a better life than they have left behind. However, over the course of time the Indonesian authorities have gathered much useful experience and have learnt how the departments responsible for transmigration activities can cooperate. During the period of the 3rd Five-Year Plan (1979/80–1983/4), the Transmigration Ministry succeeded in settling 500,000 families on the Outer Islands. If we add another 156,000 families who migrated 'spontaneously', we reach a figure of over 2.5 million people leaving the overcrowded islands within five years.[29] During the 4th Five-Year Plan (1984–5/1989–90) the authorities intend to settle about 800,000 families on the Outer Islands, up to about 4 million people;[30] however the ministries in charge believe that about one-third of this figure will be counterbalanced by immigrants to Jawa and Jakarta.[31]

Here we come to the second target of the transmigration scheme: to bring about the better utilisation of the potential of the Outer Islands. Settling people in areas which are uninhabited or which have only a very small original population posed problems right from the start. The many reports dealing with the methods, achievements and failures of resettlement projects on Sumatera, Kalimantan, Sulawesi and other islands show that some of the problems recur constantly and that others are specific for certain groups of settlers or for particular areas. However, a major handicap was that in most cases the new areas were not properly selected and prepared so that the newcomers could make a decent living. More often than not, the land was surveyed in a rudimentary way, neglecting soil and water properties indispensable for a prosperous agricultural economy.[32]

Difficulties started with the selection of transmigrants in their home villages, since this depended on obtaining information about their age, health, professional ability and family status, and the number of children and pregnant women involved. On the other hand, the administration often could not assure the interested families which place they would go to, when they would depart, and whether they would continue to be with their neighbours. For this reason many families were reluctant to register as transmigrants. Others who had registered and sold their property had already spent their savings before they were asked to leave.

In the early stages, the new settlements were conceived exactly like Javanese villages and directed towards the wet-rice cultivation that people were used to, although the new area was often quite unsuited for this kind of cultivation. Usually, the settlers were promised that irrigation facilities would be available or at least would soon be under construction. Unfortunately, these promises were rarely kept and often more than ten years passed with no irrigation water becoming available. This meant that the settlers had to shift to rain-fed cultures, the soil fertility deteriorated, and they often had to leave the land because it could not sustain them.

The resettlement schemes also brought problems of an ethnic nature. In the early days, farmers were settled in a project as they arrived. Thus neighbours were often unable to communicate with each other because they had no common language. Later, whole villages were transferred to new areas with the result that these groups formed isolated cultural units in strange surroundings, and there was hardly any assimilation with the original population of the area. A further problem arising from the relations between the original, autochthonous population and the new transmigrants was that the former did not learn a more advanced agriculture from the latter as had been hoped; rather, it was often the case that the new settlers had to learn from the original inhabitants how to produce under local conditions. On the other hand, the original farmers sometimes envied the newcomers because of the official support they received in the form of preparing the resettlement areas, building schools, social centres and so forth.

The gravest difficulty, however, arose from the modest economic returns the settlers achieved. This was because of unsuitable soils and a lack of irrigation facilities, and inadequate technical know-how to cope with the new conditions. And difficulty in marketing surplus and cash crops made many farmers turn to a subsistence economy. The problem was aggravated by the absence of economic integration and of alternative possibilities for earning a livelihood.[33] So it is not surprising that a certain percentage of transmigrants looked for other chances by clearing virgin forest, wandering to the nearest town or even migrating back to their home island.

Thus, it seems that the second aim of the transmigration scheme — a better use of the potential of the Outer Islands — is far from having been achieved. The political consequences of this twofold failure also has a dual aspect. First, it is obvious that the transmigration scheme by itself can never solve the problem of overpopulation on the densely populated islands. Therefore, additional measures have to be taken to ease the man/land ratio. The most important is family planning, but, as we have seen, it takes time before results are achieved. Another way is to make the best possible use of the available soils of the Inner Islands without forgetting protective measures; to foster afforestation, soil conservation and the improvement of waste land; and everything has to be done to create income outside the agricultural sector in order to draw the labour force away from the limited soils.

Secondly, the transmigration policy has to get away from transferring farmers to new fields where they face a fight for survival. The idea that has been envisaged since 1982 is to incorporate transmigration into the regional development being carried out on the Outer Islands to achieve a planned development of their potential. Briefly, the idea is as follows. In order to use the natural potential of the island, farmers as well as non-farmers are needed. The creation of the necessary infrastructure such as róads, ports and social institutions requires manpower; so also does the exploitation of mineral and forest resources as well as the processing industry based on them. However, the idea is not to abandon the trans-migration of great numbers of farmers to places where they can continue to farm under better conditions; nor is it to replace this programme by the transfer of limited manpower for non-agricultural jobs. Although such ideas are discussed,[34] a number of important facts are overlooked.

First of all, the great majority of potential transmigrants are farmers with little or no land of their own; hence they are interested in remaining farmers and possessing enough land. If this chance is offered to them, a sufficient number of families will be ready to leave their homes and begin a new life in the Outer Islands. Once they are settled there, some of the family members will be ready to change to non-agricultural jobs, either during the slack time of the agricultural year or because a son or daughter wants to leave agriculture for some other livelihood (if this is not successful, they can easily return to their parent's farm). Thus enough hands would be available, in the course of time, for non-agricultural activities in the same area, and a part of the economically active popula-tion could gradually move from agriculture to industry, mining, ser-vices, public works and the like, without risking unemployment far from home.

The situation would be completely different if people were encour-aged to shift to the Outer Islands for work in building roads or dams, or in some of the new industries for which the outlook is not at all secure. At the end of a time-limited job, the worker would be forced to accept any available work or return home when there is no work at all. There would not be the slightest security for him, and it is doubtful if a migration programme of this kind would succeed.

No doubt, as long as Indonesia is not in a position to secure the basic food supply for its population, all available and suitable land should be used for food production. Also for this reason the old transmigration concept offers a sound basis. But there is only a future in it if planning goes one step further: to the use of arable land should be added the use of other potentials in each area. Surplus manpower in the agricultural sector could thus be absorbed for all-round development.

It cannot be said that the transmigration officials have been one-sided or inflexible. In fact, the transmigration policy consists of six patterns, of which, however, only two have so far been given the necessary attention:

food crop farming and estate farming. The others — animal husbandary; fish-pond farming and fishery; manufacturing industry; mining — cover only a negligible proportion of the transmigrants.[35] In order that more emphasis should be placed on them, the government urgently invited foreign and domestic companies to invest in transmigration areas.[36]

However, although large areas on the Outer Islands are so far uninhabited or very thinly populated, it cannot be denied that only a certain part of them are likely to be potentially productive. In 1981 it was estimated that there was still land suitable to accommodate a population of 12.8 million. But since on the Outer Islands themselves a growing population will require land for about 2.8 million, there will only be sufficient land to settle roughly 10 million transmigrants.[37] This does not necessarily contradict statements by government officials that Irian Jaya could accommodate half of Jawa's population and that those who were against transmigration to Irian Jaya were 'separatists and enemies of the government and had to be exterminated'.[38] Such officials ignore the fact that transmigration has financial, organisational, economic, ethnic and, not least, ecological limitations throughout the Outer Islands. Estimates of the area suited for the settlement of transmigrants range from a conservative 15–20 million ha. to an optimistic 68 million ha.[39] Fortunately, the ministry in charge continuously accumulates experience. One of the latest measures, the organisation of that ministry's own technical staff independent of other ministries, is likely to improve its efficiency.[40]

The original population of the Outer Islands, as well as that of the transmigrants is growing, and the arrival of new settlers quickly pushes the population density up in certain areas. The province of East Kalimantan, for instance, had only 500,000 inhabitants in 1960, yet in the next twenty years it grew by 140% to 1.2 million. The province of Lampung on Sumatera, a traditional target for transmigrants, had 1.7 million inhabitants in 1960 and within twenty years had grown by 170% to 4.6 million. A comparison of the 1961 and 1980 censuses clearly shows the influence of transmigration on the population figures and population density in the provinces (see Tables 1 and 2). Whereas the population density in East Kalimantan grew from 2 to 6 per sq. km. in twenty years, the corresponding figures for Lampung are 50 and 139 respectively.

The transmigrant families are not distributed evenly over the Outer Islands, but are concentrated in resettlement projects. In time they attract other family members and more transmigrants from the old home areas and their numbers naturally multiply. So it is not surprising that some of the older resettlement areas show population densities similar to those on the islands of origin, and that certain provinces have already been closed to new transmigrants, at least temporarily.[41]

Seen from an environmental viewpoint, transmigration means not only intrusion by man into virgin forest, with all the ecological consequences that follow from that, but also growing population pressure on the recently denuded and cultivated soil.

NOTES

1. IBRD, 1983, p. 210.
2. Dequin, 1978, p. 22.
3. Röll, 1979, p. 34.
4. Lutz in *Allg. Deutsches Sonntagsblatt*, 4 Oct. 1981.
5. *Buku Saku Statistik Indonesia* 1980/1, p. 34.
6. Snodgrass, 1979, p. 22, fn.
7. Lutz, loc.cit.
8. *Indonesia Times*, 5 July 1982.
9. Snodgrass, 1979.
10. *Indonesia Times*, 21 April 1982.
11. Snodgrass, 1979; Lutz, loc.cit.; *Japan Times*, 19 Nov. 1980.
12. 'Persuasion from peers', 1980.
13. *Indonesia Times*, 30 April 1983.
14. Lutz, loc.cit.
15. The Indonesian dailies frequently report astonishing measures to convince people of the necessity for family planning and to keep those who participate 'faithful'.
16. Lutz, loc.cit.
17. *Indonesia Times*, 15 Jan. 1982.
18. 'Persuasion from peers', 1980.
19. Snodgrass, 1979.
20. Satoto in *Indonesia Times*, 29 Jan. 1982.
21. Loc.cit., 4 Aug. 1981, and others.
22. Thijsse, 1975a, p. 2.
23. Röll and Leemann, 1982, pp. 135–40.
24. Hardjono, 1977, p. 19.
25. Loc.cit., pp. 22–35.
26. Arndt, 1983, p. 55.
27. MacAndrews, 1978.
28. Thijsse, 1975a, p. 3.
29. *Indonesia Times*, 20 Feb. 1984.
30. Loc.cit., 14 July 1983.
31. Loc.cit., 13 March 1984.
32. The Indonesian press freely reports such shortcomings from time to time.
33. Zimmermann, 1975a; Guiness, 1977.
34. Arndt and Sundrum, 1977.
35. *Indonesia Times*, 17 Oct. 1983.

36. Loc.cit., 30 Sept. 1983.
37. *Business News*, 27 May 1981.
38. *Indonesia Times*, 26 March 1983; reference is made to internal unrest in *Südostasien Aktuell*, Hamburg, 1984, p. 118.
39. Baharsjah, 1978a, pp. 74, 82.
40. *Indonesia Times*, 17 Oct. 1983.
41. Arndt, 1983, p. 70.

Part II. JAWA: LAND USE UNDER HIGH POPULATION PRESSURE

3

IRRESISTIBLE POPULATION GROWTH

More than 150 years ago, the General Commissioner of the Dutch colonial administration, L.P.J. du Bus de Gisignies, sketched a frightening picture of Jawa's future, but we can be sure that, except for a few who shared his sentiments, nobody was too disturbed by it. One day, he wrote, the population of Jawa will crowd the whole island as it does now in a few favoured areas, namely the fertile and irrigated river plains. Millions of tenants will have to live on fractions of a hectare growing nothing but rice. Their incomes will be no more than those of poor field labourers, just covering what is absolutely necessary for survival.[1]

He was not the first to draw attention to over-population and under-employment in Jawa. In 1802, Nederburgh, a Dutch colonial official, had already described Jawa as overcrowded and its population as unemployed.[2] In 1816 Engelhard, another Dutch administrator, stated that the population along Jawa's north-east coast was far too large for the area under cultivation in a given village or district.[3]

With a population of roughly 6 million in 1830, living on more than 130,000 sq. km. of land, largely covered with dense tropical forest and presumably with extensive fertile volcanic soils, nobody seems to have taken the warnings of Nederburgh, Engelhard and du Bus seriously. Certainly, wet-rice cultivation (*sawah*) is always accompanied by a high population density since the irrigated rice fields are in the immediate surroundings of the village homesteads of the farmers. And since this traditional form of agriculture is related directly to the farming family's needs, the cultivated area is small enough to be managed by the family members themselves. Once the village population increases, the holdings often become smaller and the cultivation methods more intensive. Consequently, the population density increases further but only in the cultivated area, and thus does not enable conclusions to be drawn about the total density of the island or even the republic.[4] But beyond a certain limit, the irrigated land cannot support more people and some of them become under-employed and finally unemployed unless they migrate to a

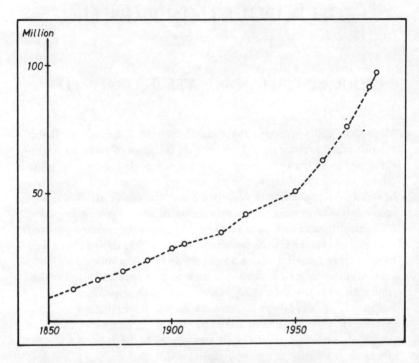

Fig. 4. GROWTH OF JAVANESE POPULATION (1850–1983)
Source: Based on Indonesian government statistics

new place where they can establish another *sawah* on unsettled land. Thus, densely populated areas spread gradually over the island raising its general population density.

Only a few decades after Du Bus' report, the trend he had envisaged became visible, and today we can see exactly what he predicted: Jawa, with an average of 690 inhabitants per sq. km., is one of the most densely populated regions of the world and has a still tangibly growing population, increasing by 2% a year. In 1980 Jawa and Madura had no less than 91.3 million inhabitants, with an absolute increase of 2.7 million every year.

However, the growth of Jawa's population came about slowly, and in the nineteenth century the Dutch colonial presence had still not affected the picture much; rather the incidence of crop failures, famines, floods, internal fights and petty wars kept the population increase low. Moreover the forced production of export crops (*cultuurstelsel*) offered labour opportunities and additonal income for the rural population, so that a relatively dense population could survive. Forced labour and the absence of men from their families also reduced fertility to some extent. It was only in the twentieth century that the Dutch colonial system began to have a tangible impact on population growth. The health services, hitherto accessible only to Europeans and to Indonesians who were in close contact with them, spread and the *Pax Neerlandica* stopped internal fighting. Flood control, irrigation and drainage were supported, and thus the danger of famine was reduced. Measures for the supply of drinking water and for sewage disposal reduced the incidence of water-borne diseases and, in consequence, the mortality rate.[5]

This trend was interrupted by the Second World War and the post-war guerrilla struggles, which caused casualities, disintegration of families, reduced health services, famine and a general economic decline and led to growing death rates and reduced birth rates. This changed during the 1950s when economic and social consolidation gained momentum; the food situation improved, public health efforts succeeded in fighting malaria and other diseases, infant mortality lessened dramatically and the birth rate remained high. In 1961, when the first post-war census was carried out, the population growth-rate for Jawa and Madura was found to be 2.24% per year, the highest ever measured or estimated in eighty years.

In the following twenty years, the Indonesian government realised the looming danger of an over-populated Jawa. To relieve the pressure of an ever-growing population there, efforts to encourage transmigration to the Outer Islands have been strengthened, as has the newly supported family planning policy. Having already dealt with these items we will recapitulate only the basic findings. Transmigration has quantitatively failed to achieve its aim of freeing Jawa from its surplus population. Against an annual population increase of 2.7 million per year, its total out-migration between 1961 and 1976 was only 1.2 million people; the same period saw half a million immigrants. Thus net out-migration was only 700,000 or roughly 47,000 persons per year.[6] Family planning has been successful but has not diminished Jawa's annual growth-rate which, according to the 1980 census, had decreased to 2.0%, that of Central Jawa to 1.7%, that of East Jawa and Madura to 1.5% and that of the special territory of Yogyakarta to 1.1%.[7] These are impressive figures, but the growth-rate of the Javanese population remains high.

Jawa in 1980 was twice as densely populated as the Netherlands, namely 690 as against 346 inhabitants per sq. km., and specialists calculate that by the year 2031 — and provided that a two-children-per-family system can be enforced — a population density of 1,300 per sq. km. of inhabitable land has to be envisaged, more than three times the Dutch level.[8]

Yet even today there are districts with such densities. In 1980, 868 people per sq. km. lived in the special territory of Yogyakarta, and in the province of Central Jawa the figure was 742 per sq. km., both above the island's average. Even in 1971, the district of Bantul near Yogyakarta counted 1,375 per sq. km. and the district of Klaten near Surakarta 1,486.[9] In Klaten the population had grown by 100% in less than 50 years with hardly any out-migration. Thus the best soils now show a density of up to 2,000 inhabitants per sq. km. and even dry farming on the slopes of the Merapi volcano sustains 500–1,200 people per sq. km.[10]

NOTES

1. Du Bus de Gisignies, 1827.
2. Quoted by van der Kroef, 1956, p. 742.
3. Loc.cit.
4. Loc.cit.
5. Lipscombe, 1972, p. 11.
6. Leiserson, 1980, p. 8.
7. *Neue Zürcher Zeitung*, 6 March 1981; *Business News*, 8 July 1981.
8. Thijsse, 1977, pp. 443–4.
9. Hardjono, 1977, p. 3.
10. Röll, 1971, p. 59.

4

LAND USE ON JAWA: THE GROWING DEMAND

Expansion and Types of Land Use

Human life on Jawa dates back to 300,000–500,000 BC or perhaps even earlier. This has been accepted opinion since the Dutch military physician E. Dubois discovered the remains of Java Man, later called *Pithecanthropus erectus*, in 1891-2 near the village of Trinil on the Solo river in East Jawa.[1] Other prehistoric discoveries followed, and by 1932 Oppenoorth had excavated many more such remains in the vicinity of Ngandong, also near the Solo river, dating back some 125,000 years.[2] Stone axes and traces of hoe cultivation dating back 10,000–12,000 years indicate that soil cultivation had begun. No doubt the cultivators were few in number and, living in extremely favourable natural surroundings, they probably caused hardly any environmental damage.

The situation of the archipelago between Asia and Australia and between the Indian and Pacific Oceans made it a typical area of transit, reached and crossed by waves of people who carried their material and spiritual cultures with them. At times there was a land bridge between the islands and between the islands and the continents, but in other periods the migrants had to use boats or rafts. After the last Ice Age the archipelago seems to have assumed its present shape.

Whereas in the beginning the newcomers were no more than hunters, fishermen and gatherers, their material culture improved as did that of the already settled peoples. And when, between 2,500 and 1,000 BC, migrants of the neolithic culture from southern China reached Indonesia, they found dry rice and millet in the fields, pigs and cattle bred for sacrificial purposes, pottery, pile-dwellings and horticulture. The soil was worked with hoes.[3] We know little of their land use technique, but can assume that they started growing mountain rice and millet on the fertile volcanic soils beyond the dense jungle, making use of the abundant rain. Following a nomadic pattern of land use and being very few in number, they had hardly any effect on the existing ecosystems.

The 'older-Indonesian' peoples were organised in tribal groups, practised a subsistence economy with dryland farming, and occasionally employed the technique of shifting cultivation, thus changing their place of residence from time to time. Their social relations were regulated by the so-called *adat* or prescriptive law which resolved all spiritual and material problems, and was at the same time flexible and varied considerably from one tribe and village to another.[4]

In the course of time, more developed immigrants settled along the coast and either mixed with their predecessors or drove them higher up into the mountains where they in turn brought new techniques and new ideas to the more backward hill-dwellers. They followed the river valleys and settled in the upper reaches and basins on the alluvial and volcanic soils; gradually, tribal and family communities evolved into village communities. Thus, the permanent village developed a more advanced form of land use, a spiritual and social culture that finally became the 'typical Indonesian village' (*desa*), largely isolated and self-reliant.[5]

Between AD 200 and 700, contacts with India and China intensified and greatly influenced life on Jawa. The main impact was the modification of the prevailing ethics and customs of the village people (*adat*). Hitherto *adat*, as part of the local culture, had been so tied to the family, the tribe or the *desa* that when laws were violated, any resulting punishment was carried out by the local society, which was so strong that any outside influence was absorbed and in consequence *adat* has to some extent survived up to the present. The most important influence, however, was the idea of the supremacy of certain persons or families outside the *desa*, religiously legitimised through Hindu and Buddhist teaching.[6]

When the Srivijaya empire of Sumatera came under strong Hindu and Buddhist influence from India, these ideas spread rapidly. But at the same time, a series of Hindu kingdoms emerged on Jawa, and, whereas Srivijaya was a maritime and trading power, the Javanese kingdoms were based on the control of irrigated rice lands.[7] It does not seem likely that Indian imperialism extended to Jawa, but rather that an existing élite imported Hindu teachings in order to justify its social position with religion.[8]

So whereas *adat* in the villages did not know a ruling class and a king who, though powerful, was relatively remote, Hinduism added this very component to the thinking of the people, who then had to accept a new way of living and working. With the development of the great central Javanese principalities which demanded tributes or taxes, people had to produce more than they needed for themselves. This led to the introduction or expansion of permanently irrigated rice fields; thus it is generally held that Hindu influence created or at least supported the new technique of land-use, *sawah*, which permitted an increase in food production and thus became a basis for population growth later.⁹

It is not quite clear to what extent *sawah* or wet-rice cultivation existed in the Javanese villages before Hindu influence arose. Dequin maintains that 'the great migrations during the thousands of years before Christ brought the *sawah* culture to Indonesia',¹⁰ and it may well be true that some time before Hinduism gained supremacy, *sawah* culture was known and practised in limited areas of Jawa that were particularly suited for this technique. But there were also other communities which did not know of kingship or any form of centralised power. These applied simple methods of land use with various forms of hoe culture and grew millet and mountain rice, but did not know of rice cultivation on irrigated fields using the plough — a feature of the *sawah* culture. This always changed with the arrival of new external influences, for instance through the Dongson or bronze culture, but particularly later through the Indian influence with ploughing, kingship and *sawah*.¹¹

It is almost certainly safe to say that Hindu influence greatly encouraged wet-rice cultivation, but it is also possible that Indian tutorship led at the same time to the shift from traditional millet to mountain rice as the staple food. This would mean that the indigenous mountain population changed from the more nomadic agriculture, growing rice on non-tilled soils, to shifting cultivation. Now upland rice was grown on dry fields, also called *huma* or *gogo*,¹² and even in those days this led to soil depletion. Some believe that large-scale soil depletion characterised the situation at the end of Hindu rule and that the introduction of Islam which followed in no way improved the deplorable system of agriculture then prevailing. According to Kievits, it was only after the power of the native kings had been severely curbed, that, thanks to Dutch influence, wet-rice culture was generally adopted.¹³ This does not, however, exclude the fact that during the period of Hindu rule *sawah* culture spread widely over Jawa.

Historically, the dominance of the Hindu kingdoms on Jawa proceeded from west to east in the course of about 1,000 years. Taruma in West Jawa was the Hindu centre from roughly the fifth to the seventh century, before the Hindu principalities shifted to Central Jawa (seventh to tenth centuries) with Mataram as the dominant state. From the tenth to the fifteenth century, the principalities of East Jawa were the most important, and of those Majapahit alone was dominant from 1293 to 1478. The first kingdom with integrated wet-rice systems as a basic economic institution was Mataram in Central Jawa, which covered the area between the volcanoes Subing, Sindoro, Merabu and Merapi and the

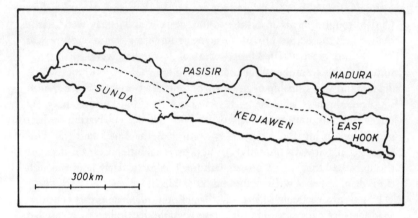

Fig. 5. THE CULTURAL REGIONS OF JAWA
Base map: Geertz, 1963, pp. 42–3.

south coast, including parts of the upper Solo river and the Serayu valley; at the height of its dominance it reached as far as East Jawa. Gradually wet-rice cultivation spread from here to the upper and middle Brantas river around Malang and Kediri, to the Madiun river around Ponogoro and to the Bondoyudo plain around Lumajang — all in East Jawa. Generally the wet-rice areas were concentrated in the inner reaches of the rivers flowing north and the upper basins of the rivers flowing south.[14]

It will prove useful in discussing the spread of wet-rice cultivation to divide Jawa into four regions,[15] of which *Kejawén* (meaning 'Jawa proper'), with Mataram, forms the core covering the southern part of the island's centre, with a good natural potential and about a quarter of the surface growing wet-rice. *Sunda* comprises the continuation towards the west, and the *East Hook* the extreme east. The whole northern coastal area from the Sunda Strait to Surabaya is called *Pasisir*, meaning strand or coastal line.

The expansion of wet-rice cultivation depended greatly on natural conditions. Around Sunda it never went far and is limited even now. By 1750 it had reached Sumedang and Tasikmalaya and then spread to the basins of Bandung and Bogor. Later, wet-rice began to be grown on the plateaux of Cianjur, Sukabumi and Garut. Today it is estimated that

15% of the total area of Sunda is under *sawah*. Although the Sunda high-lands have sufficient water and good drainage, the natural fertility of the soil is low and the relief is not very favourable to this crop. Wet-rice cultivation had already advanced into the East Hook by about 1200. However, since the east has a pronounced dry season, and river water is limited, there had to be the greatest possible reliance on rainwater. Modern irrigation did not arrive until the Dutch showed an interest early in the twentieth century, and today no more than 15% of the total area is estimated to be under *sawah*.

Although the Pasisir region has flat sedimentary soils and adequate water from the mountains, only the areas around the small estuaries became *sawahs* in the earlier days. Malaria was endemic and the open coast invited invaders who took possession of it many times in its history. The early technology was insufficient to manage the large deltas, and the only areas under wet-rice worthy of mention were those developed around Demak in the east and Cirebon in the centre. Only in modern times and using efficient irrigation and drainage techniques has the area under *sawah* spread to about 35% of the total area.[16]

It is evident that Indian-style cultural influence was limited to those areas where the rulers could establish control, and therefore territories under the rule of Hindu kings should not be regarded as monolithically organised entities. In fact the kings ruled the provinces through vassals — hence mainly over the better organised and densely populated *sawah* districts, where the plough was used and irrigation strictly organised.

There was close contact over the centuries between the *sawahs* and the ruling class, the latter requiring labour and military services from the peasants. Those who had been used to working for themselves were forced to devote their labour to a more powerful group. They had not only to construct palaces and temples but also to produce a surplus for their superiors. This socage in the form of construction work — to which the temples and other buildings so much admired today owe their existence — often compelled the peasants to neglect their own fields and led to a deterioration in their standard of living. It seems that many of them, tired of this enforced labour, migrated eastwards — a trend that became so prevalent that the rulers were forced to give up Central Jawa and follow them.[17] Thus this foreign cultural influence led to the estab-lishment of previously unknown social strata and to the initiation of a certain cultural and economic progress. This becomes clear when we consider people who lived in remote places far from *sawah* society, without knowledge of a king or any centralised organisation. They still earned their living in the most traditional way, worked dry fields, used the hoe, and produced millet and upland rice for their own consumption. Only those who cut trees, worked the timber and brought it to the markets maintained contact with the new world that was developing.

Islam slowly superseded Hindu rule on Jawa after taking over Majapahit in the early sixteenth century — long after becoming established in northern Sumatera — and a new *adat* arrangement developed. Its growing influence caused the Hindu caste system to recede, and three new social strata emerged: the masses, the Islamic teachers in schools and mosques, and the royal families with their administrators.[18] There was, however, no influence of any importance brought to bear on the economic structure in general and on land use in particular. The *sawah* area kept more or less in step with the growing population, but the *sawahs* themselves maintained their ecological equilibrium. With the intensification of the Dutch colonial system and its intervention in the autochthonous economy, things changed.

Dutch intentions in South-east Asia were not originally directed towards territorial expansion but were aimed at the accumulation of wealth through the spice trade; however, competition from Portuguese, Spanish and British merchants, often backed by armed forces, finally led to the creation of an enormous colonial empire, the 'Netherlands East Indies'. After 1602, Dutch trading interests were represented by the United East India Company (*Vereenigte Oostindische Compagnie* — V.O.C.) which was later given political and military powers; but after gaining control of Maluku in 1660, Sunda in 1684, the whole north coast of Jawa in 1743 and finally Kejawén in 1755, the Dutch traders were no longer satisfied with what was offered to them in the traditional trading ports. They now wanted to control and if possible to increase the production of the marketable goods which were so profitable for them.

In 1680 the V.O.C. started acquiring land, but instead of using it directly they leased it to local entrepreneurs, mainly Chinese, who immediately squeezed the utmost out of the land and people living on it and made a profit as middlemen. In their area of influence, the Dutch began to exercise political power — but without, for the time being, interfering with the traditional hierarchies and the structure of the village communities. They demanded tributes from the local rulers, who in turn commissioned intermediaries (*bupati*) to collect them from the villages without over-burdening the peasants in accordance with *adat* regulations. In the areas acquired by the Dutch such protective customs soon disappeared. People had to supply certain quantities of pepper, rice, yarn and other goods at fixed prices. When coffee plantations were established on the Preanger (Priangan) highlands south of Bandung, and Javanese producers voluntarily started clearing the forest and planting coffee trees, they were also forced to supply coffee, even after a decline in local coffee prices.[19] This encroachment into the natural forest, although on a moderate scale, marginal and unsystematic, was the beginning of deforestation carried out for Dutch commercial advantage rather than for the provision of food for a growing population.

At the end of 1799 the V.O.C. ceased to exist, and the newly established state of the Netherlands took over in Indonesia. But in the meantime, liberalism had entered into the thinking of Europeans, and new commercial ideas also spread among the administrators in the Netherlands Indies. Without considering the Javanese mentality or the legitimate interests of the local rulers and their people, they opened the door to ideas of free enterprise, mainly by introducing the money economy and forcing it upon the peasants with all its dubious consequences. There were two legal measures which brought a profound change to the economic as well as the ecological situation.

In 1830 the Dutch Governor-General van den Bosch replaced the land tax with a new Cultivation System (*cultuurstelsel*). Each peasant had to grow exportable crops on one-fifth of his land and sell them at fixed prices. Such crops were coffee and tea, tobacco and pepper, but also indigo, cinchona bark and cinnamon. One of the main crops, however, was sugar-cane, a plant that requires irrigation and thus competes with rice for space on the *sawahs*. Later, with the introduction of the Corporate Plantation System, the peasants were obliged to lease part of their irrigated land to Dutch sugar factories, which managed everything from planting to harvest. This led to serious changes in the social structure and the ecology of the Javanese village, particularly since it often happened that more than the prescribed one-fifth of the land was given over to export crops instead of rice, the basic food. Consequently, people tried to find a substitute by clearing jungle and growing maize and cassava on dry fields.

In 1870 a law prohibited the sale of land owned or used by Indonesians to non-Indonesians. At the same time, however, all other land was declared to be in the government domain. Any private party could lease land on a long-term basis (*erfpacht*) either from the government or from Indonesian landowners. These measures attracted private capital to Indonesia and encouraged the Dutch and other Europeans to migrate to the archipelago; numerous foreigners settled, acquired fertile land, and hired farmhands. They introduced new crops such as palm oil and rubber and cultivated them on estates of an economic size. From 1870 to 1890 exports doubled and imports quadrupled, but it did little to improve the welfare of the indigenous people.[20] This new system of land property led to large private holdings under Dutch and Chinese control, and by 1830 more than 1 million ha. belonged to 345 landlords.[21] Although the Cultivation System never involved more than 6% of the island's cultivated land and about a quarter of its population, and lasted only two decades, the new liberal regulations tended, as Geertz put it, 'to make Jawa a mammoth state plantation' with all the ecological and social consequences.[22]

The history of land use in Jawa is complicated since it depends not solely on the natural environment but also on processes of adaptation to foreign intruders over the centuries, or simply on political decisions. Therefore, West, Central and East Jawa as well as the island of Madura show different land use traditions which, to some extent, have been preserved to the present time. In Central Jawa, on rich volcanic soils and with sufficient rain, the people grow irrigated rice with hoe or plough (the *sawah* system); in fenced-in spaces around their houses they cultivate a mixture of fruit trees and vegetables, but also root crops (*pekarangan* or mixed garden system). Cassava, sweet potatoes, maize, peanuts, soya beans etc. are also grown either in fields which cannot be irrigated (*tegalan* system) or on dry *sawahs* (*polowijo* system).

The Sunda area in West Jawa, where there is plenty of rain but where the soils are of a different quality, traditionally employs what we may generalise as *ladang* systems. The dibble is used to grow upland rice on *swidden*, i.e. cleared plots of land (*ladang* system). Also a semi-permanent cultivation of dry rice (*tipar* system) is carried on using the dibble. After rice, the *swidden* may carry all kinds of annual crops in a secondary-crop system. Finally, perennial crops are generally grown around the village (*talun* system).

East Jawa belongs partly to the wet and partly to the dry area, and its soils also vary from good to poor. Originally *ladangs* together with fruit gardens (*kebunans*) prevailed, but from 1200 onwards 'Javanisation' started with the introduction of *sawahs*, which, however, often depended on rain. They were in many cases abandoned after continuous crop failures, mainly on the dry island of Madura.[23]

To understand the environmental consequences, we have to distin-

Sawah is the Indonesian term for irrigated and, if necessary, terraced fields where wet-rice is mainly grown. Irrigation can be provided from stored rainwater or via modern irrigation systems taking water from rivers or reservoirs.

guish between two basic types of land use prevailing in Jawa: *sawah* and *ladang*.

Sawah refers to irrigated and terraced fields where mainly wet-rice is grown. Irrigation can be provided from stored rainwater or via modern irrigation systems taking water from rivers or reservoirs. The rice terrace may be regarded as an integrated ecosystem when properly maintained. Experience over the centuries has shown that, once the system has been established, it can remain stable and supply food over an extremely long period without much investment. The growing of wet-rice is actually a remarkable form of cultivation because even with low natural soil fertility, food will grow if the water supply is dependable. The reason for this is that the water carries nutrients from the places of its origin, nitrogen is available through blue-green algae living in the paddies, the chemical and bacteriological decomposition of organic matter enriches water and soil, and aeration is secured by the gentle movement of the water.[24]

Obviously the establishment of *sawahs* requires level land, natural or artificial plains on gently sloping land, to hold back water. Therefore, *sawah* land has developed on the extended river terraces and flood plains as well as in the basins that often exist in the upper reaches of the river systems, and of course on the coastal plains. The crucial factor in wet-rice production is the proper supply of water both at the right time and in the right quantities, since the crop is extremely delicate in this respect. Thus two limiting factors have to be considered: first, areas where topographically it is possible for proper *sawahs* to be established have to be found; and secondly, proper management of *sawahs*, and substantial investment in soil preparation, water supply and good organisation of the water users is needed.[25]

Sawah cultivation has the advantage that once the necessary investments are made, food production can go on for centuries. We know of systems of terraces in Indonesia that are more than 1,000 years old, but we know of hardly any former working terraces which are no longer in production. The number of people who live on a piece of land can be increased if the labour input and imaginative use of land are intensified. And because the establishment of a wet-rice irrigation system requires an enormous amount of work, peasants prefer to intensify their efforts on existing *sawahs* instead of creating new ones elsewhere. Following the population growth and the colonial exploitation of the country, the Javanese *sawah* area has expanded from 1.3 million ha. in 1833, to 3.5 million ha. in 1957 and to more than 4.5 million ha. today. At the same time the *sawahs* have absorbed a great number of additional people. In places with very favourable conditions we find a rural population density of about 2,000 per sq. km. This occurs in the Adiwerna district on the northern coastal plain between Tegal and the Gunung Slamet. In this region, water rich in nutrients comes from the upper volcanic reaches to irrigate *sawahs* on alluvial soils, and thus maintains yields at a fairly high level.[26]

For a long time the peasantry of Jawa has been able to absorb the growing population by putting more and more labour into the *sawahs* and sharing the production among more and more mouths. This led to more people working on the land than was necessary for the production of the harvest. By the 1950s, it was estimated that about a quarter of the labour force in agriculture could have been withdrawn without lowering the output of the *sawahs*.[27]

We now turn to *ladang*, which is usually translated as 'shifting cultivation'. Some authors feel this to be a *scientific* interpretation, but in the Indonesian language it simply means a 'dry' or 'non-irrigated' field.[28] We should therefore distinguish between shifting cultivation (*ladang*) and dry farming (*tegalan*). Since we shall analyse shifting cultivation in more detail when dealing with the Outer Islands, only a rough explanation need be given here.

Shifting cultivation is a method of using forest soil by cutting a plot free of trees and burning it. The ground thus cleared will be cultivated for one or two harvests, until the yields drop; depending on local conditions, this may happen even after one harvest. The forest is then allowed to regenerate for ten to twenty years and to accumulate nutrients in the exhausted topsoil. The peasant then shifts to another place to start clearing the jungle again. Thus a great deal of land is required for a family to survive, but such cultivation can be practised over many generations provided enough fallow time is allowed — the longer the fallow period, the more complete is the regeneration of the forest.[29] But as soon as the population pressure increases, the system collapses, because the fallow period becomes shorter, the yields decrease, and finally the ground is so exhausted that not even new bushes can be grown. Then, *Imperata cylindrica*, a useless, strong-rooted grass, springs up, covers the area and makes the land useless for cultivation.

Since *ladang* in the sense of shifting cultivation concentrates on soils under tropical evergreen rain forest with closed-cover and without a dry season (as is prevalent on most of the Outer Islands), only the extreme western and south-western parts of Jawa (e.g. the Preanger highlands) offer the necessary environment. In fact, the growing population of this island has already shortened the fallow period in a dangerous way. In 1935, for instance, there were in the Banten (Bantam) area only 2–5 years of fallow after a single year of cultivation. Accordingly, the yields of mountain rice remained between 700 and 875 kg./ha.[30] But shifting cultivation is no longer of much importance on Jawa because the population density has grown far too high. Instead dryland cultivation prevails wherever irrigation is not possible, on hills and high plateaux, and the farther east we go in Jawa so much the more do deciduous or monsoon

Ladang (usually translated as shifting cultivation) is a method of using forest soil by cutting a plot free of trees and burning it over. Ground cleared in this way will be cultivated for one or two harvests until the yield drops.

forests take over, with the dry season becoming extended and conditions for shifting cultivation worsening.

The colonial export policy, together with a growing population, worked strongly in favour of dry farming. Since the promotion of colonial export products started early in the eighteenth century with coffee growing on the Priangan highlands, and particularly since Dutch planters and investors became active after 1870, plantations have covered an ever-growing area of the uplands; and since part of the *sawahs* was also occupied by export crops, the growing population — *sawah* as well as *ladang* farmers — had to switch to permanent dry farming (*tegalan*) to grow their food.[31] To understand this, we have to see to what extent export crops influenced the subsistence economy of the Javanese village, and at the same time see how it influenced the ecological equilibrium that had been affected only marginally till then.

Among the export cultures, there were annual crops such as sugar, indigo, tobacco and the like which could be grown in rotation with wet-rice on the *sawahs*. But sugar required irrigation and an environment similar to that of wet-rice; so it had to be grown on *sawah* land in competition with rice, and additional labour had to be hired in the appropriate season. Since the market price of this crop was high and middlemen often encouraged or even forced the peasants to devote more than the prescribed one-fifth of their land to it, occasional shortages in rice resulted, together with periods of hunger.

Other export crops such as coffee, tea, pepper, cochineal and cinnamon did not require irrigation and could thus be grown on 'waste land', land well-suited to dry land cultivation but not yet used for it. After coffee growing had started in south-west Jawa, it slowly spread over all the highlands right to the East Hook, replacing the natural forest with plantations. Such development often led to serious destruction as can be seen in the Gunung Sewu (Southern mountains), where a once sound forest was cleared to grow coffee, but the ground was abandoned after the soil proved to be unsuitable. Now, this karst area is one of the poorest districts of Jawa, where people eke out the barest living on rainfed agriculture. Plantation production in the uplands was not directly linked to the peasants' subsistence production. On the contrary, forest clearings to establish coffee plantations were often cut in sparsely inhabited highlands where there was a shortage of farm labour. In contrast to sugar plantations, coffee needs skilled and constant care; this offered a chance of permanent employment to landless or jobless villagers who settled in the new coffee plantation areas. This process led not only to a shift in population but also to a tangible impact on the forest cover of the Javanese highlands. Interestingly, it was the decline of soil fertility in the coffee plantations and consequently the fall in production which first gave rise to concern over soil problems, as can be seen from the studies by Fromberg in 1858 and Holle in 1866.[32]

The introduction and expansion of exportable crops on dry land, as well as on parts of the *sawahs*, together with the overpopulation of these *sawahs*, caused many peasants to diversify into dry land farming. They abandoned the wet-rice tradition and switched to maize and cassava and

occasionally upland rice, encroaching on the remaining forest and greatly
disturbing the ecosystem of an ever-shrinking surface. The acquisition of
land for dry farming in the remaining forest gained new impetus with
Indonesia's independence; after the exodus of the Dutch, their plantations, which had produced export crops since 1870, were occupied by
landless peasants. They established, over an area of about 80,000 ha.,
family plots of 400 to 1,500 sq. m., which were often too small to maintain a family and sometimes provided only sufficient space for a house
and garden.[33]

A growing population brings with it a growing demand for food and
other agricultural products, and this in turn requires either more land or
more intensive land use. The Dutch, during their rule, insisted strongly
on the expansion of *sawah* land wherever possible and on the permanent
use of dry fields where irrigation was out of the question. They did
everything to eradicate the Sundanese habit of *ladang* cultivation, and
tried to turn people into sedentary permanent-field (*tegalan*) tillers.[34]
However, the growing of export products finally made the production
of basic food for home consumption a problem. Since the end of the
nineteenth century, according to Pelzer, 'this has developed into a
discouraging if not nightmarish race between population growth and
the intensification of agriculture to raise yields. As a result, Jawa's
smallholders can contribute less and less to the export sector of the
economy, and the island depends more and more upon support by Outer
Indonesia in the form of food surpluses and foreign exchange to be used
for the purchase of food imports.'[35]

Expansion of Cultivated Land

Population growth usually results in a growing demand for food, wood
and other raw materials and for living space — a tendency further
exacerbated by government interest (Dutch as well as Indonesian) in
strengthening the export economy, based as it is on agricultural products. In Jawa, consequently, cultivated land spread all over the island
and du Bus's vision of 1827 has become reality: vast parts are now as
densely settled as only certain limited areas were 150 years ago.

After 1900, rice production on Jawa could not keep pace with population growth and consequently rice intake per capita dropped. To counteract this development, people took various measures to prevent a general
famine: they grew secondary crops together with the rice, and dry-land
farming spread rapidly to produce sorghum, cassava and other drought-

resistant crops. After 1930, individual farmers took over communal grazing grounds. Almost all land was taken under cultivation, and even on the bunds of rice fields and terraces food or fodder crops grew and flood-prone areas were made suitable for cropping by means of the camberbed system (where slightly elevated dykes keep the plants above flood level).[36] However, most of the expansion of the cultivated area took place at the expense of the forest. On Jawa and Bali, almost all suitable land came under cultivation, in some places exceeding the ecologically acceptable limit.[37]

The expansion of land under irrigated rice (*sawah*) is a good indicator of how a growing population will produce its basic food in the most productive way on an increased surface. For 1833 the *sawah* areas were estimated at 1.3 million ha.; by 1916, they had increased to 2.6 million and by 1970 to about 4.0 million ha. Since then, the area has stabilised between 4.0 and 4.5 million ha., indicating that there will be little further substantial growth of *sawah* land.

Our knowledge of the area under upland rice in the early days is limited, although it may have played an important part before it was superseded by the *sawah* technique; it regained importance when the *sawahs* became over-populated and dry-farming spread. In the twentieth century, the area under upland rice has fluctuated between 250,000 and 400,000 ha., but at present there is a downward trend. Still, rice accounts for about 57% of all food crops, and although the absolute area under cultivation may stagnate or decrease, the yields of both *sawah* and *tegala* rice grow by roughly 3% a year. This is the result of government efforts to intensify rice production and create self-sufficiency, and although targets are not generally attained, the results of certain programmes such as BIMAS and INMAS are encouraging.[38]

As more and more people had to make their living from dry-land farming the importance of maize and tubers as staple foods grew. 70% of all maize in Indonesia is now grown on Jawa (the area fluctuates slightly, 2.3 million ha. in 1968; 1.4 million ha. in 1975; 2.1 million ha. in 1983). The area under cassava, 75% of which is grown on the island, shrank from 1.2 million ha. in 1968 to 0.8 million ha. in 1983, but both crops attained increasing yields. This is also the case with sweet potatoes which have shown a large decrease in area from 247,000 ha. in 1968 to 117,000 in 1983. Peanuts and soyabeans, of which the lion's share is grown on Jawa (about 80%), are likewise constant in area and production, although there is a certain increase in yields. One could conclude that food production in Jawa takes place on about 4.2 million ha. of irrigated land and on about 3.6 million ha. of dry-land.

Non-food crops grown on Jawa show a very uneven development. Some of them, e.g. coffee and sugar, reach far back into colonial times, when they played a far more important part in the export economy than they do today.[39] But even in colonial times, the area under cultivation grew substantially as did the yields — partly, however, at the expense of local food production. Apart from sugar, indigo and tobacco, which compete with rice for irrigated land, most of them are dry-land cultures which contributed to the encroachment into the forest.

Coconuts, a typical smallholder crop, show a steady increase in area on Jawa (516,000 ha. in 1968; 701,000 ha. in 1975; 928,000 ha. in 1983), whereas rubber, a typical estate culture, with smallholders playing only a secondary part in Jawa, shows a decrease in cultivated area (278,000 ha. in 1968; 205,000 ha. in 1983). In 1983 there were about 45,000 ha. of coffee plantations in Jawa, the bulk in East Jawa, in addition to another 100,000 ha. of smallholder plantations, the latter figure having increased recently. Thus there were 107,000 ha. under coffee in 1968 and 145,000 ha. to date.

The production of sugar in Jawa (which can be traced back to 1635) is largely in the hands of estates, with an annual growth rate of 5.2% for the area under cultivation (about 230,000 ha. in 1983) and of 7.3% in output. A certain proportion is cultivated by smallholders, particularly in East Jawa. Tea, again a typical estate crop, is concentrated in West Jawa, but although the area under cultivation remains rather constant (about 48,000 ha. in 1983), the output has increased noticeably. The smallholders of Jawa cultivate another 48,000 ha. of tea plantations, but their production is rapidly decreasing. Although most of the tobacco estates are in Jawa, smallholders cultivate by far the largest area; the total area under tobacco fluctuates between 176,000 and 230,000 ha. (1975–83). Among the recently propagated perennial crops, cloves have gained a leading position as a crop for the smallholder; the clove-growing area in Jawa, about 32% of the national area under cloves, increases annually by no less than 28% (68,000 ha. in 1975).

A rough analysis of the development of the area under cultivation, and of the yields and the production of the main crops within the last few years enables us to draw the following conclusions:

— The expansion of cultivated areas has come to a halt, and in some cases it has even declined, particularly in dry-field production. This suggests that Jawa's capacity to cultivate more ground is exhausted or nearing exhaustion.

— There has been an intensification of agricultural production in yield per ha.; this is the result of growing inputs of labour per unit as well as of more technology such as irrigation, chemical fertiliser and the like. Of all fertiliser used in Indonesia, 70–85% was used on the fields of Jawa, and of this, 96% on food crops, particularly wet-land rice and crops subject to intensification programmes. However, a comparison of the removal rate of nutrients by plants shows that fertilisation — though substantially increased in the last few years — has not yet replaced the nutrient losses. Hence there is still quite a production reserve as long as expensive inputs are made.[40]

— If we consider population growth, the development of the land under cultivation and per capita production of food and the costs involved, we see evidence of a trend towards intensification of food production. Ecologically, a further expansion of the area under production cannot be tolerated. On the other hand, an intensification of production on the most suitable soils results in growing costs which can only be covered when producers' incomes increase proportionately. So far, only wet-land rice shows an increasing per capita production and a corresponding increase in farmers' incomes.

Because of the ever-growing population that has to be fed, Jawa's agriculture is caught in a dilemma between the ecological limits of expansion and the cost-limits of intensification. Barbara Schiller summarises the situation thus:

Jawa's population continues to grow, and its natural wealth of soil, water and vegetation continues to be eroded and polluted. Oil and other raw materials will be in increasingly short supply. New technology may provide some respite, but it is hard to see the final solution in a '*green revolution*'. *Pressure of people on resources* is a vital consideration in Jawa. However, it is not the only one, nor is it enough to solve the population problem. The increasing *pressure of people on people*, depriving the poor of access to resources, is a central problem in Jawa's — and Indonesia's — future.[41]

So it seems that Jawa has to choose between a further deterioration of the environment, preserving as far as is possible the social structure of a smallholders' society, or protecting the environment at the expense of this structure and thus fostering a society composed of a minority of large landholders and a majority of landless people; this, of course, is if the population pressure cannot be spectacularly reduced.

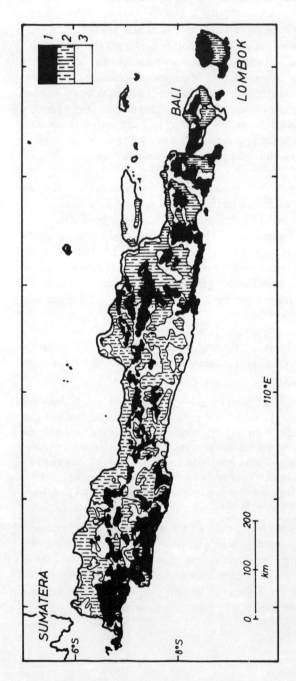

Fig. 6. LAND USE IN JAWA

1 = Forest, woods, mangroves etc.
2 = Rice fields (*sawah*)
3 = Arable land

Source: Compiled from various sources by the author

Table 3. LAND USE ON JAWA, 1980

Type of land use	Area (1,000 ha)
House compounds and surrounding areas	1,554
Bareland, garden	2,658
Shifting cultivation	264
Steppe, pasture	78
Swamps, uncultivated	47
Dyke	129
Water ponds	37
Preliminarily not used	82
Planted woodland	268
State forests	2,124
Estates	603
Irrigated land	3,491
Others	838
Total	12,173

Source: Statistical Yearbook of Indonesia, 1982, pp. 184–5.

The Social Impact

Without idealising Javanese village life in pre-colonial times, it can be said that for a long time it was characterised by economic and social subsistence regulated by a more or less stable village (*desa*) structure under customary (*adat*) law. In these conditions, social obligations ranked above economic interests. The peasant was, therefore, not necessarily inclined to maximise his income, and the increasing population, together with the shrinking *sawah* area per head, did not in any case give him much of a chance to advance economically. A commercialised agriculture did not fit into this picture or into the prevailing mentality. The peasant's only task was to survive, together with his growing offspring, on an ever-shrinking plot. He achieved this by investing more and more labour in the remaining land.[42] The village structure gave an income to everybody, both farmers with a very low level of mechanisation (so that there was enough work for the people), and those with a specialised skill such as butchers, millers, noodle-makers and sugar-palm tappers. In addition, many seasonal types of work gave a chance to women and children to supplement their families' income.[43]

In the course of time and long before the economic impact of colonialism was felt, this idyllic picture changed. With the rise of a *desa* aristocracy, certain families — often descended from those who had founded the village — acquired a leading position, interpreting the *adat* autocratically and thus accumulating power, land and wealth.[44] For a long time, however, an intricate network of patron-client relationships between the families of a village provided everyone with an income in the village economy, thus preventing outright starvation. Certainly, some scholars regard this as a thinly camouflaged system of exploitation, but it saved the disinherited from becoming completely redundant in the village economy.[45]

Things changed dramatically after colonial economic interests interfered with the social and economic structure of the village. From the nineteenth century on, the colonial system penetrated into the Javanese rural economy. *Sawah* land as well as dry land in the hills was used for colonial crops, and the growing commercialisation and monetarisation integrated the village into the world market with all its new dependencies. This created the basis for many of the problems which are still prevalent today in Jawa and in Indonesia generally.[46]

Sugar-cane competed with wet-rice, and thus reduced the *sawah* area available for food production to subsistence levels and led to periodic food shortages. The growing imports of cheap industrial goods paralysed the local manual labouring class and thereby systematically destroyed the social equilibrium in the Javanese village and severely threatened its ability to subsist. The plantations emerging in the uplands were short of labour and thus offered an income to villagers with too little land or no land at all, and although these privately-owned estates upset the traditional system and created considerable discontent among the defenders of the traditional methods of social and economic survival, so many people offered themselves as farm-hands that the owners had ample opportunity to exploit them. Thus unrest emerged in this sector too, and since the traditional leadership was unable to cope with these novel problems, radical groups often spurred the discontented to take action. By the end of the nineteenth century, the peasants' chances of survival had declined so badly in many areas that unrest and rebellion flared up: e.g. the Ciomas rebellion near Bogor of 1886 and the Banten revolt of 1888, both in West Jawa.[47]

In addition, the increasing population pressure on the remaining cultivable land led to a number of changes in ancient structures. Those who practised shifting cultivation (*ladang*) had to become sedentary and switch to dry farming (*tegalan*) since there was no longer enough land to maintain a proper bush fallow; intensely cultivated backyard gardens increased food outputs, but often the upland was used indiscriminately, thus leading to serious erosion; in irrigable areas the *sawahs* were expanded by reshaping the land with improved water and land use, but the suitable area was limited, and communal land was often taken over by private parties for economic exploitation. Certainly, rice production fields were intensively cultivated, but those which yielded two harvests a year and perhaps an additional secondary crop were still limited.[48]

The Dutch, finally acknowledging their responsibility for the people of Indonesia, in 1901 instituted what they called an 'ethical policy' which later led to the promotion of modern irrigation systems and the construction of schools and health centres.[49] But since the politicians continued to think in terms of commercialisation and world market prices, colonialism remained for the rural masses what it had always been: a threat to their survival. This was especially so after 1930, when all arable land in Jawa was taken under cultivation. Various riots and disturbances accompanied the decline of the peasants' situation between 1912 and 1919. During the Second World War the Indonesian economy was cut off from the world market and the food supply crises eased, but after 1965, when the new strategy of the 'green revolution' was initiated, the old problems reappeared in an even more serious form because of the steep population growth which had taken place in the meantime.[50]

The 'green revolution' involved a new rice technology, and this decisively changed the rural structure. This technology is based on specially bred varieties that give high yields compared with native ones, and are thus generally called 'high-yielding varieties' (HYV). This term, however, is somewhat misleading, because to produce extremely high yields these seeds require enormous inputs in the form of fertilisers, irrigation, plant protection and additional labour. If this is available, they may well respond with super yields — hence the more correct name 'high-responding varieties' (HRV) has been proposed. Nobody doubts that such varieties have led to substantial production increases in all parts of the world, and that many more jobs have been created — directly in the fields and indirectly in the supply sector which provides the inputs. It is also true that the 'green revolution' has bought us some time to prepare for the nutritional future of mankind. But unfortunately the social impact and technology of the new varieties have caused catastrophic changes in the producing villages.

This is not the place to discuss at length the 'green revolution', which is already the subject of innumerable studies. However, we must understand that the rural masses are less and less interested in environmental problems because they are increasingly being deprived of their subsistence basis. The introduction of the new technology required, first of all, considerable investment by the peasants, which only few of them could afford. The rest were unable to benefit from it, and those who tried and failed usually forfeited their property to village moneylenders in the absence of any formal credit system. In the densely settled areas of Jawa, only 10–20% of the peasants were able to participate in the government's new programmes; they were the ones with access to formal credit which gave them the chance to accumulate money and acquire fields from their less fortunate fellow villagers, so that more and more land came into fewer hands. Polarisation in village society became ever more pronounced, and the smallholders' traditional production was further endangered. Moreover, there is evidence that there were also educational and psychological barriers which prevented many of the farmers from participating in the new technologies.[51]

Furthermore, many traditional activities which had formerly given women and children a certain income during the season were lost: transplanting, weeding, harvesting, rice pounding and so forth. Even moderate mechanisation (e.g. the replacement of the traditional rice-cutting knife [*ani-ani*] by the sickle), together with the loss of land suffered by former small farmers, had severe socio-economic consequences: there was a decrease in the amount of seasonal or 'side' work available, while the number of people competing for such work increased because of the continuous loss of property.[52] This growing supply of available hands even led to a substantial reduction in animal traction and its replacement by human labour — naturally, at the expense of crop productivity.[53] In the traditional village, the whole population had participated in the harvest and thus gained part of their living; with the new rice technology the harvest is often sold while still standing in the field and the buyer hires limited labour (*tebasan* system).[54]

The impact of the new technology policy has been described thus: 'The history of the "green revolution" means that when a minority of the rural population succeed in becoming modern farmers, the success is theirs alone. For the majority it means a growing deterioration of their standard of living.'[55] Also:

A perusal of available case studies of agrarian and other social-economic developments in rural Java would point out the following changes in recent years; unequal distribution of the direct and indirect benefits of new biological and chemical technologies in rice production; new technologies in cultivation, weeding, harvesting and processing which cut costs for the larger farmer but reduce the employment and income opportunities of labourers; more frequent harvest failures resulting from the new varieties' vulnerability to drought, flood and particularly to pests, which have affected the income of small farmers more seriously than those of large farmers; declining real agricultural wages; unequal access to agricultural and other forms of subsidised government credit, while informal interest rates remain high for small farmers and the landless; unequal access to other government services: differential impact of inflation on large farmers compared to small farmers/labourers; shifts in the market system with larger traders taking over the role of small traders in the bulking process of rural produce; increasing landlessness and an acceleration in the purchase of agricultural land by wealthy villagers and the urban élites; decline of many traditional labour-intensive handicrafts and home industries in competition with more capital-intensive substitute products.[56]

Various scientific efforts have been made to analyse and to ease the lot of the rural masses, but it is clear that political decisions and actions are required which are unlikely to be welcomed by the well-to-do. So far, no government has been strong enough or sufficiently highly motivated to take the necessary steps. J.H. Boeke, who described the 'dualistic economy' of Indonesia (the co-existence of a modern, world-market-oriented economy alongside a more or less self-sufficient subsistence economy in the Javanese village), did not see how the two parts could be merged. The main reasons he gives are the rapid overcrowding of villages on the one hand, and the impossibility of absorbing this surplus population into industry on the other. As an alternative he suggests keeping the village traditional and self-sufficient in the framework of a basic subsistence economy: 'It is the programme professed and embodied by Gandhi who has formulated it in five words: plain living and high thinking.'[57]

No wonder that other writers strongly oppose the ideas of the 'simple life'. Fruin, for instance, rightly criticises a number of contradictions in Boeke's concept, and proposes solutions which hardly go beyond the absorption of surplus manpower from the villages by the industrialised towns or the Outer Islands. A modest degree of prosperity would eventually lead to a more rational way of life with reduced population growth.[58] He carries some conviction when he writes: 'The population of Java has been banished for good from the "paradise" of a village economy with practically no dualistic money transactions. How could a situation ever be attained in which the man of the *desa* would go back to making his own tools and clothing, replacing kerosene by vegetable oil he himself has refined, smoking cigarettes only from his own tobacco, and never travelling any more?. . . How are the traders and the money-lenders to be kept out of the *desa*?'[59]

Yet a remarkable development can be observed in the Javanese villages. Everybody thought that in the course of modernisation the subsistence economy would disappear. Yet, for the mass of poor people in the villages (and to some extent even in the towns) its importance is growing in the form of 'survival production'. Peasants with small lots see that they cannot keep abreast of market requirements, and often take refuge in production for their own subsistence on their remaining parcel of land. Similarly, trades and crafts have reappeared in the village, because many can no longer afford to buy imported goods. This new petty production of articles for daily use is no longer the traditional handicraft but rather production for survival.[60] It seems that Boeke's ideas are being realised, not as the result of a purposive policy for the betterment of the rural poor, but as a result of spreading misery and the determination of the poor to survive.

Care is needed in attempting to depict the social consequences of colonial penetration for the Javanese village. Geertz, whose studies centre on villages in East Jawa, finds that the reaction of the peasants was 'involution' and 'shared poverty':

Tenure systems grew more intricate; tenancy relationships more complicated; co-operative labour arrangements more complex — all in an effort to provide everyone with some niche, however small, in the overall system. [. . .] Under the pressure of increasing numbers and limited resources, Javanese village society did not bifurcate . . . into a group of large landlords and a group of oppressed near-serfs. Rather it maintained a comparatively high degree of social and economic homogeneity by dividing the economic pie into a steadily increasing number of minute pieces, a process [which I call] 'shared poverty'.[61]

Obviously other scholars in those days shared this opinion: '. . . . thanks to various measures of the Dutch colonial administration, there are hardly any big landowners in Indonesia and there is, in contrast to nearly all parts of South-east Asia, no landless rural population.'[62] However, we must regard as a 'big landowner' any man who owns or controls an area that is substantially greater than the average holding of 0.5 ha. Thus in view of the population pressure in the village and the shortage of cultivable land (particularly sawah), we have to include owners of as little as 2–3 ha. in this group.[63]

Yet there are other analyses proving the opposite. In several areas around Bandung there was a rapid concentration of land in the hands of absentees or members of the village élite; half the arable land fell into their hands, and indebtedness led to the feudal servitude of those who lost their property: 'It is indisputable that a polarisation of classes based on the widening distinction between the land-owning and the landless is . . . a social dynamic of major importance in contemporary Indonesia.'[64] An intensive field study in the village of Cibodas, near Bandung, showed about 8% belonging to the 'commanding' elements and the remaining 92% to the 'serving' elements — smallholders, owners of backyards and the landless.[65] This is certainly an extreme example, but the general tendency pointed out in these studies is manifest in many parts of Jawa.

These few examples show that any generalisation is risky, since some areas are more tradition-bound than others, and the impact of the colonial economy, monetarisation and modernisation did not affect the whole island equally. Moreover, the colonial impact alone cannot be blamed for the structural changes in the Javanese village since there is evidence that after independence the *status quo* was disrupted in many ways, even in places where tradition and harmony seemed strong. Certainly, the history of Jawa shows quite a number of upheavals since the second half of the nineteenth century, and social harmony in the villages seems to proceed mainly from the attitudes of the lower strata. These people have often given up hope of ever improving their lot; they are passive, take little interest in community problems and, when decisions are required, often follow their patrons rather than those from outside who claim to have their interests at heart.[66]

This should not suggest that the revolutionary potential is lost. But resistance has hitherto come from those who feel that any changes would curtail their privileges. When, following the land reform of 1960, the first attempts were made to redistribute land above 5 ha. and to humanise the payment of rent, a massacre took place in 1965 (in the guise of an anti-Communist pogrom) which cost the lives of around 1 million poor farmers, while representatives of the big landowners, radical Islamic youth and the army did everything in their power to block a policy of development in rural society.[67] Unfortunately, the problem continues, and the government has hitherto been able to do little to solve it. An analysis made by the Ford Foundation as early as 1969 may be cited, with slight modification, to summarise the present situation:

The issue of land reform in Java has proved to be particularly attractive for exploitation [by any social movement]. Nearly 90% of the villagers in Central and Southern Java do not own the land they cultivate. Peasants work on leased land, retaining only 40% of the harvest, the rest constituting rent payment. Although a reform of this situation has been proposed for many years, it has never been carried out. Neither has the Sukarno government taken up the issue in earnest.[68]

Instead everything in this direction was blamed as 'communistic' — the easiest way for the ruling class to defend their interests. And while nothing is done, the population continues to grow, the cultivated area per head shrinks further, incomes decline and fragmentation and indebtedness lead to ever more landless families being forced to sell their land in order to pay their debts. Regrettably more and more unsuitable land is turned over to cultivation and therefore is bound to be lost in the near future.[69]

NOTES

1. Dubois, 1894.
2. Oppenoorth, 1932; Kälin, 1952.
3. Von Heine-Geldern, 1932.
4. Uhlig, in Kötter *et al.*, 1977, p. 263; Soemardjan in Kötter *et al.*, 1977, pp. 183–4; Röll, 1979, p. 30.
5. Bernet Kempers, 1959, pp. 89–93; Uhlig in Kötter *et al.*, 1977, pp. 263–4.
6. Dahm in Kötter *et al.*, 1977, p. 72.
7. Speed, 1971, p. 25.
8. Dahm in Kötter *et al.*, 1977, pp. 69–74.
9. Smits, 1938, p. 500; Terra, 1958, p. 164.
10. Dequin, 1963, p. 38.
11. Bernet Kempers, 1959, p. 104. Scholars are still debating this question at the present time; van Setten van der Meer, who reviews a multiplicity of opinions, concludes that '*sawah* cultivation, directed by the ruler or by religious bodies, was based on a foundation of purely indigenous irrigation organisation already established before the arrival of Indian influence' (van Setten van der Meer, 1979, p. 133).
12. Terra, 1958, p. 161.
13. Kievits, 1892, p. 79.
14. Geertz, 1963, pp. 41–2.
15. Loc.cit., pp. 42–6.
16. Loc.cit., pp. 38–46.
17. Bernet Kempers, 1959, pp. 104–5.
18. Subardjo in Kötter *et al.*, 1979, pp. 116–23.
19. Dahm in Kötter *et al.*, 1977, pp. 81–4.
20. Speed, 1971, p. 34. ˋ
21. Dahm in Kötter *et al.*, 1979, p. 85.
22. Geertz, 1963, p. 53.
23. Terra, 1958, pp. 157–66.
24. Geertz, 1963, p. 31.
25. Wittfogel, 1957.
26. Geertz, 1963, pp. 28–37.
27. Horstmann, 1958/9, p. 415.
28. Uhlig in Kötter *et al.*, 1977, pp. 267–8.
29. Blaut, 1960, pp. 185–98.

30. 'Bebossing van geerodeerde gronden op Java', 1973, p. 105.
31. Pelzer, 1971, pp. 261–75.
32. Fromberg, 1858; Holle 1866a.
33. Röll, 1971, p. 63.
34. F.de Haan, 1910–12.
35. Pelzer in MacVey, 1963, pp. 130–2.
36. Blom, 1979, p. 11 (on Jawa called the *surjan* system).
37. Baharsjah, 1978b, p. 132.
38. This and the next paragraph is largely based on information from Woelke, 1978.
39. Boeke, 1948, pp. 27–9, 31–6; Geertz, 1963, pp. 66–80.
40. Woelke, 1979, pp. 214–22.
41. Schiller, 1980, p. 93.
42. Blom, 1979, pp. 9–10.
43. Speed, 1971, pp. 47–8.
44. Huizer, 1972, p. 3; Dahm, in Kötter *et al.*, 1977, pp. 67–8.
45. Palmer, 1980, p. 222.
46. Evers and Hartmann, 1981, p. 10.
47. Huizer, 1972, p. 1.
48. Blom, 1979, p. 10.
49. This new policy has its origin in a newspaper article of 1899 by the Dutch liberal, Conrad Th. van Deventer: 'Een Ereschuld' ('A Debt of Honour'). In it the author calculated that the Netherlands had hitherto drawn a net profit of 187 million Guilders from Jawa and claimed that it would be necessary to pay this debt back by helping the impoverished Javanese to develop their island (Dahm, in Kötter *et al.*, 1979, pp. 89–91).
50. Clauss and Hartmann, 1981.
51. A. Hartmann, 1973.
52. Speare, 1978, p. 93.
53. FAO/IBRD, 1979; quoted by Blom, 1979, p. 12.
54. Palmer, 1980, pp. 223–6; Evers and Hartmann, 1981, pp. 10–11; *Indonesia Times*, 29 Sept. 1981.
55. Clauss and Hartmann, 1981, p. 18.
56. B. White, 1979, pp. 95–6.
57. An extended bibliography of Boeke's works on this subject is given by Fruin, 1961, pp. 406–8.
58. Fruin, 1961, pp. 342–3.
59. Loc.cit., p. 338.
60. Clauss and Hartmann, 1981, p. 18. In this context we should note that, to some extent, Boeke's ideas have become part of the actual rural development policy of the Indonesian government. In 1983, the Director General for Rural Development explained that he aimed at a stage where all Indonesian villages might be self-sufficient in supplying their basic needs. He envisaged a development from *swadaya*, the fully traditional village, via *swakarya*, the village in transition, towards *swasembada*, the village 'in which development will not disturb the customs, having rational relations among its people and will apply technology to increase the people's productivity' (*Indonesia Times*, 11 Nov. 1983). Similarly, the mutual help between poor farmers and poor villages is encouraged. This would further not only the spread of simple applied technologies but also create a kind of self-reliance

among those who belong to the lowest economic class. (Djajanto, *Indonesia Times*, 2 Nov. 1983).

61. Geertz, 1963, pp. 82, 97.
62. Horstmann, 1958/9, p. 418.
63. See also Huizer, 1972, pp. 33–8.
64. Van der Kroef, 1965, p. 194.
65. Ten Dam, 1961, p. 349.
66. Huizer, 1972, pp. 8–9.
67. Loc. cit., p. 52.
68. Hatzfeld, 1969, p. 69, quoted by Huizer, 1972, p. 53.
69. See also Röll, 1973, pp. 305–21; Burger, 1975, pp. 151–61; Schlereth, 1983, pp. 75–6. In 1983, the President of Indonesia admitted that landlessness was one of the severest problems of the agricultural sector and its development. Of some 25.5 million farmer families, only 17.5 million owned land, the rest being tenant farmers without land of their own to till. But the majority of those who own land possess less than one hectare of land, and 8 million farmer families do not own even a small piece of land at all. (*Indonesia Times*, 30 July 1983).

5

THE ENVIRONMENTAL IMPACT

Even today the view is widespread that the tropics are the most fertile region on earth, where food can be produced in almost unlimited quantity. There are authors who estimate that the tropics alone could feed more than the present population of the earth.[1] They maintain that the capacity of our planet to feed people is much higher than is often contended.

This opinion is fostered by the impression received by anyone who looks at the tropical vegetation or biomass of an undisturbed area of rain forest. Its luxuriance suggests that the soil must be extremely fertile, so that after the natural vegetation has been removed, useful plants like foodgrains and vegetables would likewise grow luxuriantly. Similarly, an abundance of rain seems to provide the necessary moisture for the growing process. Unfortunately neither view can lead to sound generalisations. We will show that the soils of tropical Indonesia are by no means universally fertile and that the climate, though favourable in some parts of the archipelago, often adds to the difficulties already presented by the soils. The island of Jawa is a special case in so far as its soils are to a great extent fertile because of their volcanic origin. As for climate, the island shows a differentiated picture.

The Climate of Jawa

Situated between the fifth and the tenth parallel south of the equator and stretching roughly 1,000 km. from west to east, Jawa shares with the rest of the archipelago a location between the Indian and the Pacific oceans as well as between the Asian and Australian land masses. Thus the island is exposed to the great movements of moist air called the monsoons. However, its extremely mountainous terrain (reaching a maximum height of around 3,700 m.) makes climatic averages of little value because microclimatic conditions often change from one valley and one plateau to another.

Jawa's situation between the Australian and the Asian continents has a tangible influence on the monsoon winds and, correspondingly, on rainfall. From June to August pressure is low over Asia and high over Australia, and warm, dry air flows from Australia over Jawa on its way

The luxuriance of the tropical forest suggests an extremely fertile soil.

to the Asian continent; this 'east monsoon' brings with it a dry season. Between December and February, the situation is reversed: the air flowing from Asia to Australia passes over long stretches of sea and is full of moisture. This 'west monsoon' brings much rain to Jawa — the wet season. In the transitional periods between the wet and dry seasons the inter-tropical convergence zone crosses the archipelago. This causes a heavy collision of air masses by strong convection and is often supplemented by the ascent of air masses following strong local winds. Therefore, inter-monsoon rains may be stronger than the monsoon rains proper.[2]

Due to the pronounced relief of the island, local winds are of great importance. Meeting mountain barriers that cause orographic uplift, they lose their moisture and produce rain showers for long periods. After passing the crest of the barrier, however, they descend, warming up and drying out the surrounding area on the lee side. This phenomenon is called the föhn effect, and is particularly impressive when the dry easterly monsoon meets the mountain chain of Jawa from the south. Then, dessiccating winds are observed on the northern coastal plain in the areas of Cirebon, West Jawa, or Tegal on the northern slope of Central Jawa, and in the areas of Pasuran and Probolinggo in East Jawa north of the Tengger mountains.

Heating over land areas during daytime may also produce convective showers by attracting sea winds. However, sea winds off the south coast of Jawa are more pronounced, and thus the south of the island is generally moister than the north all the year round. Hence there is an appreciable rainfall in Jawa even during the dry season.

Finally, zenithal rains play an important part; extremely heavy rain falls shortly after the sun has passed its zenith. They are connected with the movement of the inter-tropical convergence and produce an enormous amount of water. In Bogor, for instance, 60 km. south of Jakarta and 266 m.a.m.s.l. (metres above mean sea-level) zenithal rains fall on 322 days of the year, and a rainfall of 0.5 mm. per minute is not unusual. The same is true of the frequency of thunderstorms, which is probably the highest in the world.[3] Up to a certain altitude above sea level rainfall increases, but thereafter it decreases again; however, at those higher altitudes dew and mist take its place.

According to Köppen's classification, the western two-thirds of Jawa belong to Am (tropical rain forest climate despite a short dry season according to the monsoon-type precipitation cycle), only some parts at the highest altitude were classified as Cfa (warm temperate climate, moist, with precipitation in all months, with a hot summer, warmest month over 22°C). In the eastern one-third, however, Jawa belongs to the Aw climate (tropical but with a pronounced dry season). This shows that an important climatic division of the archipelago runs right through Jawa: west of the line we have high rainfall all the year round, an average of 2,360 mm. in West Jawa and 2,400 mm. in Central Jawa; and east of the line a pronounced dry season and consequently a lower rainfall, namely an average of 1,660 mm. in East Jawa. This has an important impact not only on the natural vegetation and soil fertility, but also on the carrying capacity of the different landscapes.

Naturally, these average figures can only convey a very rough idea of conditions. For the potential of a specific area to be assessed, more detailed observations are needed since the variations from place to place

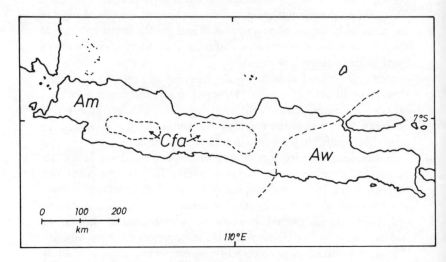

Fig. 7. CLIMATE OF JAWA

Am = Tropical climate, rainforest climate despite a short dry season (monsoon type)
Aw = Tropical climate, tropical savanna climate with a pronounced dry season
Cfa = Temperate rainy climate, wet in all seasons, with hot summers

Base map: Meyers Grosser Weltatlas, 1979, pp. 246–7.

are substantial.[4] The highest mean annual rainfall is 7,069 mm. per year
in Tenjo, Central Jawa;[5] the absolute maximum was measured in 1909 in
Sirah Kencong (Kediri) at 10,112 mm.; the highest daily rainfall at 511
mm. was observed in Besokor (Semarang) and 613 mm. on the island of
Madura. Most stations record a daily average of 200 mm. On the other
hand, critical rainfall also occurs on Jawa, i.e. rainfall too low to permit
optimal forest growth. At 910 to 950 mm. per year we find savannah
vegetation or transitional vegetation forms in the north-east.

Although Jawa generally receives high rainfall evenly distributed
throughout the whole year (an evermoist monsoon climate), differ-
entiation in space and time can be substantial. We have rainfall figures
for Jakarta for 70 consecutive years which show an absolute annual
minimum of 1,177 mm. and an absolute annual maximum of 2,397 mm.

The annual mean of 1,766 mm. was exceeded in 38 years and not attained in 39, and that often tangibly by 33–36%.

Jawa's location not far from the equator and surrounded by a vast area of water results in comparatively high and stable air temperatures. The mean annual air temperature at sea-level remains between 26° and 27°C, the difference between the mean maximum and minimum being no more than 1–2°C. Greater differences are brought about by altitude, since each 100 m. is equivalent to 0.57°C difference in the mean temperature:[6]

	Mean annual temperature
Jakarta (7 m.)	25.9°C
Bandung (768 m.)	22.1°C
Pangerango (3,023 m.)	8.9°C

The daily range in temperature is small but somewhat greater in the dry season (5–10°C). Frost appears on Jawa at high altitudes, but this is often the result of nocturnal radiation after which cold air collects in depressions. This rather infrequent phenomenon has an adverse effect on agriculture. Frost has been observed in the highlands of Pangalengan (1,500 m.) in West Jawa in July–September, on the Dieng Plateau (2,100 m.) in Central Jawa and in the uplands of Kalisat (1,100 m.) in East Jawa.[7]

Soil temperatures differ somewhat from the atmospheric temperature; they depend on the temperature of the surrounding air, radiation, and the colour and water content of the ground. It is an important factor in the assessment of the potential of a certain area since it influences the underground minerals, the weathering of the parent material, the development of humus, the germination of seeds, plant health and development, and so on.[8]

Under forest cover the temperature of the soil at a depth of 60–100 cm. is almost constant (25–27°C). In the open, however, it is 3–4°C higher, and as altitude increases the soil temperature decreases by 0.57°C every 100 m. At a depth of 10 cm. the day temperature varies by 6°C, at the surface between 15°C and 45°C or more. At the highest altitudes where the soil is unprotected and exposed to radiation and insolation, minima of – 15°C and maxima of 70°C occur. Thus, at least under forest cover and in moderate altitudes, the soil temperature is favourable for the growth of vegetation.

Despite the high rainfall, hours of sunshine reach 55 to 75% because much of the rain falls during the night. But the solar energy which is so important to plant production is partly absorbed and dispersed because of the high humidity of the atmosphere; a substantial 38% does not reach the ground. However, the abundance of vegetation proves that insolation supplies sufficient energy, particularly in the coastal lowlands which have the longest period of sunshine, since cloudiness increases with altitude.

Climatologists analysing Jawa's climate with special reference to its influence on agricultural land use found that Köppen's system has no practical purpose since it is based on mean values of precipitation over many years. In fact, precipitation in Indonesia undergoes considerable variations from year to year, so that the number of dry months — a most important datum in the estimation or determination of a region's agricultural potential — is not clearly revealed. Therefore, in 1950, a national working group was established to construct a climatic map suitable for practical application, and the result was the new Schmidt and Ferguson concept.[9] This system considers the number of dry and wet months year by year and then takes the average value over many years. The climate type is determined by the quotient of the average number of dry and wet months. In this way eight climatic types are determined (A to H), with no dry months at one extreme and no wet months at the other. The stages between the extremes indicate climatic zones with a growing share of dry months from B to G. On this basis, the authors developed an impressive map of rainfall types for Jawa (see Fig. 8), giving not only a detailed picture of rainfall distribution, but also showing the distribution of natural vegetation and the potential for growing certain cultures, provided the soil is suitable.[10]

This climatic classification subdivides the islands in such a way that the progressive dryness from west to east becomes clearly visible. In West Jawa, the north coast belongs to the D and the C type. The very wet A type prevails in the mountainous central part, surrounded by the B type reaching right down to the south coast. In Central Jawa, the B and C types prevail with moderate rainfall, whereas the wet A type is confined to small areas around the Dieng mountains and Mt Slamet. Dryer areas (D type) appear along the north coast and in the south. Towards the east of the island, the dry D type plays the most important part, and the still dryer E type appears along the north-east coast and on Madura. The mountainous part shows the C type whereas the higher altitudes, such as the Tengger and Ijen mountains, show types B and even A. The map shows clearly the increase of dry months from west to east.

The climate, as briefly presented here, has a direct influence on the

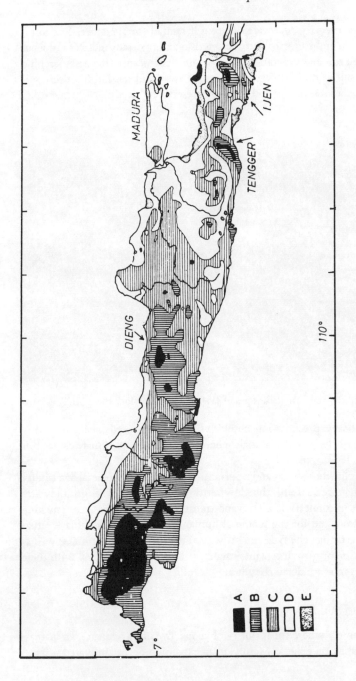

Fig. 8. RAINFALL MAP OF JAWA (after Schmidt and Ferguson, 1951)

Climatic types A to E with progressive dryness.

carrying capacity of an area, through rainfall, temperature, humidity, insolation and so forth. But there is also an important indirect influence through soil and vegetation. Heavy rain, for instance, increases leaching of the soil so that nutrients are washed away and soil fertility decreases, whereas a long dry period helps to preserve soil fertility. On the other

The terraced and irrigated slopes of Jawa, admired all over the world.

hand, heavy precipitation combined with heat speeds the weathering process of the parent material, whereas a dry and cool climate causes this process to slacken.

The climate also has an important influence on the general availability of water, since rain, the downflow of rivers and the groundwater reserves in connection with evapo-transpiration are the basis for any kind of land use and for the whole of human life. Finally, the climatic situation determines the type and growth of the natural vegetation, as well as the limits for growing plants useful for man.[11] We shall deal with these subjects in more detail elsewhere.

Soils and Soil Reserves

In a country where the majority of people live on agriculture, an increasing population generally results in an expansion of the cultivated surface.

As we have seen, this has been happening in Jawa to such an extent that today the island is one of the most densely populated areas on earth. While this population growth continues, it has to be asked: to what extent can additional land be cultivated? In view of environmental damage in many places, it should also be asked how much of the land that is being used for agriculture should be taken out of production and reforested.

The carrying capacity of a country as large as Jawa is very difficult to judge and to quantify; it depends greatly on the investments that can be made and the technology that can be applied. A hilly area which is well terraced and irrigated can obviously feed many more people than the same area left in its original state with, say, scattered forest and wild pasture. The Javanese have proved this all over the island, to the admiration of the whole world. However, the growing population pressure has driven people into areas less and less suited to cultivation, particularly since the necessary investment to protect the soil and to increase its productivity was unavailable.

The population of Jawa is nearing the 100 million mark, with an average population density of nearly 700 people per sq. km., but regional differences in density clearly indicate corresponding variations in natural potential. Mohr, who analysed this phenomenon in the 1930s, found a clear correlation between population distribution and the pattern of volcanic eruptions.[12] No matter whether the wind spread the fertile material or rivers washed it away and deposited it in distant locations, people followed it and settled there. In Jawa, unlike other countries where the rural population accumulated near the cities where they found markets for their products, the soil quality has been the criterion for centuries.

Let us take, for example, the Mt Slamet volcano in Central Jawa. The Gung river carries the ejectamenta towards the northern coastal plain, and the Banjaran and other rivers carry it towards the south. Thus extremely fertile plains and valleys have developed on both sides, which have been partly irrigated by the settlers. By the 1930s, rural population densities of between 700 and 1,600 people per sq. km. were recorded. In contrast, cultivated areas on old volcanic ground 100 km to the east had densities of less than 400, and those on Tertiary soils, far from the nearest volcano, less than 300. A similar picture can be found in the zone of influence of the Dieng volcanoes and the Merapi volcano north of Yogyakarta, or of Mt Kelud in East Jawa south of Surabaya. There, wind and water deposited fertile volcanic material, the population densities

topped 500 and even 1,000, whereas the less favoured neighbouring areas
remained below 500 or even 200. The soils are therefore a key factor in
judging the potential of a country, and in Jawa decreasing soil fertility
attracted the interest of researchers in the middle of the nineteenth
century when the declining yields of export crops became generally
apparent.[13] It seems that the first paper ever written on soil fertility and
fertilisation in Indonesia was one by Fromberg who reported in 1854 on
his experiments with guano, ashes and other fertilising materials on
tobacco fields which had suffered from decreased yields.[14] When the
fertiliser industry in Europe and particularly in Germany started to
expand and the use of chemical fertiliser became popular, many Indonesian plantations made liberal use of it on tropical crops such as coffee,
tobacco and sugar-cane, although the expected success was not always
attained.

The growing demand not only for export crops but also for food to
sustain the indigenous population led to a more systematic examination
of the phenomenon of soil exhaustion and the possibility of counter-
action through fertilisation. Van Hall, who published an early study in
this field in 1873, found that fighting soil deterioration can only be
successful if it is supported by the rural population; and this again can
only be expected when the individual farmer can freely dispose of his land
— in the sense that he alone is responsible for it. This, however, would
not be the case as long as communal possession of the soil and the dominant position of the village chieftain prevailed.[15] It is perhaps the earliest
instance of a Dutch colonial officer or technician realising the interdependence between land tenure and environment.

The more forests were cleared and forest soils taken under cultivation,
the more farmers and agricultural technicians became aware that the
fertility of these soils, originally regarded as extremely high, rapidly
decreased as soon as the land was used for annual crops. In particular, the
decline of tobacco yields prompted the then famous Dutch soil scientist
van Bemmelen and the well-known forester S.M. Koorders to look into
the matter.[16] Their papers are among the oldest dealing with the fertility
of tropical soils in relation to their natural vegetation — a controversial
topic even today. Farmers and scientists alike learned that the lessons of
the developing soil science in Europe could not be applied in Indonesia
without consideration of the special tropical conditions prevailing there.
This was the case especially with liming, which at that time was very
popular in Europe, particularly in Germany.[17]

Gradually, Dutch and Indonesian soil scientists abandoned European practices and based their work on Indonesian conditions such as volcanic soils, the soils underlying the tropical rain forests, alluvial sedimentary plains under local conditions, and so forth. After 1910, pedology in Indonesia developed rapidly, and no aspect of soil science and soil management, from microscopic and chemical analysis to soil mapping, was neglected. Thus we are in a position to describe the soils of Jawa, referring to a number of studies produced over the last few decades, despite the lack of uniformity in pedologic nomenclature. In 1975, the Soil Research Institute of Bogor issued an estimate of the soil types of Jawa and their quantitative distribution over the island (see Table 4).

The soil sketch map of Jawa shows the broad belt of volcanic soils stretching from east to west, and the grumosols and lateritic soils covering the slopes on the north and south. The soil types covering most of the island can be divided in three: volcanic, alluvial and limestone soils covering respectively 50, 20 and 12% of the surface. The most important, particularly in the sense of productivity, are the *volcanic soils*.

The feature that dominates Jawa, physically as well as pedologically, is the chain of young Quaternary volcanoes, several of which are active. They are mainly strato-volcanoes, some with enormous calderas, reaching up to 2,000 m. and sometimes more than 3,000 m.[18] The most significant difference between Jawa and Sumatera is that the volcanoes of Jawa have mainly been producing andesite which is basic to neutral in nature. This makes an important contribution to the fertility of the island's soils. About 23% of Jawa's surface is covered by young volcanic material, and the whole history of the island is marked by catastrophic eruptions resulting in colossal destruction. But even though they have fled, the people always return because of the favourable soils.[19]

The material ejected from the volcano settles in the surrounding area, unless driven away by the wind or washed down by the rains, and the first soil which develops from it is called volcanic ash soil or regosol. Due to the influence of the environment, it develops into andosols and later into latosols or grumosols, depending on local conditions. Each of these soils has both advantages and disadvantages as regards suitability for cultivation. Regosols cover 10–11% of Jawa's surface and consist of unweathered or barely weathered volcanic material and therefore form the initial state of the transformation of raw materials such as andesite, augite and hypersthene into structured soil. They are still sandy and lack a developed profile although organic matter occasionally accumulates on the surface. Texture and chemical composition differ from place to place, but regosols are generally regarded as the most fertile soils of Indonesia,

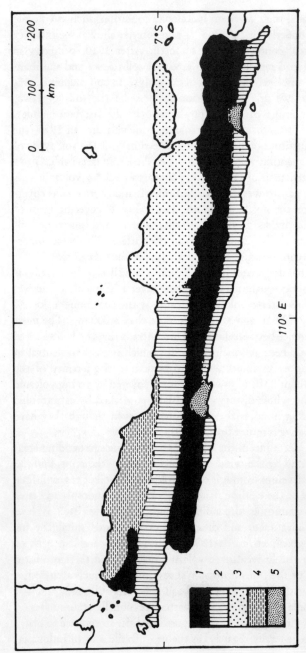

Fig. 9. THE SOILS OF JAWA

1 = Andosols (young soils from volcanic ashes, unclassified) in mountains, often associated with meadow and meadow steppe soils.
2 = Mountain soils of humid and sub-humid tropics, unclassified. 3 = Dark, heavy soils of the temporary humid tropics and subtropics (vertisols, grumosols, tirs, regurs, etc.). 4 = Lateritic soils, partly bleached (latosols, kaolisols, ferralitic soils, ferralsols and ferrisols), unclassified, of the humid and sub-humid tropics. 5 = Mineral hydromorphic soils and alluvial soils, unclassified.

Base map: Ganssen and Hadrich, 1965.

Table 4. ESTIMATED AREA OF THE VARIOUS SOIL TYPES OF
JAWA

Soil type	1,000 ha.	%
Latosols, lateritic soils (inceptisols, oxisols)	2,831	21.4
Alluvial soils (entisols)	2,550	19.3
Podzolized soils and complexes (ultisols)	2,394	18.1
Limestone soils, red Mediterranean-type soils (altisols)	1,625	12.3
Grumosols, black earths, margalitic soils (vertisols)	1,481	11.2
Regosols (entisols, inceptisols)	1,431	10.8
Andosols (inceptisols)	844	6.4
Others	63	0.5
Total	13,219	100.0

Source: Soil Research Institute, Bogor, 1975 (quoted by Baharsjah, 1978, p. 83).

having reserves of phosphorus, potassium and other minerals and plant nutrients. They weather rather quickly under humid tropical conditions so that the nutrients become available.[20]

Although the existence of regosols is not bound to specific climatic conditions, the latter determine their development into andosols or other soils. Most of them are found in Central and East Jawa, where there is a dry season of 3–5 months and where rainfall reaches 1,200–2,000 mm. But since precipitation is greater in mountainous regions, annual rainfall may be as high as 5,000 mm. at higher altitudes. Under such conditions regosols, particularly when covered with mountain forest, serve as water reservoirs, attracting the water, together with nutrients, to the lower layers of the mountain system. Unfortunately many regosols are found on extremely steep slopes and are thus subject to severe erosion under heavy rainfall. This often leads to the formation of so-called *lahars* (torrential streams of mud) rushing down upon the less pronounced slopes and plains, devastating fields, irrigation works and villages.[21] To some extent this danger can be reduced by the planting of trees or dense grass mountain pasture or, under favourable conditions, even by the construction of *sawah* terraces. But since the gradient increases sharply with altitude and the slopes are extensively cut by ravines, the upper part of the volcanoes consists only of naked ash, fully exposed to wind and rain. This can be seen on Merapi, Kelud, Raung, Bromo and Ijen, to name only a few.

Under the influence of the climate, young regosols slowly mature into andosols, grumosols and, after a long period of weathering, into latosols. A tropical climate with a pronounced dry season favours the formation of latosols; in a more temperate climate, as at high altitudes, andosols are the final stage of soil development; and grumosols develop where there are extreme quantities of water.[22] Accordingly, andosols are persistent at high altitudes with a wet and cool climate. On higher slopes they are often used to grow commercial trees such as pines (*Pinus merkusii*), eucalyptus and spruce (*Agathis alba*), whereas on high plateaux commercial crops are grown. In the wetter parts of Jawa, e.g. near the Puncak Pass south of Bogor, extended tea plantations exist whereas further east, with decreasing precipitation, coffee (*Coffea arabica*) takes over. More recently, with the development of a road network and growing markets in the large towns, the production of vegetables, potatoes and temperate fruits has spread on these soils: this can be seen around the Puncak Pass, in the area of Penggalengan south of Bandung, and on the Dieng Plateau in Central Jawa where horticultural production continues up to 2,250 m.

The lower parts of the volcanoes often slope far into the plains or basins. They offer excellent conditions for growing various crops, particularly wet-rice, once the water supply is secured. Here the famous terraced landscapes have been developed, and in Jawa they extend up to 1,800 m. and, in the Tengger mountains of East Jawa, are found as high as the climatic cultivation line at 2,250 m. The irrigation of these terraces depends on the existence of wells to supply water. Sometimes we find clearly defined spring lines between 600 and 900 m. forming a distinct boundary for irrigated land. At a higher level, rain-fed agriculture with bushes and tree culture prevails, and the terraces become steeper and cruder in construction.[23] The upper limit for cultivation is set by the gradient as well as the climate. Rice cultivation, even where irrigation is possible, does not occur above 1,480 m. in Jawa — a relatively low climatic limit for a tropical country. But the temperatures are comparatively low due to extreme daytime rainfall and cloud. When the night sky is clear, frost may even occur locally in depressions above 1,500 m. between July and September, a serious limitation for tea estates situated mainly at such altitudes.[24]

Finally, as the last stages of soil development from volcanic parent material, the grumosols and latosols have to be mentioned. Grumosols, also called black earths or margalitic soils, have a very high clay content since their parent material is no longer exclusively volcanic. It also comprises marl of marine origin and old alluvial sediments. Therefore grumosols in the north of Central and East Jawa are rarely found above

600 m., and hill grumosols as found in the Kendeng and Rambang mountains are severely eroded and their material mainly transported into the valley of the Lusi river. Here, as well as in the lower valley of the Solo river and on the plains of Demak, are the most important deposits of grumosols: this means that they are bound to lower zones, undulating as well as flat, and that they often face flood problems.

Hilly grumosols, because of their propensity to erosion, are best suited to afforestation, preferably teak, but they are less favourable for cultivation. Despite a reasonable natural fertility, there is a lack of nitrogen, phosphorus and potassium, and green manuring is recommended as a counter-measure. The physical properties are bad. Because of a high clay content, grumosols are too hard and dry in the dry season, and too wet and heavy to be worked in the rainy season. Thus it is only for a short period in the year that the humidity is exactly right to work the soil — provided this coincides with the sowing or planting period. As soon as the soil dries out, however, it tends to shrink, crack and break the hair-roots of the plants. In August and September large cracks as much as 4.50 m. deep may be found.[25]

Latosols, finally, occur in areas with constantly high soil temperatures, abundant rainfall and good drainage. They are not homogeneous, and there are many intermediate forms between andosols and latosols. Brown latosols cover the higher reaches of the volcanic slopes where temperatures vary between 18° and 24°C and there is no dry season. Dark-red latosols, however, are the final state of a pedogenetic development that started with unweathered volcanic ash. They prevail in areas under 100 m. and cover more than 10% of the island. The agricultural value of latosols differs from place to place. Although their fertility is regarded as low to medium, they are fully integrated into agricultural production; their main problem however, is that *padas* or hardpans are formed under the topsoil, particularly in conditions of permanent irrigation where hard crusts develop, containing iron and aluminium, and preventing water from circulating and roots from penetrating into the deeper layers of the soil. Since the soils are generally deep, they are well suited to the construction of *sawah* terraces.[26]

We must complete this brief survey of the volcanic soils by mentioning that the volcanoes in the west of Jawa are closely related to those of Sumatera, which means that they produce acid material, dacitic and liparitic ashes and lavas. This leads to the formation of less fertile soils in West Jawa; here too, as in Sumatera, the distribution of fertile and less fertile soils is uneven. There is the fertile plain between Jakarta and

Bogor, and there are the inter-montane basins of Garut, Bandung and Cianjur where good soils support a dense population. But to the north-west, in the region of Banten and south of Bandung in the Priangan district, the population is much lower. The different volcanic ejectamenta clearly result in different soil qualities and thus carrying capacities, expressed in population density.

The second important soil group consists of the alluvial soils. Alluvium is the result of the erosion and deposition of other soil material that is often transported by water over considerable distances. It is usually rivers that transport the material, and broad alluvial flood-plains have been formed along them. Examples among the rivers of Jawa are the Serayu, Madiun and above all the Solo and Brantas (in Central and East Jawa respectively) where the famous remains of the *Meganthropus* and the *Pithecanthropus* were found. Thus from the earliest days of mankind they gave easy access to hunting grounds, and later they offered a soil that was easy to work and grow the first crops on.[27]

Many of the large rivers have built up natural levees by depositing the coarser material first along the river banks while carrying the finer silt further away. Similarly, silt is also transported through irrigation canals and ditches. But apart from the activities of the rivers, it can be said that a gigantic process of sedimentation has been going on on Jawa ever since the island has existed. In particular, it has built up the northern coastal plain and is extending it further into the Java Sea. Finally, in some of the inter-montane basins, deep alluvial layers have accumulated, consisting partly of lacustrine sediments, as near Bandung and Malang. Altogether alluvial soils account for about 20% of Jawa's surface.

The fertility of alluvial soils, often regarded as high, depends on the parent material from which it is eroded, i.e. on the chemicals transported to the plains. If the plains are bordered by volcanic highlands where basic to neutral ejectamenta are in various states of pedogenetic development, the mud deposited in the lowlands is rich in plant nutrients. Such soils are generally not only fertile, but also easy to work since the original crude material has already been broken down elsewhere.[28] In Jawa and Madura part of the alluvium is formed by marine sedimentary marls and is less fertile than that coming from igneous parent material.

There are several problems connected with the alluvium. The first is salinity which occurs on the lowlands of Demak, Kudus and Kajen near the north coast. This can be counteracted by using the *surjan* method. It consists of making alternating broad furrows and hills which permit the salt and superfluous water to collect in the furrows, whence it can be drained while salt-tolerant cultures can be grown. The hill crops on the

Areas prone to flooding are made suitable for cropping by means of the camberbed or *surjan* system.

other hand receive water with only a minor salt content.[29] The second handicap of the alluvium is a tendency towards laterisation. However, the loss of fertility of the topsoil that is part of the process is often reversed by the arrival of fresh sediment with the next flood. Floods, caused by too shallow ground water, poor internal drainage and the influence of tidal movements near the coast are another problem facing alluvial soils, often hampering a proper and profitable use of the land.[30]

It is evident that the agricultural value of alluvial soils not only depends on the chemical and physical properties but also, and mainly, on the availability of water. Particularly in Jawa, alluvium is a good rice soil, and if there is enough water at the right time, two to three harvests are possible; however, 20–25% of the *sawah* farmers use only rainwater for irrigation and can thus harvest rice no more than once a year. But even if there are irrigation systems, the water supply is often insufficient for a second or third rice harvest. The decreasing rainfall from west to east is clearly reflected by the following figures: in West Jawa, about 42% of the *sawahs* produce a second harvest, but in East Jawa only 35%.[31]

The residual humidity in the soil after the rice harvest is used to grow cassava, sweet potatoes, maize, peanuts or soya beans, mainly for the farmer's own consumption, or tobacco and sugar-cane for the market. If the dry season is very pronounced, the terraces may be used as pasture for water-buffaloes.[32] The population density in the districts where alluvial soils prevail is high and may well exceed 1,000 inhabitants per sq. km.; accordingly the holdings are very small, seldom exceeding 0.3 ha.

Notohadiprawiro concludes from his studies of alluvial soils that the major problems of this soil type are linked with the area's hydrology. Consequently, there should be a strong emphasis on water management including irrigation and drainage, salinity control, flood control and regulation, and land reclamation in tidal areas. He rightly feels that the introduction of better plant varieties and fertilisation means little unless proper water management can be established first.[33]

Apart from soils produced from igneous rocks and ashes and those accumulated by sedimentation, which also partly originate from volcanic material, a substantial area of Jawa (about 12%) is covered with limestone soils. These consist of Tertiary sediments and derive from limestone rocks or conglomerates. We find limestone plateaux along the southern coast, and in the north the Northern Chalk Range (Pengunungan Kapur Utara) stretches as far as the island of Madura. In these areas and on some of the coastal plains limestone soils prevail. They are relatively infertile but can be enriched when covered by wind-carried material rich in nutrients. Often the soil is mixed with sand, but the presence of alkaline water prevents humus from collecting.[34] The major disadvantage of these soils is their porosity; the water they are unable to hold instead percolates underground where only plants with long roots can reach the water level. An example is the teak tree (*Tectona grandis*) which grows in the eastern and still dryer part of the limestone massifs, whereas in the more humid west, rubber trees grow on these soils. In the wet season the rain often washes the soil into depressions where it accumulates and permits agriculture on relatively deep soils. Drought on limestone soils is therefore mainly caused by the physical properties of the soils themselves and less by low rainfall.[35] It is only the fact that soil moisture rises during the dry season and so partly restores the nutrients in the impoverished soil that secures a minimum fertility.[36]

For a long time the limestone areas were under-populated, for not only does the moderate fertility of the soil and its permeability make profitable agriculture almost impossible, but in addition there is a lack of water, even for drinking. The farmers work hard, but they hardly ever fertilise their fields and consequently the yields are low. Maize barely brings in 700 kg. per ha. and cassava no more than 7 tonnes.[37] Thus the farmers who cultivate limestone soils are among the poorest on Jawa.

The areas with tropical karst, in particular, show the problems of limestone soils. The best known example is the Gunung Kidul or South-ern mountains south-east of Yogyakarta, and especially the Gunung Sewu or Thousand mountains, covering an area of roughly 1,220 sq. km. As the name indicates, it is dotted with limestone outcrops, beehive-like cupolas of porous rock material where limestone soils accumulate in the depressions and in every cavity in the rocks. More than 100 years ago, the Gunung Sewu was still largely forested, and the botanist Junghuhn described it as the most beautiful landscape in Jawa;[38] and Daneš, who visited the mountains in 1915, reported a densely forested area.[39] But today the forests have disappeared and scrub, coarse grass, bamboo, acacias and wild bananas have replaced them. Consequently an enormous process of soil erosion has taken place and all the mobile material from the karst mountains and slopes has been carried into the depressions that have formed in the karst material; these are known as poljes or *karst-wannen*. This transported material consists of volcanic tuffs, sand and debris from the slopes but above all terra rossa, a weathering product from the limestone that can accumulate to a thickness of 30 m. Hence islands of intensive agricultural cultivation develop in these depressions, and since terra rossa fills all the cavities, a relatively multifarious and dense flora covers the ground in the wet season. In addition, the soil that still covers the slopes is cleared by cutting the scrub, and it is then pro-tected by innumerable small terraces and low stone-walls so that perma-

nently cultivable fields of different sizes are formed (see picture).

In the Gunung Sewu, surface water has completely disappeared underground and there is a system of subterraneous rivers with a water-table perhaps between 10 and 50 m. deep depending on the rainfall and underground conditions. This water may reappear through karst-wells

Wells dug several metres deep dry up quickly once the rain ceases.

either in the depressions we have mentioned or at the end of the karst-massif, thus watering the neighbouring plains. In these rare conditions, in or next to a karst landscape, wet-rice can be cultivated on *sawahs*. Rain usually disappears underground as soon as it falls, but if there are small depressions with a ground well, impermeable because of a layer of clay, natural lakes come into existence and are often created or enlarged by man-made walls or earth dams. Of these lakes (known as *telagas*) Escher found 433 in the Gunung Sewu in 1931;[40] only 372 were found in 1965,[41] which possibly indicates a drying-up process in the region.

Such *telagas* are often the only source of water for many months, since even wells several metres deep dry out once the rain stops. When this happens, water has to be carried from as far as 8 km. away. The *telagas* are also used for washing, bathing, irrigating and providing the animals and people with drinking water and are thus a dangerous source of infection.

The water percolating through the karst often reappears in the form of sub-marine sources along the southern coast 1 metre or so below the mean sea-level. In such cases, people often use these pools of brackish water for washing purposes. Apart from the few cases where the existence of karst-wells permits irrigation, the fields in the Gunung Sewu depend entirely on rain; therefore, maize and cassava are the crops providing the basic food. In addition, partly inter-cropped second harvests are grown in the same field and in the same season, e.g. peanuts, soya beans and tobacco, but also dry-land rice.

The future of the areas with limestone soils and particularly those with tropical karst appears bleak since the process of deterioration of the soils is progressing swiftly. Although limestone and karst areas tend to be sparsely populated, the general population increase on Jawa has not spared them. With 350 inhabitants per sq. km. the population of the Gunung Sewu is relatively high,[42] but there are flat areas on the Monosari Plateau where the population reaches 423 per sq. km., an enormous increase compared with the 1930 Census when the figure was 253.[43] The permanent use of the soil, which lies fallow only during the dry season, and its exposure to sun, rain and wind causes a high degree of denudation, so that the naked rock becomes more and more visible and not a single plant can grow. While this picture does not convey the impression that the soils all over Jawa have excellent physical and chemical properties, with high fertility and productivity, qualification should be made according to topography (gradient and altitude), climate etc.

If, following the example of the Soil Research Institute at Bogor in 1975, we divide the soils into eight classes, we find that 'suitable' soils constitute only 12% of Jawa's surface, whereas 20% are classed as 'unsuitable' soils. The 68% inbetween are classified as 'conditionally suitable', which means that they are suitable for certain crops, in certain climatic or topographic conditions or after substantial investments have been made.[44] The fact that nearly 100 million people live on the land only proves that they have invested much labour in the soils to increase their carrying capacity.

However there are two facts which have to be remembered in this context. First, there are zones in Jawa where the suitability of the soils is worse than the island average would suggest. Dames, who analysed the eastern part of Central Jawa, found that the productivity of the soils is 'very high to high' in 18% of the region, and 'low to very low' in 49%, while 33% showed a moderate productivity inviting the necessary investment.[45] Secondly, the soil varieties which prevail on the island and offer a certain fertility are in danger of becoming unproductive because of poor cultivation methods. Even in the nineteenth century, the Dutch technicians were alarmed by the fact that large tracts of soil had become unproductive waste land which they called *stervend* (dying) *land*. Although there are no accurate data as to the size of waste land, it is agreed that it has reached serious proportions and that poor forest management has also contributed to expanding deterioration of soils.[46]

Jawa's potential can also be judged on the basis of the terrain. When the World Bank undertook an agricultural sector survey in Indonesia in 1974, the territory was subdivided into three types of terrain: (1) mountain land at least 200 m. above sea-level; (2) almost level to undulating and hilly land less than 200 m. above sea-level and not belonging to type 3; and (3) lowland including swamps. Of the island of Jawa together with Madura and Bali, with a total area of 14 million ha., 38% belongs to type 1, 29% to type 2 and 33% to type 3. However, only 7.8, 22.9 and 26.4% respectively are suited to agricultural use. Only 57.1% of the total area was regarded as suitable for agricultural land use.[47] If we remember that even in 1937 the total area harvested on Jawa and Madura was given as 8.7 million ha. or 65.8% of the total area, we may presume that much unsuitable land has already been taken under cultivation.

At present we can only offer rather crude estimates. It would be necessary to analyse watershed after watershed to discover the potentials provided by nature for the growing of food or other crops including forest and to estimate the magnitude of investments necessary to increase the carrying capacity. This, however, would require more time, qualified field staff and funds than will ever be available in the near future. We have therefore to make the most conservative estimates, since experience shows that much of the land so far taken under cultivation is unsuitable, yields little, and will in due course become completely unproductive. In general, the target for a land use policy can only be to use suitable soils to the utmost (which is probably already the case), to direct the necessary investments towards the conditionally suitable soils so as to increase production in a protective way, and finally to take the unsuitable soils out of crop production and give them a new use through afforestation.

Forests and Deforestation

As we have seen, the demographic and economic development of Jawa resulted in a rapid expansion of the area under cultivation at the expense of the natural forest cover. We can assume that the original natural vegetation consisted of primeval forest, except on young volcanic ashes, dunes and the like. But now, except for a few inaccessible areas in the higher volcanic mountains, for instance on the summit of Mt Muria,[48] little of the original vegetation remains.

Our knowledge of the interdependence of climate and altitude enables us to reconstruct the vegetation that once covered the island. In general, there are four prevalent forest types, distributed roughly as follows. In the west and south-west are tropical evergreen rain forests; the east is dominated by deciduous monsoon forest; the central mountain spine carries tropical evergreen montane forest; and there are a few coastal areas with mangrove stands. The distribution clearly reflects the decreasing rainfall from west to east as well as the growing altitude from all the coasts towards the central mountains; but of course the rather

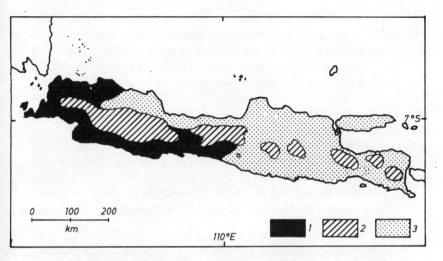

Fig. 10. THE FOREST TYPES OF JAWA

1 = Tropical evergreen rain forest
2 = Tropical upland rain forest
3 = Tropical monsoon (raingreen) forest.

Source: compiled from various sources by the author.

rugged relief of the island creates innumerable microclimates which in turn support a great many micro-vegetation zones in valleys and basins, on slopes and the like.

The tropical evergreen rain forest, also called lowland forest, is composed of a multiplicity of species due to the abundant humidity and high temperatures all the year round. The trees form three different tiers and the crowns of the tallest trees reach 50 m., whereas the undergrowth is less dense due to lack of light. Among the trees, the Asian element prevails, represented by legumes as well as species of the *Dipterocarpus* and *Bursera* families.[49]

The deciduous monsoon forest stretches from northern Central Jawa through the Lesser Sunda islands to Timor, an area in which aridity increases towards the east. It may be regarded as the only deciduous monsoon forest area south of the equator. This forest type consists of fewer species than the rain forest, but includes extended teak (*Tectona grandis*) forests in the north-eastern part of Central Jawa. Foresters regard the north and east of Central Jawa as the natural environment of this tree.[50]

The tropical evergreen montane forest, which grows from about 1,500 m. upwards, is determined by decreasing temperatures together with sufficient rain, and its leading species are oak varieties, tree fern and *Asplenium nidus*. The number of species is small and the giant trees of the lowland rain forest are absent. Between 2,000 m. and 2,200 m. cloud forest takes over and a constantly cool climate and permanent moisture leads to a further reduction of species — which are often covered with moss and bearded usneas. Locally, where the climatic conditions cause a very humid microclimate, mosses (*Bryophytes*) are so extensive that the forest is known as 'mossy forest'.[51] The forest line runs approximately along the 3,500 m. isohypse.[52]

Although historical data on the deforestation of Jawa are scarce, experts in this field maintain that large-scale clearance only took place after 1800:

Prior to 1800, agriculture was confined to the fertile young volcano plains, the lower volcano slopes and a few scattered settlements in the other regions. As the population in Jawa only amounted to some 4 million there was never any need to clear the forest on less suitable land. Moreover, the forests surrounding the villages provided adequate protection against attack from without. Only in the immediate vicinity of such settlements was any wood-felling necessary — on the one hand to provide firewood and timber and on the other hand to keep out wild animals. With the increasing population since 1800 — in 1850 the population of Jawa and Madura had reached the 10 million figure; in 1900 it was about 25 million and in 1950 roughly 50 million — and with the coming of coffee and other plantations, radical changes will inevitably have taken place.[53]

Although we may presume that man's use of forest goes back to the first settlers on the island, we can accept too that for centuries the local people used it only to satisfy their personal needs. The manual nature of their agriculture and the limited accessibility of the forests prevented serious destruction. Their activity was most likely confined to a·strip of forest along the rivers, perhaps 1–2 km. deep. With their primitive tools they cut no more than was neccessary and for the rest they were hunters and gatherers.[54]

Increased population and technological developments brought changes. When man used fire, he started the work of degrading the natural forest: he used it to support his hunting activities, he freed areas as pasture for his cattle, and finally made clearings for planting food crops. Extended savannahs spread in the eastern part of the island where cattle took over the land, and in West Jawa *ladang* farmers established their cultivation techniques to the detriment of the forest. Certainly, the forest occasionally reconquered its terrain, but usually this was second-ary bushland (*belukar*), poorer in species and less productive than the primeval forest. Deforestation led to a deterioration in environmental conditions and consequently the succeeding vegetation cover was poorer and more homogeneous, as the teak and *Casuarina* forests in East Jawa show.[55]

Originally, the Dutch bought only what was offered to them and did not interfere with locally developed land use. But when they fully established their colonial system, they forced the farmers to grow what they wanted for export, and finally they took over the production themselves in forming the plantation system. This rapidly changed the picture of land use. The peasants were now obliged to look for additional land where they could grow the necessary amounts of food, and they increasingly found it outside the favourable *sawahs* on still forested ground. It was not long before the foreigners too wanted to clear addi-tional land for their plantations and demanded help from the local people. Subsequently they hired farm hands to tend these plantations, and the new labourers came with their families into a formerly forest-clad area, settled near the plantations and cut further trees to gain space to grow their food. In this way, more and more forest was encroached upon and turned into dry farming and plantation land.

The export cultures of sugar, tobacco and indigo only indirectly resulted in deforestation because they usually occupied *sawah* land and thus forced the peasants to clear the jungle for their own food. However, coffee, tea and rubber had a direct effect, for they were grown on the uplands that had hitherto been forested. The Priangar highlands in the west and extensive areas in East Jawa were thus deforested. The process was often interpreted in the early days as 'cultivation of fallow land', but this is misleading for two reasons. First, much of this land was an impor-tant reserve of the village economy before it was turned into plantation, supplying the people with firewood and timber as well as with land for dry-farming. Dry farming was practised either in addition to the *sawahs* or in case of a bad harvest. In short, forest land was used for field produc-tion when the *sawahs* could not fully cover the villagers' needs. Now it belonged to foreigners, which often endangered the farmers' food situation.[56]

Secondly, forest is not just 'land not yet in agricultural use', it also has an important ecological value. In the beginning however, these ecological functions were neglected because the newly cultivated forest land constituted only a moderate expansion. But it soon covered a substantial part of Jawa and the Dutch farmers became alarmed. Reports

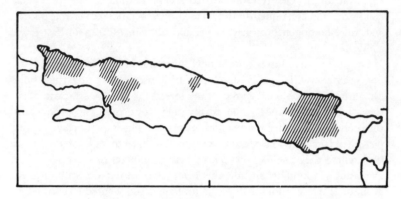

Fig. 11. THE MOST IMPORTANT AREAS OF PLANTATION
ECONOMY IN JAWA

Source: Hartmann, 1981, p. 95.

of catastrophic deforestation, usually in connection with soil erosion and
hydrology, reach back to the middle of the nineteenth century.[57]

Some of the Javanese regions have been carefully observed as far as
deforestation is concerned. One of them is the limestone and karst area of
the Gunung Sewu that stretches along the south coast not far from
Yogyakarta. According to many travellers, the whole mountain complex was densely forested and of exceptional beauty as late as 1850.[58]
Then Dutch planters partly cleared the hills and planted coffee, but the
crop did not thrive. However, although the coffee scheme was finally
abandoned, the peasants continued felling timber and settled on steep
slopes from which the shallow soil had now been cleared, so that soil
erosion inevitably began and grew worse, particularly as the felling
continued during the Japanese occupation of 1941–5.

Although the authorities have made various efforts at reforestation
with teak and mahogany in the last 100 years, it has had little impact.
But even this insufficient cover is threatened by the peasants who extract
timber and firewood and burn the undergrowth between the trees. Thus
large areas are now classified as 'dead land'. In 1960 the forest cover of
this 'regency' was officially given as 7.7%, but in fact it is only about
3%.[59] The Gunung Sewu is a model of the consequences of growing
population pressure, deforestation and mismanaged plantation economy,

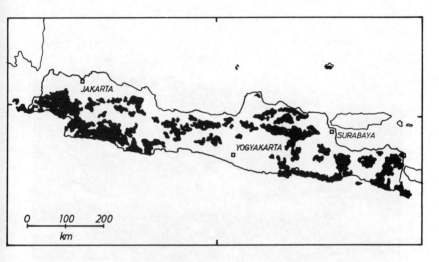

Fig. 12. THE FOREST COVER OF JAWA IN THE LAST QUARTER
OF THE TWENTIETH CENTURY

Source: compiled from various sources by the author.

and shows how soil erosion, over-population and hunger oedema follow
man's activities.

Towards the end of the colonial period, before the Second World
War interrupted any systematic silviculture in Indonesia, the discussion
of the problem of deforestation and its consequences had reached its
climax. Many experts blamed the Dutch administration for not being
active enough in this field. G. Gonggrijp, then an outstanding Dutch
forestry expert, thought it absurd that the Netherlands spent vast sums
on reclaiming land from the Zuiderzee while they remained indifferent
to the fact that vast areas of fertile tropical soils were being degraded and
becoming unproductive.[60]

There were various reasons for this deplorable development. The
main one was certainly the defective administration of the forest areas —
although most had been declared government property long before.
Gonggrijp, also aware of the need for conservation, maintained that in
1937 about 3 million ha. of forest land were properly administered and
another 120 million ha. practically neglected. Furthermore, the local
authorities did not support the fight against the misuse of forests, since
they were corrupted by the short-term income drawn from forest areas

Despite various efforts, the deforestation process on Jawa continued until most of the accessible forest was cleared. These slopes, near Malang, were densely forested only twenty years ago.

and did not understand the ecological background.[61] This is perhaps understandable since scientific considerations had hardly any influence on the decisions of civil servants.[62]

It is true that from 1924 onwards good laws existed to protect the remaining forest area — laws which have constantly been amended and improved. However, their application and enforcement has suffered from a lack of funds and trained personnel as well as from the unwillingness of the local populations to support the efforts of the administration. Many of the Europeans were temporary residents (*niet-blijvers*) who wanted to make a quick fortune and did not care what happened to the land. The indigenous population had no understanding of the protective measures outside the *sawahs* that might have restrained their urge to clear land and exploit the natural environment. Both groups were also fond of hunting — without giving a thought for the conservation of natural resources.[63]

Thus, despite various efforts, deforestation continued until most of the accessible forest was cleared. The population continued to grow, the plantation and dry-farming areas continued to expand, and when the Indonesian peasants took possession of the abandoned Dutch plantations after independence and established their smallholdings on them barely any time remained to look for a solution. More and more land had not only lost its forest but also its topsoil.

At present the official reports state that the remaining forest area amounts to 2.9 million ha. or 22% of the island's surface, subdivided as follows:

	Ha.
Protective forest	642,000
Productive forest	1,845,000
Mixed forest	2,000
Natural conservation	317,000
Not designated	85,000
Total forested area	2,891,000

In fact, only 30,000 ha. have been surveyed and experience shows that official forest figures often merely reflect the areas under the control of the ministry or department in charge rather than the area actually covered with forest. J.P. Thijsse, an extremely knowledgeable specialist, maintains that the forested area had dwindled by 1973 to a mere 11% of the island's surface area.[64]

A study of project documents dealing with forest areas shows that the surface actually covered with forest vegetation is much smaller than officially reported. When, for instance, the Dutch-Indonesian Kali Konto Project near Malang in East Jawa was inaugurated in 1979, the total official forest area within the project area was given as about 15,400 ha., 5,000 ha. of which were considered 'productive forest'. In fact, only 750 ha. were found to be of such quality.[65] Yet despite the fact that Jawa is largely deprived of its forest, the local people as well as the timber economy still rely on it. Of course, Jawa's contribution to the timber production of Indonesia is negligible and the output of roundwood is not important apart from some teak production. Between 1971 and 1975, the annual average of the island's wood production amounted to 581,420 cu.m. log equivalent, which was no more than 3% of the national output.[66] With 66,000 cu.m., the island did not even contribute 1% to national wood exports.[67]

More important, however, is the part that wood still plays in national consumption. In 1975, for instance, only about 12% of the production was exported, 88% being used locally. This is easily explained by the fact that timber is not only needed for construction purposes, particularly in rural areas, but that 70% of all energy consumed in Indonesia is supplied by wood. If there is an energy crisis in rural Indonesia, and particularly in

Jawa with its degraded woodlands, it is a crisis of firewood. This is typical for many countries of the Third World. The island's consumption of wood amounts to 0.75 cu.m per head each year, 0.05 cu.m of which goes into the wood industry and construction whereas 0.70 cu.m is consumed as firewood.[68]

Rural areas need firewood for cooking and heating and for cottage industries such as brick and tile making, the production of soyabean cakes (*tahu* and *tempé*) etc. But firewood is not only the most important energy supplier for the rural population, its collection and sale are also a source of income for villagers who lack sufficient land or employment, so that its prohibition for conservation reasons would mean hardship for such people unless other sources of income were made available.[69] Sources of firewood are usually the undergrowth of forests, dead or damaged trees and the stems of fast-growing shrubs. In areas where there are house or coffee gardens, shade trees are often lopped and used for the same purpose. *Leucaena leucocephala* or lead tree, locally called *lamtoro* is of outstanding importance, but firewood plantations have not yet gained the importance they deserve.[70]

There is a close relationship between the prices of firewood and of kerosene, and where wood is close at hand, it is the cheaper fuel. But if transport is needed, as where the deforestation has already destroyed the sources of wood in the vicinity of the village, then kerosene may be cheaper because its price has long been subsidised. The Indonesian government has, over many years, artificially lowered the kerosene price to help poor people and at the same time protect the forests from further exploitation. At first glance, this seems reasonable; people would naturally prefer a fuel that is ready at hand, easy to handle and cheap in contrast to firewood that is costly and has to be carried over a long distance. However, if we look a few decades ahead, we have to admit that such a policy is shortsighted, since it encourages the consumption of a non-renewable fuel (oil) and discourages the use of a renewable source of energy (wood). In addition such a policy deprives rural people of a desperately needed supplement to their meagre incomes. Howard Dick, who studied the problem in detail, comes to the following conclusion:

If oil prices are allowed to rise steadily until the prices of all fuels reflect their true scarcity, it may become clearer that the denuded hills and mountains of Java, which are now such a human tragedy and an environmental disaster, represent a great latent potential for development. Java's natural forest, like the tiger, has almost disappeared. The challenge is to replace natural with man-made forests which, carefully harvested like a vast house compound, can provide a perpetual supply not only of timber, firewood, and charcoal, but also of leaves, grass and even fruits, nuts, oils, and honey. For the time being wood is scarce and oil is abundant, but the abundance will not last. Any attempt to overcome deforestation by increasing dependence upon oil will, in the long-run, be like jumping out of the pan into the fire.[71]

Before we consider the ways and means of stopping and reversing the deforestation process in Jawa, we have to look at its environmental consequences: 'soil erosion'.

NOTES

1. Penck, 1924; Carol, 1973.
2. Domrös in Kötter *et al.*, 1979, pp. 24–37.
3. Sukanto in Arakawa, 1969, p. 215.
4. Boerema, 1926, divided Jawa into 69 zones, each with a characteristic monthly rainfall distribution. The studies undertaken in the Serayu valley (e.g. van der Linden, 1978*a*, pp. 11–16) show substantial climatic differences in a comparatively limited area due to altitude, exposure and so forth.
5. Domrös in Kötter *et al.*, 1979, p. 26.
6. 'Bebossing . . .' 1973, p. 4.
7. Sukanto in Arakawa, 1969, p. 218; Domrös, in Kötter *et al.*, 1979, pp. 32–3.
8. Pitty, 1978, pp. 136–43.
9. Schmidt and Ferguson, 1951.
10. Sukanto in Arakawa, 1969, pp. 222–7.
11. Bakker and van Wijk, 1951; Mock, 1973; Uhlig, in Imber and Uhlig, 1973, pp. 41–2.
12. Mohr, 1938, pp. 479–83.
13. Van Baren, 1950, pp. 3–4.
14. Fromberg, 1854.
15. Van Hall, 1873.
16. Van Bemmelen, 1890; Koorders, 1893.
17. Dubois, 1899/1900, pp. 410–18.
18. Röll, 1979, p. 116.
19. This is true even today. When Mt Galungung in West Jawa erupted in 1982, thousands of people lost their property and fled. Two years later, 21,000 victims had already returned to the most dangerous area where the lava flow still threatened the settlers (*Indonesia Times*, 16 Jan. 1984).
20. Dames, 1955, p. 28; 'Bebossing . . .' 1973, pp. 31–2.
21. Uhlig in Kötter *et al.*, 1979, pp. 44–9.
22. 'Bebossing . . .', 1973, pp. 25–6.
23. Birowo *et al.* in Kötter *et al.*, 1979; Uhlig, in Kötter *et al.*, 1979.
24. Domrös, 1976, pp. 97–107; Uhlig, in Kötter *et al.*, 1979, pp. 44–9.
25. Hardjono, 1971, pp. 100–2; 'Bebossing . . .', 1973, pp. 29–31.
26. Hardjono, 1971, pp. 97–8; 'Bebossing . . .', 1973, pp. 34–8; Uhlig, in Kötter *et al.*, 1979, pp.51–2.
27. Notohadiprawiro, 1972, p. 248.

28. Hardjono, 1971, pp. 94–5.
29. Soemarwoto, 1981, p. 21.
30. Notohadiprawiro, 1972.
31. Birowo *et al.* in Kötter *et al.*, 1979, pp. 405–6.
32. 'Bebossing . . .', 1973, pp. 27–8.
33. Notohadiprawiro, 1972, p. 247.
34. Hardjono, 1971, p. 94.
35. Birowo *et al.* in Kötter *et al.*, 1979, pp. 406–7.
36. Uhlig, 1976, p. 321.
37. Birowo *et al.* in Kötter *et al.*, 1979, p. 407.
38. Junghuhn, 1845.
39. Daneš, 1915.
40. Escher, 1931, p. 259.
41. Uhlig, 1976, (suppl.).
42. Flathe and Pfeiffer, 1965, p. 535.
43. Pelzer in MacVey, 1963, p. 18; Uhlig, 1976, p. 325.
44. Baharsjah in Ensminger, 1978, p. 83.
45. Dames, 1955, p. 127.
46. Baharsjah in Ensminger, 1978, p. 81.
47. Woelke, 1978, pp. 1–2.
48. Dames, 1955, p. 20.
49. U.Scholz in Kötter *et al.*, 1979, p. 55.
50. Hesmer, 1970 (vol. II), p. 55.
51. Dames, 1955, p. 22.
52. Van Steenis, 1935 (quoted by Dames, 1955, p. 22).
53. P.W. Milaan's contribution to Dames, 1955, p. 21.
54. Sudiono and Daryadi, 1978, p. 121.
55. Van Steenis, 1937, p. 640.
56. Hartmann, 1981, p. 12.
57. *See* van Baren, 1950, pp. 6–13, with an extended bibliography.
58. Junghuhn, 1852/4.
59. de Haan, 1942; Bailey and Bailey, 1960, pp. 266–78.
60. G. Gonggrijp, 1937, p. 682.
61. Loc. cit.
62. Thorenaar, 1937.
63. Appelman, 1937.
64. Thijsse, 1974, p. 5. The Indonesian Forestry Service feels that Thijsse exaggerates when he maintains that between 1940 and 1973 two-thirds of the forest area was lost; they say 'only half' of it was lost. But even if this is true, the process of deforestation is still going on and if it continues at a rate of about 50% in 33 years, the Javanese forest will have disappeared by the year 2010. See also Thijsse, 1975*b*.
65. 'Kali Konto Project', 1979, p. 7.
66. Woelke, 1978, p. 212.
67. Manning, 1971, p. 31.
68. Wiersum, 1978*a*, pp. 307–8.
69. Palte, 1980, pp. 79–81.
70. Loč. cit., pp. 57–9.
71. Dick, 1980, p. 60.

6

THE FINAL STAGE: SOIL EROSION

Soil Erosion: Facts and Figures

Soil erosion is a geomorphological phenomenon which is present for as long as there is any material to be eroded. The present configuration of vast areas of the earth has resulted from this process, taking place during geological eras. To understand its importance for present-day agricultural cultivation, we only have to look at the alluvial plains worldwide. Even geographers belittle the warnings of today's soil erosion; without it, they say, there would hardly be enough level and fertile land to feed the world's population. Defenders of the 'beneficial effects of erosion' maintain that it is a good thing for soil particles, spread over the mountain slopes, to be blown away by the wind or washed off by rainwater because they are then accumulated elsewhere in the form of thick layers of fertile loess or as alluvial plains. All these highly productive zones would not exist if wind and water had not done their work.

This statement can easily be verified in Jawa, particularly in the northern part. There the terrain consists of alluvial plains, the material of which has been brought down from the mountains and accumulated in the lower reaches. This process probably started some 5,000 years ago when the maximum post-glacial sea-level was reached, but it has still not yet come to an end. If we calculate the sedimentation and accumulation process on the basis of the actual annual formation of new alluvial land, the whole plain of northern Jawa would have been built up in no more than 2,000 years. However, some 5,000 years have passed since the process began, and we can therefore assume that the speed of accumulation increased, as the result of erosion, by 2.5 times and that it is still increasing.[1]

The extent to which the northern coast of Jawa is advancing into the sea is indeed remarkable. The coastline is actually being pushed out by 10–50 m. annually, the amount depending on the extent of erosion.[2] Soil accumulation occurs, as would be expected, particularly around the river deltas where most silt is deposited. By contrast, delta formations are few along the southern coast because the heavy breakers, tide and currents of the Indian Ocean remove all the mud as soon as the rivers carry it into

117

Table 5. INCREASE OF RIVER DELTAS ALONG THE NORTHERN
COAST OF JAWA (in sq. km.)

| River system | Period | Watershed | Delta growth | |
			Total	Annual
Punagara	1869–1934	1,450	18.4	0.28308
	1934–46		10.7	0.89167
Manuk	1857–1917	3,650	26.0	0.43333
	1917–35		20.1	1.11667
Bangkaderes	1853–1922	250	8.8	0.12754
	1922–46		3.0	0.12500
Sanggarung/Bosok	1866–1922	940	18.0	0.32143
	1922–46		18.1	0.75417
Pemali	1865–1920	1,200	7.1	0.12909
	1920–46		10.8	0.41538
Comal	1870–1920	710	2.2	0.04400
	1920–46		6.2	0.23846
Bodri	1864–1910	640	5.9	0.12826
	1910–46		15.2	0.42222

Source: Hollerwöger, 1966, p. 347.

the sea. Along the northern coast the tide is less high and the currents are comparatively weak.[3]

Thanks to topographical maps and, more recently, aerial photographs, it is possible to reconstruct and measure the process of delta formation since the middle of the nineteenth century. This gives us a good insight into the erosion and sedimentation processes, and allows us to conclude that soil erosion is expanding rapidly because of the unremitting population growth, deforestation, the acquisition of land and the increasing cultivation of dry-land with annual crops. Within the past 100 years, its speed has increased by three to four times. For instance, between 1869 and 1934 the delta surface of the Punagara river (West Jawa) grew annually by about 0.28 sq. km.; but between 1934 and 1946, it grew at the rate of 0.89 sq. km. per year. The delta surface of the Bodri river (Central Jawa) grew by 0.13 sq. km. a year between 1864 and 1910; and between 1910 and 1946 by 0.42 sq. km. Table 5 gives more examples that were computed by Ali Djojoadinoto of Bandung on the basis of Franz Hollerwöger's maps.

But we have to remember that the picture we see of the transport and deposit of sediment is not complete, since with the growing depth of the coastal waters, more sediment is needed to build up a particular delta surface, and of course some of the sediment is carried away by the sea.[4]

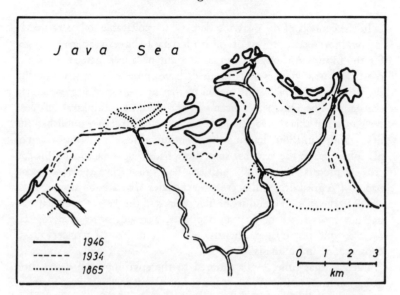

Fig. 13. GROWTH OF THE CI PUNAGARA DELTA (WEST JAWA)
Source: Hollerwöger, 1966, p. 349

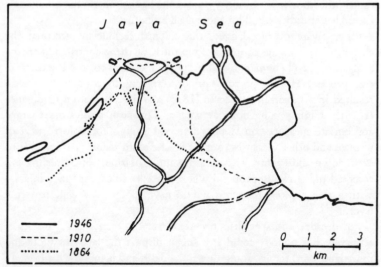

Fig. 14. GROWTH OF THE KALI BODRI DELTA (CENTRAL JAWA)
Source: Hollerwöger, 1966, p. 353.

In the context of the growing demand for cultivable soil, particularly in Jawa, this recent process of soil erosion has to be seen as a great danger for the future. Although a certain amount of land is gained along the coast, and new villages and rice-fields have come into existence during the past century, the loss of topsoil further up the slopes is greater than the gains. The Dutch technicians in the colonial administration were well aware of this development. It seems that K.F. Holle published the first warning in 1866, drawing the attention of the administration and of his colleagues to the fact that something had to be done to stop the soil erosion process. His study, entitled 'Een groot gevaar, dat sluipend nadert' ('A great danger that creeps ever nearer'⁵) became an often-quoted classic in the erosion literature of Indonesia, and the danger he anticipated has never ceased to occupy agronomists, agro-engineers, pedologists, foresters and not least politicians. He still remains the pioneer of soil conservation in Indonesia.

Even at that time, Holle referred to the environmental destruction going on in America: 'Who does not know that in the United States of America whole regions are brought to hopeless infertility through a long-continued system of reckless and exhaustive cultivation?' This was eighty years before the publication of William Vogt's *Road to Survival*⁶ and ninety years before Rachel Carson's *Silent Spring*.⁷ Most significantly, the knowledge of what is *today* the basic problem for Indonesia's future has already existed for more than a century.

Once aware of the danger, the colonial technicians accepted the challenge, and in the course of time most of them contributed to the analysis of the phenomenon and proposed measures to stop the destructive process. Holle's propagation of terracing to counteract erosion resulted in a government decree (1873) and a Reclamation Ordinance (1874), obliging indigenous as well as European cultivators to take appropriate measures on sloping fields. Through the construction of terraces and other measures to meet local circumstances, erosion was to slow down and eventually cease. But not all farmers or all technicians accepted the terracing proposal wholeheartedly since, especially in areas with a thin topsoil, terracing would have spoiled this delicate fertile layer.

For decades, soil specialists disputed over the merits and demerits of terracing, but nobody could any longer dispute the fact that more and more land was literally going down the drain and being lost for ever. In 1933 the soil devastation had become so impressive in some areas that, after a comprehensive press campaign, public opinion became alarmed.

Land is continuously washed from the deforested slopes and literally 'goes down the drain'.

The campaign referred mainly to the Sindang-palai area, a hill tract south of Majalengka in West Jawa belonging to the catchment area of the Lutung river. This area, built up of highly erodible marlstone and also containing sandstone and limestone, was being intensively used as pasture for cattle, partly also for dry-land cultivation (*tegala*); it was only used to a minor extent for the production of wet-rice. In particular, overgrazing led to a heavy violation of the protective grass cover so that rain and wind could easily find weak points. And since the annual precipitation is 3,000–3,500 mm. and storms are frequent, the area was soon violated by innumerable deep ravines or gullies visibly losing their fertile topsoil. Thus the term '*het stervend land*' ('the dying land') was introduced and accepted as a technical term.[8] However, *het stervend land* was by no means confined to the area mentioned. For example, in the neglected Indonesian tea-gardens that stretch along the southern slopes of Mt Salak, about 25 km. south of Bogor in West Jawa, Coster observed that about 50 cm. of topsoil was washed away so that the roots of the tea-bushes and the sterile subsoil were visible. *Alang-alang* and ferns had invaded the area.[9]

The Keruh river, like the Lutung, is a tributary of the Ci Manuk. Rising on the slopes of the Ciremay volcano, it flows northwards over an area built of marl soils that are very much prone to erosion. According to Junghuhn and Verbeek, who travelled there in 1837 and 1890 respectively, the ground was well covered with natural forest and they found no sign of erosion — which, according to the oldest inhabitants of the watershed, started at the beginning of the twentieth century when pasture land expanded visibly. The catastrophe came when there was a great food shortage in 1917–18 and the government encouraged the peasants to clear the forest and grow as much food as possible. From this moment on, the erosion of the soil, which was completely unsuited for annual crops, increased unavoidably. The smallest creek or tributary developed into a steep gully often 20 to 25 m. deep that could no longer be secured by artificial walls, dams or the like. Thus this area too developed into a typical *stervend land*.

Coster described another form of soil devastation as a consequence of inexpedient land use in East Jawa. Here the fertile alluvial plains, on which the sugar factory of Besuka near Buduan grew its cane, were surrounded by forested hills. These hills were eventually offered for clearing and settlement with the result that enormous gully erosion developed and the rains carried a mass of mud, gravel and stones down and spread it over the sugar-cane fields. Hence not only did the protective forest of the hills and the forest soils disappear, but the fertility of the plain was also tangibly reduced.[10] This did not actually come as a surprise to those who had studied and observed the development scientifically and had warned of the threatening catastrophe. As early as 1882, the hydrological aspect came under discussion and the disastrous consequences of unrestrained deforestation were spelled out. Thousands of people cultivating the alluvial plains would be endangered by floods and deposits of debris caused by the few who had cleared the hills.[11] In these early discussions, the protective role of the tropical forest and the limited fertility of its soils also came into the picture. Hitherto everyone had seen the soil underlying the tropical tree cover as an inexhaustible source of fertility, and Indonesians as well as foreigners took full advantage of this alleged free gift of nature. But already in 1899, European coffee planters were aware of the negative consequences of deforestation:

The soil, formerly protected against the sun by a canopy of leaves is now an easy prey as regards destruction of organic matter, solution of plant nutrients and the carrying off of the same by rains. The brooks, formerly containing clear water, are now loaded with suspended soils, part of the most fertile soil being transported to the lowlands. It has to be acknowledged that this loss cannot be avoided if we reclaim virgin land for coffee cultivation.[12]

Indeed it seems that the coffee planters slowly realised that they had not taken enough care in protecting the soil from which they earned their livelihood, and that their techniques might destroy in a few days what nature had built up over hundreds of years.

In the following decades, soil specialists looked for techniques that would allow soil cultivation without soil destruction. A proper drainage of the rainwater was felt necessary, and level ditches were recommended to reduce surface run-off and to enrich the ground water. Already in 1920, covering plants such as *Leucaena glauca* were proposed to protect the soil against rain and run-off, and in 1930 investigations became more and more detailed as erosion was also found to be prevalent under a vegetative cover. Unfortunately, while investigations and technical efforts went on to this very date, the population continued to increase, thus nullifying all the efforts.

If we consider the history of soil conservation in Indonesia and the experiments, discussions and recommendations produced over a century,[13] we can only share the frustration of the few responsible specialists fighting against an ever-growing flood of millions of people who spread over the land, cut down the trees and till the ground, watching helplessly as it is washed away leaving the ground bare.

We can now understand very well the difference between 'natural' erosion that took place over geological eras and mostly in places where there was hardly any population worth mentioning, an erosion that built up the land which is used today to feed people; and the rapid 'man-made' erosion of densely populated areas, an erosion that destroys the land from which people earn their living. These facts are today more or less known to the interested public, because since the 1970s they have found their way from scientific journals to the press. However, the deforestation and soil degradation that are occurring all over the world are often either exaggerated or underestimated, sometimes according to the writer's interests. We will therefore try to quantify the process as far as possible, drawing attention to the many factors which, by their interaction, lead to the destruction of soils. This is a complex process and accordingly measures to stop it require a complex approach. If we quote data concerning, for example, suspended sediment in certain rivers or soil erosion in a certain area, we refer to experiments made by specialists in various parts of the country, mainly in Jawa. Some of these experiments were carried out many decades ago when sophisticated measuring methods were not yet known; and even today all experiments include assumptions that may be valid for the place of the experiment but do not necessarily allow the findings to be generalised. Soil is by no means a homogeneous material and the landscape from where it is washed away is by no means uniform. If it is stated that from a given watershed a soil layer of x mm. is removed, this can only be an average value; there are areas where much more soil is carried away and others where some may even be deposited. Similarly, if it is observed that at a given point on a river a certain quantity of suspended material is being carried down, this quantity is not necessarily uniform from source to mouth.

As we have seen, soil erosion in Jawa is by no means a recent phenomenon. In 1952 de Haan estimated that 30–40% of the land surface had been to some extent eroded,[14] and a number of Dutch and Indonesian specialists have provided details of erosion research and control.[15] Nowadays, practically all agricultural projects, often carried out with foreign assistance, include an environmental component dealing with soil erosion[16] so that agriculturists, politicians and indeed ordinary people become conscious of the imminent danger. For example, studies of the Lutung river basin revealed that the annual loss of topsoil had doubled between 1917 and 1948 due to deforestation, overgrazing and cultivation of unsuitable soils. Under the prevailing conditions, an average loss of 10 cm. of topsoil for the whole catchment area would occur within 50 years. In 1911/12 this river removed a total silt load of 821,000 tonnes, but in 1934/5 the figure had risen to 1,790,000 tonnes. Assuming a specific gravity of 1.5, this would equal a soil removal in tonnes per hectare of 13.2 in 1911/12 and 28.9 in 1934/5, or a yearly erosion of 0.9 and 1.9 mm. respectively.[17] A more recent investigation found that in 1973 the soil removal from the same watershed amounted to 75 tonnes per hectare and the annual erosion had risen to 5.0 mm.[18] The increasing speed of erosion has been proved in various catchment areas: the watershed of the Serayu river in Central Jawa had an average annual erosion of 2.5 mm. in 1911, which had increased to 3.9 mm. by 1973.

A particularly serious case is that of the upper Solo river watershed. Here, about 50% of the surface has been severely eroded and visual reconnaissance suggests that the process is accelerating. However, the catchment area is tangibly heterogeneous. Whereas 60,000–100,000 ha. belonging to the Surakarta regency are critically eroded, the soils of the Madiun regency are less affected because of extended National Forest Estate land. Measurement of the suspended sediment load of the Bengawan Solo river over a 5 month period (rainy season) at Surakarta in

1970/1 showed a total of 7.3 million tonnes from a catchment area of 289,000 ha. The two sets of data have to be adjusted, however. First, if one bears in mind the eroded soil that was not sampled, whether it was in the bed load of the river or had been deposited upstream, and the fact that 15% is coarse sand probably deriving from the Merapi volcano (thus not part of the eroded topsoil), then the total sediment comes to 8.6 million tonnes. Secondly, it is believed that most of the sediment comes from about 162,000 ha. of the Southern mountains in Wonogiri county and from the slopes of the Lawu volcano in the Wonogiri and Karanganyar counties. Thus there are specific parts of the catchment area with particularly strong erosion.

Assuming that one hectare of soil with a depth of 17 cm. — the so-called plough sole — weighs 2,200 tonnes, the equivalent erosion is 3,923 ha. losing a topsoil layer of 17 cm. in one rainy season. If we take instead a depth of 2.5 cm., this would affect a surface of 26,245 ha. If these figures are applied only to the steep lands of the entire upper Solo watershed, we come to an estimate of 10,500 ha. eroded by 17 cm. in a 5-month rainy season. If, finally, we assume an erosion equivalent of 26,245 ha. per year to a depth of 2.5 cm. and most of this comes from 162,000 ha., then only 6.2 years would be required to erode all that land to this depth, and less than 41 years to erode it to a depth of 17 cm. Of course, erosion is not uniform over large areas, but the figures proclaim the seriousness of the problem, particularly since the area in question is among the most densely settled on Jawa: the average in 1970 was 700 persons per sq.km., with a minimum of 250 and a maximum of 1,500. And already more than a decade ago, large areas produced only one-third to one-tenth of their former output. Productivity and potential are further reduced and much land has been abandoned as useless.[19] Table 6 gives the erosion rates of a number of Javanese catchment areas.

A comparison of the specific degradation, expressed in the annual loss of soil in tonnes per sq. km., with other rivers of continental Asia shows that the Javanese data generally exceed those of the continental rivers. The Kosi coming from the highly eroded Himalayas of Nepal has a specific degradation similar to most Javanese rivers, and the two Huang tributaries, the Lo and the Ching, come from a quasi-desert area where use of unsuitable land seems to have caused extremely severe erosion.[20]

Comparison of the measured data and of changes observed in the field of soil degradation is extremely difficult and sometimes impossible, and one cannot pretend to be offering mathematically exact figures. The methods used may well be accurate, but the materials — soil, water,

Table 6. SOIL EROSION IN SELECTED CATCHMENT AREAS
OF JAWA

River (gauging station)	Catchment area (sq. km.)	Soil type	Annual erosion rate (mm.)	Source
Ci Manuk (Parakankondang)	1,480	mainly volcanic	3.7	NEDECO/SMEC, 1973
Serayu (Cilacap)	n.a.	volcanic, marl, lateritic	1.6	Hollerwöger, 1964
Serayu (Banyumas)	2,665	as above	3.8	SMEC, 1974
Ci Lutung (Kadipaten)	620	breccias, sandstones, clay st., marly sediments	1.0 (1910) 2.0 (1934)	Van Dijk and Vogelzang, 1948
Ci Lutung (Kamun Weir)	600	as above	8.0	NEDECO/SMEC, 1973
Pekacangan (Brengkel)	129	white clays	5.6	SMEC, 1974
Cacaban (n.a.)	79	marly	4.5	Van Dijk and Ehrencron, 1972
Pengaren (Semarang)	n.a.	marl, limestone, lateritic	4.3	Hollerwöger, 1964
Lusi (Purwodadi)	1,966	n.a.	2.5	SMEC, 1978
Konto (Selorejo Dam)	185	n.a.	3.0	Brabben, 1979
Brantas (n.a.)	2,050	n.a.	3.4	Brabben, 1979

Source: Summary of results of erosion studies.

Table 7. SPECIFIC DEGRADATION OF CATCHMENT AREAS OF
JAWA, COMPARED WITH OTHER ASIAN RIVERS

River	Catchment area (sq. km.)	Specific degradation (annual soil loss in tonnes per sq. km.)
Huang Ho, Yellow river (China)	673,000	2,804
— Lo tributary	26,000	7,308
— Ching tributary	57,000	7,158
Ci Lutung (Jawa)	600	5,714
Pekacangan (Jawa)	129	4,000
Cacaban (Jawa)	79	3,214
Kosi (Nepal)	62,000	2,774
Serayu (Jawa)	2,665	2,714
Ci Manuk (Jawa)	1,480	2,643
Kali Brantas (Jawa)	2,050	2,429
Kali Konto (Jawa)	185	2,143
Yüan Hong, Red river (Vietnam)	119,000	1,092
Brahmaputra (Bangladesh)	666,000	1,090
Irrawaddy (Burma)	430,000	695
Yangtze Kiang (China)	1,942,000	257
Mekong (Laos)	795,000	214

Source: Stoddart, 1969; various Indonesian studies.

vegetation, rain and wind — are by no means homogeneous; besides, the measuring is not done under laboratory conditions but in the open.[21] Moreover it often depends on the judgement of the researcher whether or not an area is categorised as, say, 'severely eroded' or 'moderately eroded', as 'bare land' or 'critical land'.[22] For instance, the estimates of critical and eroded soils of the 639,000 ha. watershed of the Ci Tarum range between 86,000 ha. or 13% and 185,000 ha. or 30%.[23]

When in 1955 Dames analysed the soils of eastern Central Jawa over an area of 1.6 million ha., he found 49.0% not eroded, but 36.0% severely, 10.5% moderately and 4.5% slightly eroded.[24] When, in 1970, an Indonesian survey team analysed the erosion of six watersheds in West Jawa covering together an area of 2.1 million ha., only 9% were classified as slightly, moderately and severely eroded.[25] Data and classifications should thus be regarded as trends or approximations rather than as mathematically exact.

Erosion mapping in Indonesia is still in its early stages. Dames made the first efforts in this field (see Fig. 15) which includes the severely eroded watershed of the upper Solo river. Another map, showing the

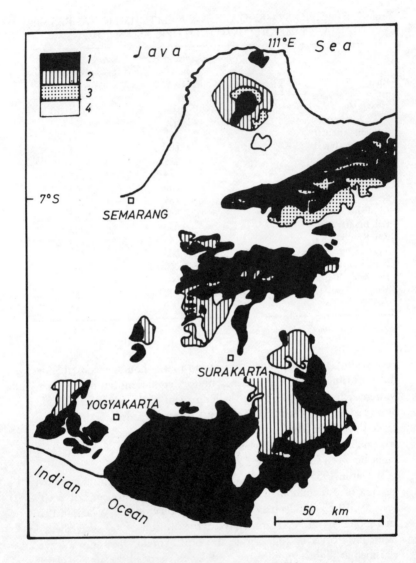

Fig. 15. EROSION IN EASTERN CENTRAL JAWA

1 = severely eroded
2 = moderately eroded
3 = slightly eroded
4 = not eroded

Base map: Dames, 1955, p. 130.

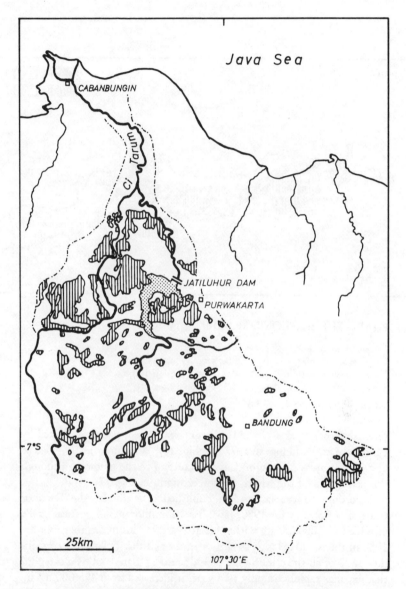

Fig. 16. AREAS OF SEVERE EROSION IN THE CI TARUM
WATERSHED (WEST JAWA)

Source: 'Bebossing . . .', 1973, map VI.

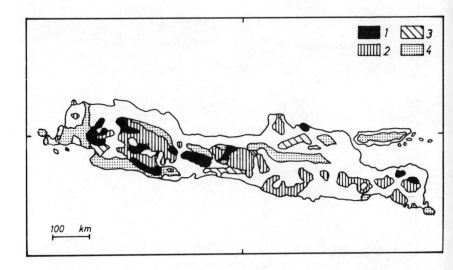

Fig. 17. THE EROSION DANGER IN JAWA

1 = elevation > 500 m; slope > 30%; very serious erosion danger
2 = elevation > 500 m; slope > 30%; serious erosion danger
3 = elevation < 500 m; slope > 30%; serious erosion danger
4 = elevation < 500 m; slope > 15%; erosion danger

Source: Wiersum, 1980*b*; and FAO.

seriously eroded parts of the Ci Tarum watershed, was contributed by
Pengusahaan, Bandung, in 1972 (see Fig. 16). Wiersum, finally, tried to
identify the areas of erosion danger according to the gradient, although
this is only one of the criteria for soil erosion (see Fig. 17).

According to the latest data published officially by the Director
General of Forestry for 1980, Jawa has 1.2 million ha. of 'critical land',
of which 167,000 ha. are within forested areas. This means that roughly
10% of the island belongs to this category and that Jawa alone contributes 18% of all the critical land of Indonesia.[26] 'Bare land' is a category
that expands rapidly; in only two years it increased from 495,000 ha. by
62% to 800,000 ha. in Jawa alone.[27]

In many areas with dry-field agriculture where heavy rainfall can
easily attack the topsoil, there is a regular annual loss of 2–4 cm., so that

in some places the bedrock is already visible and the population has left.[28] According to Thijsse, the total area of dry soils (*stervend land*) in the whole of Indonesia may be about 100 million ha., 40% of which are already in a degraded state.[29] Conservative estimates come to about 80 million ha. of derelict land in the year 2000,[30] and many of the rugged valleys of Jawa will have lost their humus, together with the population that once tilled their soil. [31] The march to catastrophe is accelerated by the growing population and the shrinking surface of arable land.

Two more questions arise: What exactly leads to erosion, and what follows it?

The Factors of Erosion

Vegetation, Land Use and Erosion. We have seen that a certain 'natural erosion' is taking place everywhere in the world, and that in geological times it was often of catastrophic dimensions. Nowadays, however, most of the severe cases of soil erosion and other forms of soil devastation are man-made. Apart from absolute deserts, the soils are generally covered with some natural vegetation, and where the climate and fertility permit, such cover usually protects the underlying soil from destruction. But as soon as man is no longer content with what nature offers him and interferes with the process of natural production, the balance is disturbed. Since research has discovered and described this imbalance — and in Indonesia this goes back to the mid-nineteenth century[32] — many studies and experiments were carried out to establish a new balance that would satisfy both the ecological equilibrium and man's demand for higher land productivity. The population pressure in Jawa has already reached such a level that there is hardly any chance of implementing the necessary techniques to establish a new equilibrium. Nevertheless, research is going on that will provide us with the basis for a better understanding of the process of soil erosion, its causes and consequences.

It is generally believed that dense forest cover is the best natural protection against soil erosion, and there is no doubt that the lack of vegetative cover and certain land use techniques may increase erosion by a hundred times.[33] But different types of forest have a different protective value. Teak forests (*jati*), for instance, give only moderate protection since they have little undergrowth, which is often collected or even burnt. And in natural forests the burning and the collection of litter makes the ground bare and hard, and helps advance the rainwash of soil and the formation of gullies. Grassland used as pasture becomes susceptible to erosion when it is overgrazed and trampled by animals so that the cover is violated and the topsoil becomes impermeable.

Many experiments have proved that while vegetation is of the greatest importance for the control of erosion, various plants have very different effects. This is also true of perennial cultures, i.e. trees or bushes, even within the same genus. For example, in the case of coffee, *Coffea robusta* is less protective than *Coffea arabica* due to their different roots. Rubber plantations (*Hevea brasiliensis*) are believed to give little protection because the activities of the planters press the soil together so that the rainwater runs away quickly. The same is true of cocoa plantations. Generally, rubber plantations are more prone to erosion than coffee gardens, but in principle perennial plants protect better than annual ones, provided they are carefully kept.[34]

Forests protect the soil in two principal ways: by influencing the rainfall and by influencing the soil (see Fig. 18). The forest canopy intercepts the rainfall and thus decreases the amount of water that reaches the soil directly by splitting it into a larger throughfall and a smaller stemflow. The latter is of no importance in erosion, but the throughfall has a higher erosive power than rainfall drops, according to the type of tree.[35] By its production of a substantial yield of litter (fallen leaves, rotting organic matter) and by its undergrowth, the forest influences the soil. The litter, however, is the more important anti-erosion agent.[36]

In 1938 Coster published the results of extended experiments dealing with the downflow of rainwater and with the suspended sediment load of the water.[37] Both are good indicators for the magnitude of erosion. If a large part of the rain over a certain area penetrates into the ground, the water is stored subterraneously and will feed wells; it will not run off taking soil particles with it. The percentage of run-off, i.e. the percentage of the total rainfall flowing off over the surface, permits a comparison of the anti-erosive impact of certain vegetative covers or land use techniques. Anyway, the crucial question is the extent to which the mineral soil is exposed to rain, sun and wind.

On a good forest ground well covered with litter and undergrowth the downflow will hardly exceed 1–2% of the rainfall, and there is no erosion whatever. When the forest is cleared, the downflow immediately rises to 25–50% of the rainfall and the annual erosion to 5–12 kg. per sq. m., or 30 kg. on loose and sandy ground.[38] However, the quality of the vegetative cover rises with the presence of litter accumulating on the ground, or a dense cover of herbs or grass — though bunchy-growing grasses such as *Andropogon*, *Festuca* and *Themeda* (also *Imperata cylindrica*)

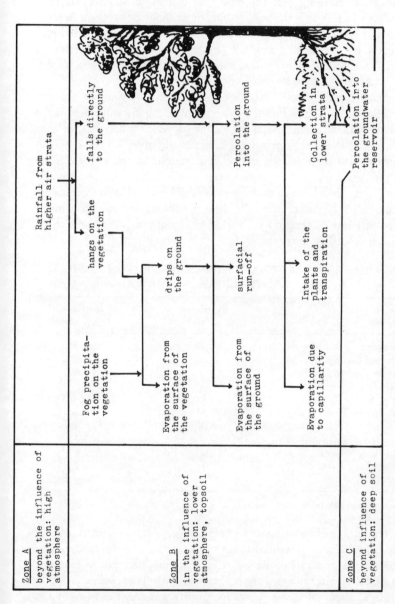

Fig. 18. THE HYDROLOGICAL IMPACT OF FORESTS

Source: de Haan, 1935, p. 1114; for a more recent interpretation see also Bruijnzeel, 1983, p. 93.

leave so much bare soil between the bunches that rain may initiate gully formation. Burning over the grass will destroy what litter has accumulated and expose the soil, leaving it without protection. *Imperata* recovers well after burning, and fresh sprouts will bring some soil protection, but repeated burning will also destroy litter and impoverish the soil. Bamboo forests have a luxuriant undergrowth, or at least a thick layer of litter, and are therefore good soil protectors.

According to Coster, there are two extremely favourable types of vegetative cover. The first is a mixed luxuriant grass wilderness in the mountains, with much wild sugar-cane growing in between, in clear contrast to the bunchy-growing grasses; usually there is a thick layer of litter so that there is little run-off and erosion. Secondly, there is the mixed rain forest with lower temperatures where litter slowly turns into humus, an abundance of undergrowth, and loose soil with microbe life and a high humic content. In these conditions the run-off remains moderate even on steep slopes. Table 8 gives a summary of Coster's findings in more than thirty test plots.

Table 8. RUN-OFF DATA FROM SMALL PLOTS UNDER
DIFFERENT VEGETATION (JAWA)

Vegetation or land use	Number of test plots	Run-off as % of average rainfall
Bare soil	6	25–55
Dry-land farming	5	2.0–16.2
Young forest plantation (under 2.5 years old)	3	2.8–9.5
Grass, mainly *alang-alang*	6	0.28–5.4
Bamboo and grass	4	5–20[a]
Jungle	3	1.5–2.6
Teak forest	1	8.0[a]
Mahogany forest	1	3.6
Thinned pine forest	1	3.2
Rain forest	4	1.3–6.2

Source: Coster, 1938 (summarised).

The change of the vegetative cover is closely connected with land use and its techniques. If we look at the two main types — dry-field cultivation and irrigated agriculture — we find tangible differences with regard to erosion. Coster analysed dry-field cultivation in Jawa, *ladang* as well as *tegala*, in 1938, when the population density was only half of what it is now:

The fact that inundation and erosion in the extremely densely populated island of Jawa (316 souls per sq. km.) are fairly small compared with those in other countries, is due to the bare soil being rapidly covered with herbaceous vegetation and shrubs. Intermittent agriculture (*ladang*) with fallow periods of many years duration, is not, therefore, very harmful in the rainy tropical climate with its luxurious vegetable growth. The regular burning off, often inherent on this form of native agriculture, destroys however the fertility of the soil.[39]

But as we have seen, population pressure results in shrinking fallow periods and more and more dry land has been taken under cultivation after the clearing of the jungle. Observations have shown that during the first two years after forest cultivation, the downflow is not very severe, but when the annual cultivation is continued, the upper soil is washed away very soon and the ground becomes barren. Especially in the mountain forest, where the layer of humus soil is no thicker than 10–15 cm., the activities connected with primitive agriculture may prove disastrous.[40] Gonggrijp, who verified Coster's findings in the Ci Widey watershed, reported that in the first year after deforestation there was no serious erosion, but that it increased after the humus layer was washed away. In the second year after deforestation, the erosion from non-terraced ground amounted to about 5 kg. per sq. m. a year, or 5,000 tonnes per sq. km.[41]

Similar to the vegetative cover, the treatment of the cultivated soil determines the degree of erosion. Terracing has long been recognised as a protective technique, and *sawah* terraces are especially useful since they can store several centimetres of water behind their earth dams as long as the terraces are dry. However, in the wet season, when they are filled with irrigation water, they cannot hold additional rain, and the downfall of 30% of the precipitation is as high as on bare ground.[42] Other writers have given even higher downflow figures, but information on the impact of *sawah* terraces on the rainwater run-off in the wet season is scarce, particularly since such terraces and the cropping cycles are not uniform. It has even been maintained that *sawahs* sometimes increase the run-off during the wet season when intensive rainstorms occur.[43] In contrast, terracing of dry-land tangibly reduces the downflow of rainwater because it facilitates its penetration into the ground. One particular area, monitored four years after deforestation, showed 25% run-off on non-terraced and only 7% on terraced slopes.[44]

This may indicate that the forest is not the most important anti-erosive agent but that the way the soil is prepared for cultivation has the greater protective effect. A recent study of the watershed of the Tarum river, near the Jatiluhur reservoir, concluded: 'Erosion is not caused by deforestation, but by bad land use practices. Activities in the forests whereby the litter layer is destroyed will result in erosion, while agriculture with good soil conservation practices will not give much erosion. Consequently, in erosion control projects emphasis has to be laid on good overall land management rather than only on planting trees.'[45] This should not be misunderstood: forests have many indispensable functions, the anti-erosion effect being only one of them, but it is not the least important. The same study showed that 'it is the forest's influence *on soil* rather than on rainfall which gives it its erosion protective value. This influence is exerted by the normal ecological process operating in a forest, in particular by the process of litter production and decomposition. If these processes are disturbed, erosion can be heavy in spite of the presence of trees.'[46]

Climate and Erosion. The discussion is still going on today as to whether or not forests influence the climate and particularly the rainfall of an area. Already in the 1930s, observers in British-ruled India had found that such interrelations cannot be excluded. From the tea plantations of the Preanger highlands in West Jawa it was reported that the downpours and consequently the floods (*banjirs*) have increased since more forests have been cut back. Similarly, in the area south of Bandung, the daily maximum rainfall has risen with advancing denudation. However, the extent to which forests influence the rainfall is difficult to determine since other factors obviously come into play. We know that deforestation in many parts of the world furthers the drying-up of an area, leading to desertification, yet under different conditions the cutting back of forests may lead to more disastrous rainfall. It is well known that forests can 'comb out' rains from the clouds drifting by, often at the expense of the hinterland which then dries out.[47] Anyway it is safe to say that a dense vegetative cover can be expected to 'smooth' the climate; the rainfall and hence the run-off on the ground may become more balanced.

We have seen that soil erosion in Jawa is mainly caused by rainfall, whereas wind erosion is quantitatively less important. We have already noted that Jawa, on account of its relief and its position in the monsoon tracks and in the intertropical convergence zone, receives an enormous annual rainfall. This can reach catastrophic proportions through its concentration in only a few months. Yet, on the other hand, the average annual precipitation decreases from west to east.

One of the worst catastrophes ever recorded in Jawa in historical times was the flood of 1861. Following a long dry season towards the end of 1860, a cyclonic storm began on 19 February 1861, and there followed continuous rain from 19 to 23 February between 109° and 114° East, thus covering the whole of Central and East Jawa. The clouds remained, hanging between the mountains and precipitating an estimated 1,000 mm of water. The consequences were described by Klinkert:[48] some of the rivers rose 12–18 m. above their mean level, some were silted up and changed their course, and rocks came thundering down the valleys. There were innumerable landslides. At the end, vast areas were flooded, including 70 sq. km. alone around the town of Surakarta. Plantations and gardens were destroyed, *sawahs* washed away and fields and pastures in the plains covered with mud and rubble, in some places 1 m. deep. Lakes and reservoirs accumulated sediment to a depth of more than 3 m. Whole villages were flooded and some disappeared into the sea. The total number of deaths was never reported. Indeed this was so great a natural disaster that coping with it seemed to be almost beyond human ingenuity. Yet the consequences of many minor disasters could be limited if the protective forests and the cultivated land were well kept. Once the forest is reduced to a mere memento and the fields are not arranged in an anti-erosive way, then even a medium downpour can do great damage.

It is obvious that the first thing that leads to erosion is the impact of raindrops upon the soil. If the rainwater can percolate into the ground, it does no damage; if it cannot, it flows off over the surface taking along parts of the topsoil. It seems, however, that — within certain limits — the strength of rainfall does not necessarily correlate with the dimension of erosion. In the Lutung watershed, the conditions between 1911–12 and 1934–5 were compared and it was found that whereas the rainfall had increased by 8%, the sediment yield had done so by 111%. This cannot be merely explained by a coincidentally higher rainfall in 1934–5, but only by a growing deforestation, over-grazing and the misuse of the soil in the watershed.[49]

Nevertheless, rain can cause damage in two ways. As well as carrying away the topsoil (erosion), it washes the nutrients off or into the underground (leaching). The widespread view that the Indonesian soils are extremely fertile and that there is sufficient rain for almost all crops to be grown is erroneous: 'The very fact of the high rainfall explains, to some extent, the infertility of soils. Rapid leaching occurs in areas where rainfall is excessive; the fact that giant dipterocarps have been growing for centuries in the rain forests of the archipelago is no indication that soils are fertile in themselves. With an almost constant movement of water downwards through the soil, plant nutrients are removed from the upper layers and the reddish clay known as laterite frequently forms.'[50] In tropical karst conditions where limestone soils prevail, the rain washes the ground into depressions and cavities where it accumulates and then is often cultivated or supports wild vegetation. The rainwater itself quickly penetrates into the underground where it gathers in subterranean streams and lakes, often at such a depth that wells up to 50 m. deep dry up shortly after the rain stops.[51]

The rainfall changes daily, seasonally, and geographically, and thus its impact is very unequal, but erosion, measured as the sediment load of the rivers, indicates the extent to which the precipitation actually has an impact on the soil. Rivers that are reached by the east monsoon generally seem to have a lower sedimentation than those hit by the west monsoon. Similarly, the chemical content of the river water can differ according to the type of monsoon.[52]

Finally, it should be remembered that rain is not the only erosive agent. Soils on steep slopes crumble away in quantity when left unprotected during the dry season. This observation was made in coffee plantations as early as 1904. On pronounced slopes the loss of soil was greater in the dry season due to the so-called 'dry-wash' than during the west monsoon.[53] High annual temperatures combined with heavy rainfall results in a rapid weathering of the rocks. In East Jawa, a lengthy dry season and strong insolation furthers the breaking-up of the rocky marls and volcanic tuffs — which then weather quickly. And if the vegetation cover is thin, climatic factors can easily erode the ground: sunshine, heat, wind and, last but not least, rain.[54]

On the other hand, climatic conditions play an important part in the regeneration of natural vegetation, determining a natural anti-erosive development for the protection of the soil. In most parts of Indonesia, the climate, expressed in terms of temperature and humidity, supports such regeneration by furthering growth once the area is left to itself and is well protected from fire, cattle, and, of course, man. Such natural reforestation had been observed and described before the end of the nineteenth century, where coffee plantations as well as the indigenous population's subsistence plots (gogo-fields) had been abandoned 15–20 years earlier: the process had been accelerated by the favourable climate and some shadow trees left over from the former plantations.[55]

Specialists do not recommend rainfall data as an indicator of the dangers of erosion, but instead base their decisions on the amount of sediment in the rivers. As we have seen, there is little correlation between rainfall and sediment, but there is a link between the soil type and the river load.

Soil Properties and Erosion. 'It is a well-known fact', write van Dijk and Ehrencron,[56] 'that many factors such as soil type, rainfall, slope and vegetation influence the degree of soil erosion caused by surface run-off from an area. Human influence too, either for good by judicious managing and planting, or evil by improper use of land, may either delay or accelerate the erosion.' When they analysed the sedimentation of two rivers running north from the Central Javanese volcano Mt Slamet in 1938–9, the two authors found a very good example of the influence which soil properties have on erosion. The catchment areas of the two rivers are adjacent, the annual rainfall lies roughly in the same magnitude, viz. between 3,000–4,000 mm. and the types of land use are quite comparable, namely some forest, some *sawahs* and some dry land agriculture (*tegala*). The main difference, however, is the geological character and consequently the type of soil prevailing in the catchment area. The Rambut river runs almost entirely over young volcanic deposits of Mt Slamet, consisting of olivine containing pyroxene-andesite and basaltic efflata. The underground of the Cacaban river, however, is a middle-miocene formation of marlstone, marly sandstones, sandstone breccias, and local limestones. The velocity of the total erosion during the year of observation amounted to 0.42 mm. on the volcanic material and 4.51 mm. in the marly region. This means that the erosion on marly material exceeds that on volcanic soil at least tenfold.

This was not the first investigation into the influence of soil properties on the rate of erosion. As early as 1917, L. Rutten had submitted a study based on a number of Javanese rivers on different geological ground (see Table 9). He concluded that the denudation velocity depends to a great extent on the geological situation of the watershed and that the velocity is smallest in volcanic areas, increasing with the growing content of miocene clays and marls in the ground. According to his findings, it seems that the influence of rainfall is very moderate. A catchment area with 4,000 mm. rainfall had a smaller annual denudation than one with 2,000 mm. Similarly, the influence of relief seems to be smaller than expected. Catchment areas with a pronounced relief showed a tangibly smaller denudation than those with a more balanced relief.[57]

Table 9. CONNECTION BETWEEN GEOLOGY AND EROSION OF
SELECTED CATCHMENT AREAS IN JAWA

Catchment areas by geological type	Annual denudation (in mm.)
Totally, or almost totally volcanic rocks: — Ci Liwung river — Brantas river — Banjuputhi river	0.25–0.38
Volcanic rocks prevail: — Ci Manuk river — Tajun river	0.36–0.60
Volcanic and clayey-marly rocks are in balance: — Ci Lamaja river	0.1–0.4
Clayey-marly rocks prevail: — Jragung river — Pengaron river	2.65–3.75
Mixed geological basis: — Lusi river — Serayu river	1.20–1.60

Source: Rutten, 1917 (summarised).

Another example is the Ci Lutung. Two-thirds of its watershed lies on miocene deposits. Van Dijk and Vogelzang estimated that 90% of the sediment load came from these layers and that a loss of topsoil of 10 cm. could be expected in 35 years.[58] It may be remembered here that in a similar way to vegetation cover, soil properties also tangibly influence the run-off of rainwater. And since this is the most important erosive agent, low run-off means a high degree of percolation and consequently less erosive power. Recent studies in the catchment area of the Serayu river revealed that river basins situated mainly on volcanic rocks and their soils had a mean annual discharge of 57%, those on sedimentary rocks and their soils 95% and those on a mixed underground 79%.[59]

Thorenaar[60] compiled the following information concerning soil types, their properties and their erodibility:

Ash soils, a very young and loose material, strongly affected by erosion. On steep slopes following heavy rainfall they often form mud streams (*lahars*).

Laterite soils usually resist the denudation process quite well, but on slopes the brown and red topsoil is often so eroded that the bedrock shows through.

Limestone soils are easily eroded especially in areas of pronounced relief. On the upper parts, the soils are often thin, or have already disappeared and accumulated in thick layers in the valleys.

Marly soils are likewise soft and very prone to erosion. In areas where lime/marl soils prevail, concretisations of iron and lime are sometimes left behind on the surface.

Quartz sands form soils that are rather resistant to erosion, except for the red topsoil which weathers easily.

Old forest soils in the high mountains are easily washed away, and sandy *dune soils* are usually displaced by the wind.

As regards soil properties, forest areas exhibit different characteristics. It is possible for densely forested areas on marly ground to produce higher floods and lower minimum downflows than deforested areas of volcanic soils. This does not mean, however, that there is no protective function afforded by the forest, but simply that the volcanic soil types are hydrologically more favourable than marly soils even under forest cover. To see the difference between the two soil types properly we have to imagine the marl area without forest cover.[61] According to its outward shape we can distinguish various forms of erosion:[62]

Splash erosion starts under the immediate impact of the raindrops upon the soil, breaking up loose soil particles and washing them down the slope. Eventually, this leads to clogging of the soil's pores so that percolation is no longer possible and the total rainfall is forced into run-off. If the surface is smooth, the running of water will uniformly erode large areas causing *sheet erosion*, which prevails mainly on a less pronounced relief. This type of erosion is not very spectacular, but it can cause a significant loss of topsoil from a given area. Two basic processes are involved in sheet erosion: the breaking loose of the soil particles (splash erosion) and the transport of the same (sheet flow). Results of sheet erosion often show up as light-coloured patches of soil on hillsides. The dark-coloured surface soil, with its characteristic organic content, has been lost, exposing the organically deficient lighter-coloured soil, indicating a loss of productivity.[63] Sheet erosion in the proper sense presumes a rather smooth surface that hardly exists in nature. Thus, once the water attacks a weak point in the soil such as a depression, a violated grass cover or the like, the next stage — *microchannel rill erosion* — begins. The erosive agent is no longer the raindrop but the flowing water that digs into the soil and, depending on its velocity, washes away more and more soil particles. Intense storms may even cause the water to dig into the subsoil, whereas sheet erosion remains on the top layer. However, in theory rill erosion can be done away with by ploughing.

The next stage is *gully erosion*, which often follows rill erosion or is initiated by natural depressions and man or animal-made incisions such as tracks and ruts up and down the hill. Field boundary gullies are one example. Great quantities of water coming down the slope at great speed dig depressions or incisions often many metres deep, destroying the cultivated surface of the landscape. The amount of water available is closely related to the size and run-off producing characteristics of the area involved. The rate of gully erosion is further conditioned by the soil characteristics, the size and shape of the gully and the slope in the channel. In many parts of Indonesia, these types of erosion can be seen and their annual progress observed. Ploughing can no longer heal such wounds.

Under certain conditions, the conservationist has to face a type of soil devastation that is grouped with others under the collective term of *mass wasting*. Soils and overburdens, being poorly held together, are very susceptible to gravity movement, which means that the top layer of a whole slope may slowly slide down. This process requires a subterraneous stratum slippery enough, through water and its physical properties, to permit such movements, also called *soil creep*.[64] In the northern part of the Serayu valley, whole slopes have been lost together with the rice terraces built on them and the tree vegetation growing on the creeping layer. A motorable road had already been covered and a bridge had slipped with the soil about 20 m. downhill.[65] Soil erosion does not simply denude the hills and destroy cultivable land, it also leads to other serious consequences.

The Consequences of Erosion

Erosion and Hydrology. The hydrological effects of erosion fall into two categories, namely the surface run-off and the underground reaction. Once the protective function of the forest or any other vegetation cover is destroyed, the hydrological consequences are visible. The regulative function is lost and that means that when it rains the downflow data of every creek, rivulet or river rises substantially, as does the run-off over the slopes that we have already described. When the rain stops, the downflow of the watercourses decreases rapidly, and in many places where river water formerly permitted a second crop to be grown, the fields have to remain fallow and barren because of lack of water. The difference between the wet and the dry seasons becomes more pronounced. Whereas in the rainy season catastrophic floods (*banjirs*) often destroy the crops, in the dry season droughts halt their growth.

The hydrological behaviour of the river largely depends on its geological and pedological basis. Rivers on sedimentary underground show a violent reaction to rainfall. They rise and fall with the rainfall level, so that fluctuations between high and low flow occur very often, sometimes even daily. Infiltration is small and so is the groundwater discharge. Such rivers transport large amounts of sediment, ranging from clay to boulders, which settle as soon as the flow slows down, helping to form gravel and sandbanks. In this way they evolve an instable bed and often change their course.[66] Such hydrological behaviour has severe consequences for the farmers who settle in the watershed, even if it is not an area of pronounced relief. If we take, for instance, the watershed of the Lamongan river in East Jawa that emerges from the edge of the Kendeng hilltracts and flows over an extended plain to debouch near Surabaya, we can illustrate the disastrous hydrological results of denudation and erosion. Though the soils of the area (andosols, regosols, Mediterranean-type soils and grumosols) are very prone to erosion, local patterns of land use neglected any form of soil conservation. After very heavy rainfall in 1968, extensive *banjirs* breached the dykes and large numbers of people and animals were killed. Damage to *sawahs* was estimated at 1,100 million Rp. and to dykes and roads at 300 million Rp.[67] There are many other examples recorded in the literature. Since most of the watersheds of Jawa are small, floods do not generally reach dangerous dimensions,[68] but for those who are hit by such a disaster it may well be catastrophic. Every year large areas of cultivated land are destroyed by flooding.

In contrast, rivers flowing over volcanic rock behave in a different way. Their discharge is more even, which means that the peak flows are lower and the low flows higher and longer sustained. Similarly, the reaction to daily variations is slower and the groundwater storage higher. Consequently the sediment transport is lower and more even.[69]

However sediment transport goes hand in hand with erosion, no matter from what type of soil it is derived. The most important economic result of sedimentation is the silting-up of irrigation channels, river mouths, coastal areas and — last but not least — water reservoirs.

The sediment that does not settle in the river itself by forming sand and gravel banks — often forcing the river to change its course and thus destroying cultivable land — will be deposited near the coastline where the river usually becomes sluggish. Here, it often clogs the mouth of the river, dams the water and leads to flooding of coastal stretches that had been drained and reclaimed earlier.[70] In this way much cultivable land is lost.

Often there is a port at the mouth of the river, and its silting-up is usually well recorded. It has been estimated that 16–20 million cubic m. of sediment clogs the Indonesian ports every year, of which only a minor part can be removed by outdated dredgers.[71] In principle, however, the hydrological consequences of erosion in the form of sedimentation cause permanent problems through the silting-up of channels and by innundation. It is estimated that annually about 65 million cubic m. of silt had to be dredged out of the rivers of Indonesia to maintain their courses, apart from the construction of dykes and other regulatory measures.[72] A particularly serious matter is the sedimentation of artificially built reservoirs. These expensive constructions are meant to regulate the downflow of a river, store water for irrigation in the dry season and are also used to generate hydro-electric power.

Apart from sedimentary problems however, the construction of heavy dams in earthquake-prone areas — such as Jawa — can be potentially catastrophic for two reasons. First, damage caused by earthquakes, either directly or indirectly, for instance by triggering landslides, could be disastrous. Secondly, the history of seismic activity in the past suggests that regional stresses in the rocks underlying some of the proposed dam sites and the surrounding areas are often high. Thus any mechanism, for instance the filling of a reservoir, that triggers a failure in the subsurface rocks, can induce the release of geophysical energy and thus cause an earthquake.[73]

Often the hopes and finance invested in building a dam are destroyed by sedimentation, but most engineers take into account a reservoir's limited lifespan due to siltation at the design stage. This is the case all over the world since to a certain degree, erosion goes on everywhere. But when the Selorejo dam in the Kali Konto watershed near Malang in East Jawa was put into operation, it turned out that the sedimentation level during the first seven years was five times higher than anticipated, that it is increasing, and that so far erosion control measures have not brought the expected results.[74]

When in the course of the development of the Serayu river watershed the damming of the Merawu river (Maung reservoir) was proposed, the preparatory studies revealed that this river would carry a very high load of suspended sediment into the reservoir. The very fine material would seal its bottom within one year with a thick layer of impermeable deposits. According to the Australian consulting engineers, the Snowy Mountains Engineering Company (SMEC), the anticipated lifespan of the proposed Maung reservoir is less than fifteen years, probably an optimistic estimate.[75] In the karst areas where people are highly dependent on the limited surface water, reserves are kept in small artificial lakes (*telagas*). These ponds become shallower every year due to the erosion of the surrounding hills, from where rainwater washes much of the residual soils down into the lakes. Thus the life-basis of people is slowly being destroyed.[76]

The underground behaviour of water following denudation and erosion is another pertinent phenomenon, and it seems that it is even more important to look at the consequences of erosion from this viewpoint than with regard to sedimentation and floods.[77] A great number of villages draw their drinking water from natural springs or artificial wells dug into a water-table. Moreover many *sawahs* to which river water cannot be diverted likewise depend on springs or wells, unless the farmers restrict themselves to rainwater alone. Springs and water-tables however, depend on rainwater percolating into the ground, and as soon as this percolation is reduced or even stopped the flow of springs and wells reacts accordingly. It is evident that deforestation and erosion prevent large amounts of water from percolating into and enriching the groundwater table.

Various observations[78] have shown that several factors determine the extent of possible groundwater reserves and their accumulation: the nature of the soil, the vegetation cover and the type of terrain and climate (particularly rainfall). The relation between direct run-off and conservation is largely determined by the extent to which the basin is saturated with water. The very small groundwater reserves of a denuded marl area are quickly filled after a considerable amount of water has been absorbed by the hydrophile soil itself. In young volcanic areas however, the filling of the groundwater reserves is limited by the rainfall. If such areas are afforested or terraced, the percolation of rainwater underground can be increased. The increased infiltration in such forested areas amply counterbalances that of the transpiration losses caused by the vegetation cover. The discharge from wells and springs in dry periods is thus larger in forested, young volcanic areas. Cinder cones which are well forested can also function as a water reservoir for the *sawah* terraces below, but once these forests are destroyed, the *sawahs* may easily become worthless. Experimental basins located on young volcanic material in the Ci Widey area of West Jawa showed that deforestation reduces both infiltration and the volume of groundwater reserves to less than half of the original amount after a few years. If terracing is applied after deforestation, this reduction is much less serious.[79]

We have only scanty evidence about the long-term development of springs to help illustrate the connection between deforestation and their water-discharge. In the eighteenth century the tiny kingdom of Mangku Negara covered about half of the watershed of the upper Solo river. Around 1850 most of its mountainous area was still covered with forest, but by 1915 a great deal had been cut down to establish coffee plantations and to provide agricultural land for the growing population. Due to disease however, many of these plantations were abandoned and the land was left bare. The resulting disturbances in the hydrology of the area were such that a growing percentage of the 2,000 or so local springs dried up or diminished their discharge. In 1915, 85 of the springs in Mangku Negara had dried up while an additional 289 had suffered a reduced discharge. By 1940 the situation had further deteriorated: 175 sources had ceased to flow and 606 had reduced their discharge.[80]

It has to be admitted that most of the quoted data must be regarded with a certain reservation. A watershed, even a relatively small one, is a heterogeneous entity. This means that soil types are many and varied; the slopes have different gradients; rainfall varies from place to place; forest cover and cultivated areas are dispersed; anti-erosive measures may have taken place and meteorological data change from year to year.

The process of sedimentation depends very much on conditions in the rainfall area. Divergence from a theoretical relation often occurs. Thus at the beginning of the wet period much more sediment is moved by the water than in the wet season. The reason for this is that the fine, dry material has less resistance to the water than when it is already soaked and adheres to the ground more easily. To measure all this in a catchment area it would be necessary to establish a great number of checkpoints and samples would have to be taken every 30 minutes.[81] It seems, though, that this would only satisfy academic curiosity rather than contribute to the solution of vital problems. Most data refer to and are valid for only a specific point at a specific time, and must not be generalised. However it does reveal a general tendency or average that should be sufficient to base the necessary political decisions on. A century of experiments, research, and technical propositions should finally lead to action to alleviate the suffering of the simple peasant.

Erosion and Land Capability. Since the middle of the nineteenth century, the process of soil destruction has become more widespread and visible. The indigenous population grew rapidly and required more and more land. Thus, shifting cultivation (*ladang*) and dry-land agriculture (*tegala*) spread over forested and mountainous areas. The *sawah* cultivators shifted to a type of land use to which they were not accustomed. With growing population pressure, the indispensable fallow periods became shorter and finally the land was exhausted. More and more unsuitable ground was cultivated simply to be abandoned again after depletion or erosion. The Dutch terms *uitboeren* (depriving the peasants) and *afspoelen* (being washed away) became synonymous with the problem.[82] At the same time, the Dutch colonial government's coffee cultivation (*koffiecultuur op hoog gezag*) penetrated the upland forests and reduced them substantially. An enormous area was cleared for this purpose in the course of time.

Opinion differs as to whether or not conservation work was undertaken during the Dutch colonial period. Edelman maintains that anti-erosive measures were very modest and that extended areas had been deprived of their fertile soils — though the consequences of deforestation in the form of soil destruction had not been catastrophic everywhere. Yet, he concludes, the present and the following generation must spend much to restore the land if they are to leave behind soils of some value for the generations to come.[83] Vink, on the other hand, reports that as early as 1842 practical advice was being given on soil conservation in the tea estates of Jawa. This continued until the Second World War, so that 'the mountain estates in Indonesia had in general a well balanced system of soil conservation measures adapted to their crops . . . and also to local conditions of terrain, soil, climate and natural vegetation.'[84] This mirrors Edelman's opinon that the situation in the Dutch holdings under hereditary tenure (*erfpacht*) was better than in the Indonesian units, and that much conservation work had been done in the early twentieth century. However it would be futile to try to apportion blame for the mistakes of the past. Growing population pressure is continually reducing the land's natural capability so that the soils can carry — i.e. feed — less and less people. This does not contradict the fact that actually more and more people are now sharing the produce of one piece of land and that expensive inputs are used to replace what nature has ceased to give.

We have tried to describe the interaction of forest, water and soil and explained why a certain forest cover is indispensable. Specialists maintain that no tropical country should have less than 30% of its surface covered with forest, but on Jawa this percentage has already fallen below the 20% mark and is still decreasing. We have also seen that substantial areas which once supported a dense, rich tropical jungle or, after clearance, flourishing fields or plantations, are now covered with secondary bush, growing only as long as the residual soil fertility permits. Quite often, however, the soil has deteriorated to such an extent that only useless grasses can find enough plant nutrients to thrive: 'The consequences of the insufficient forest area are clearly demonstrated on Jawa where many parts of the forests have been turned into eroded, lixiviated areas free for erosion and insolation, or which are overgrown by noxious combustible grasses. The accumulation of water is not possible in such areas. Every rainstorm will cause a direct flow-off and consequently dangerous big floods.'[85] Apart from this, such grass-covered areas are generally lost for cultivation because they develop such a dense root-system that the local people with their simple tools cannot break it.[86]

These grass areas are generally called *alang-alang*, which strictly speaking refers to *Imperata cylindrica* (cogon grass or satintail), but in fact all giant grasses have long been classified under this term. It seems, however, that *Imperata* proper forms only a minor and fragile component of the grass savannahs, especially because it is prone to flood damage and is easily trampled down. The so-called *alang-alang* areas consist in large part of *Themeda gigantea* and many other grasses.[87] The present area of Jawa under *alang-alang* is not known exactly, but estimates put the figure at half a million hectares though this is little compared to the Outer Islands.

At first, population growth went hand in hand with the growth of cultivated areas and there was a tendency towards the conversion of a tropical island into a 'tropical Netherlands', where every square metre was intensively cultivated to support the growing population. However as the population continued to grow, much of the cultivated land suffered from a decline in fertility or even became sterile. In some areas and under certain conditions — for instance on the *sawahs* — output increased substantially thanks to a growing input of labour and energy, but in other places the topsoil was completely lost and the peasants had to move.

How was this dilemma to be resolved? Ramsay and Wiersum suggested one possibility:

Clearly the only real solution to the ecological problems caused by over-population is to determine the safe carrying capacity of each class of land in terms of human occupancy, and to develop a programme to achieve this in terms of actual occupancy. Only in this way will it be possible to manage the natural resources of land and water for sustained yields, and to improve the socio-economic conditions of the farmers.[88]

This simply means evaluating the potential of each area — preferably a watershed — and planning its utilisation accordingly. It may turn out that the soils are generally so bad or the watering so limited that only a certain type of forest can grow. In these cases the population has to be shifted elsewhere and the area afforested. Usually, though, a watershed has sub-areas of different capability. Some are best suited for forest, others for pasture and some for irrigated agriculture on terraced land. Dry-land cultivation mixed with scattered trees on terraced land, provided with enough rainfall, can be productive if it is done in the form of horticulture, i.e. with intensive labour. Alternatively, plains or high plateaux may offer a potential that is best utilised in the form of large-scale plantations.

Certainly, determining the carrying capacity of precise areas is rather difficult. It depends on what we think is technically possible and desirable. Let us take, for example, a sloping dryland area on which a few families can live by cattle farming. If the area is terraced and green manure and improved seed are introduced, many more families can be supported by rain-fed agriculture. If the terraces are irrigated all the year round, even more families can live on the same area, growing subsistence as well as cash crops and gathering several harvests. This may explain why determining an area's carrying capacity is so difficult. We have to both evaluate the theoretical potential and decide politically which of the necessary investments can be afforded.

Apart from this, the theoretical carrying capacity fluctuates in accordance with the different levels of security which we want to achieve. In other words, to what extent do we want to utilise a given potential (expressed always in different population densities)? Thus, we can make the following distinctions:[89]

— The *subsistence density* is the maximum density of safe human occupation which does not allow for anything more than a subsistence level of living. This is an economic situation similar to that in most parts of Jawa.

The subsistence density cannot be exceeded without adverse ecological results nor can the population be increased; in fact in many places it even has to be reduced.

— The *security density* is lower than the subsistence density and the difference between the two serves as a cushion against natural calamities. It generally means that an area is self-sufficient and in most years probably a small net exporter of produce. Its subsistence income remains untouched in cases of catastrophe.

— The *optimum density* allows a sufficient surplus production under most circumstances to make the area a net exporter, to provide the farmers with an income above their basic needs, and to ensure a necessary security in case of natural calamity.

To increase the land capability — and that means the carrying capacity for agricultural land use — various investments in the soil have to be made. Reforestation of protective stands is indispensable and terracing and other anti-erosive measures have to be expanded. The construction of irrigation and drainage projects requires both finance and land. To some extent, such measures can be taken without moving the local population and can even involve their participation. But if we compare the population pressure and the carrying capacity which can be achieved after all the necessary investments have been made, there is no doubt that at least the *annual* population accretion has to be moved elsewhere.

No matter how the carrying capacity is determined or envisaged, in the interest of Jawa's ecological and economic future, quite a number of families will have to be dispossessed of the land which they are tilling. 'In human terms', write Ramsay and Wiersum, 'this is an agonising decision for the government . . . but in the national interest of maintaining the soil and its productivity for future generations it is inescapable.'[90]

How, therefore, are the Indonesian people and the Indonesian government going to solve this problem? There is no doubt that both groups know what they are facing. We have seen that the government encouraged transmigration, and that hundreds of thousands of families followed the call to the Outer Islands. We have seen how the government promotes family planning and, as a result, millions of families are producing less children. Yet we have also seen that it is beyond the financial and organisational capacity of the country to absorb the annual demographic accretion by these means. At the very least, efforts must be continued to try to stop the on-going deterioration of the soils, to re-establish Jawa's ecological equilibrium, to increase the carrying capacity of the different geographical regions and ecological zones, and to settle the peasants accordingly.

How has Indonesia responded to these challenge so far?

NOTES

1. Hollerwöger, 1966 (Verstappen's contribution, p. 355).
2. 'Bebossing . . .' 1973, pp. 85–6.
3. Hollerwöger, 1966, p. 348.
4. See Verstappen's contribution to Hollerwöger, loc. cit.; Hadisumarno (1979, pp. 45–52) reconstructed the process of coastal accretion of an inland sea or brackish water estuary in Central Jawa, observing the immediate colonisation and stabilisation of the new land by mangrove vegetation.
5. Holle, 1866.
6. Vogt, 1948.
7. Carson, 1962.
8. Van Baren, 1950, p. 9.
9. 'Typen van stervend land in den Nederlandsch Indischen Archipel', *Tectona*, vols. 29 and 30 (1936 and 1937).
10. The number of areas with severely degenerated soils, the so-called *stervend land* or *gestorven land* (dying land or dead land), increased in the course of time and spread over the whole of Jawa. These include: the marl soils of Maja in West Jawa, the high plateau of Malang in East Jawa and parts of Madura (Mohr, 1933, 1938); the tea-gardens of Priangar in West Jawa, the sugar-cane production area of the Ranus mountains in East Jawa and the pastures of the upper Keruh river (Coster, 1936, 1937); the devasted karst region of the Gunung Sewu (Leeuwen, 1941); parts of eastern Central Jawa (Dames, 1955, pp. 129–31) and the watershed of the Ci Tarum not too far from Jakarta ('Bebossing . . .' 1973, p. 86), just to name some of the best documented cases. De Haan (1935, p. 1120) contributed to the debate thus: 'It is mainly the Tertiary marl and tuff areas which suffer from erosion. I include here the hilly land between Semarang, Ngawi and Jombang, the Serang and Lusi rivers, the marl terrain of South-Gundih, now completely washed away and useless for proper cultivation. Also the tuff-marl areas near Cirebon and Majalengka, the ill-famed '*stervend land*' of the Ci Lutung, the completely destroyed Ci Keruh watershed and the Pengaroa area south of Semarang. Finally, the boulder-clay area of Banjanegara should be mentioned and, as an example of an eroded volcanic zone, north-western Yang.'
11. Mounier, 1882, pp. 42–8.
12. *Koffiegids*, 1899–1900 (quoted by van Baren, 1950, p. 8).
13. Van Baren, 1950.
14. De Haan, 1952, pp. 179–84.
15. See, for instance, the bibliographies and references given by van Baren (1950, pp. 11–13), van der Linden (1978, pp. 105–10), and other reports of the Serayu Valley Project, and the older technical journals published in Indonesia (*Tectona, OSR News, Landbouw, Bergcultures, Teysmannia* etc.)
16. To cite some examples: the Upper Solo Watershed Management and Upland Development Project (FAO/UNDP); the Kali Konto Project; Development of

the Upper Brantas Watershed (The Netherlands); the Land Resources Evaluation Project (FAO); the Serayu Valley Project (The Netherlands); Area Development Project West Pasaman (F.R. of Germany); Transmigration Area Development Project, East Kalimantan (F.R. of Germany).

17. Van Dijk and Vogelzang, 1948, pp. 3–9.
18. NEDECO & SMEC 1973, Soemarwoto. 1981, p. 17.
19. MacComb and Zakaria, 1971, pp. 20–2; Soemarwoto, 1974, p. 363.
20. Kinzelbach, 1981, p. 754; Hendry, 1984, pp. 20–4. — To offer to the reader a standard of erosion magnitudes, we use the scale that Meijerink (1978, p. 27) proposes:

	Tonnes per sq. km. per year	
Extreme	over	25,000
High	over	5,000
Moderate	over	1,500
Low	under	250

21. Two examples may illustrate the large seasonal and annual fluctuations. The sediment transport of the Pekacangan river (Serayu watershed) amounted to 5,800 tonnes in January 1907 and to 120 tonnes in August of the same year (Mohr, 1908). The silt of the Ci Tarum (northern West Jawa) in 1972 was three times that of the preceding year, without any catastrophic events (NEDECO, 1973).
22. 'The critical situation is understood to be such, that those areas are neither able to yield elementary production, to act as regulators of the water system, or to fulfil protection functions. The hydrological situation is such that inundation and erosion occur' (Wölke, 1978, p. 211).
23. Wiersum, 1980*b*, p. 1.
24. Dames, 1955, p. 129.
25. 'Bebossing . . .' 1973, p. 107.
26. *Statistical Pocket Book*, Indonesia, 1980/1, p. 148.
27. *Statistical Yearbook*, Indonesia, 1975.
28. Wiersum, loc. cit., p. 2.
29. Thijsse, 1975*b*.
30. Van der Linden, 1978, p. 51.
31. Brown, 1978, p. 1.
32. Fromberg, 1854; de Sturler, 1863; Holle, 1866, 1873, for bibliographic details see Edelman, 1947, pp. 311–20.
33. Wiersum, 1978*b*, p. 1.
34. Edelman, 1947, pp. 287–8.
35. This is due to an alteration of the diameter, the distribution, and the velocity of the falling drops with a variation between the species. Compared with undisturbed rainfall, the erosive power of the throughfall amounted under *Albizzia* to 102, under *Acacia* to 119 and under *Anthocephalus* to 147 (Wiersum and Ambar, 1981, p. 36).
36. The removal of litter and undergrowth increased the erosion rates by 17 to 22 times. On undisturbed plots, the annual erosion amounted to 0.3 mm.; after removal of litter and undergrowth it rose to 7.0 mm. (loc. cit.).
37. Coster, 1938, pp. 613–728.
38. Though Coster has warned of the danger of applying his figures (gained on small

plots) to the whole watershed, other authors have produced similar results. Bakker and van Wijk (1951) observed the run-off on sloping young volcanic terrain as well as on marly ground during the highest rainfall of the observation period. They found 1.8 to 2.0% under undisturbed rainforest and 25% on non-terraced ground deforested four years earlier. Roessel (1940) concluded: 'with favourable vegetation the surface flow during maximum run-off is less than 5% of the rainfall and with an unfavourable vegetation does not exceed 30%' — the latter data being probably too low.

39. Coster, 1938, p. 728.
40. Loc. cit.
41. L. Gonggrijp, 1941.
42. Roessel, 1940.
43. Meijerink, 1978, p. 39.
44. Bakker and van Wijk, 1951.
45. Wiersum and Ambar, 1981, p. 15. — It should be added that a very specific type of surface violation caused by improper land use occurs in parts of Jawa, for instance in the Serayu watershed. Here, peasants often demarcate their land by using ditches as field boundaries, in most cases parallel to the slope. A study of an area of 910 ha. showed that these field boundaries develop into scar features, then into gullies which become deep and wide, and through small mass movements develop into large land scars which cover 1–2% of the surface area. (Bergsma, 1978, pp. 75—91).
46. Wiersum *et al.*, 1979.
47. De Haan, 1935, p. 1115.
48. Klinkert, 1937, pp. 731–6.
49. Van Dijk and Vogelzang, 1948, pp. 6–9.
50. Hardjono, 1977, p. 9.
51. Bailey and Bailey, 1960, p. 275.
52. Den Berger and Weber, 1919, pp. 35–9.
53. Kramers, 1904 (quoted by Edelman, 1947, pp. 284–5).
54. Rutten, 1917, pp. 925–6.
55. For details about natural reforestation (or *spontane reboisatie*) as well as the success of early trials to recover destroyed forests, see Koorders (1894a, 1894b, 1898) and Tobi (1894).
56. Van Dijk and Ehrencron, 1949, p. 3.
57. Rutten, 1917, pp. 927–8.
58. Van Dijk and Vogelzang, 1948, p. 9.
59. Faber and Karmono, 1977, pp. 13–15.
60. Thorenaar, 1933 (quoted by Edelman, 1947, p. 286).
61. De Haan, 1935, p. 1113.
62. F.A.O., 1965, pp. 23–7; see also Pitty, 1978.
63. F.A.O., 1965, p. 23.
64. Strahler, 1969, pp. 402–8.
65. Meijerink (1978, p. 46) comments on this specific area: 'The worst case of mass wasting and associated gullying occurs in the hydrologic land units underlain by the Merawu formation. This formation consists mainly of blueish to greyish claystone, marls and shales which are crushed by flooding. Part of the outcrop area, according to van Bemmelen . . . is subject to large-scale gravitation-tectonic

movements with a speed that can be measured in a decade . . . Gullies are often choked by the sliding debris, *sawah* systems destroyed, and the rivers are not only capable of transporting the debris supplied, but are also actively eroding the toes of unstable slidden masses or outcrops of crushed rocks. . . . The sediment yield from this area seems to be more than 30,000 tonnes/sq. km./year, a truly exorbitant figure.'

66. Faber and Karmono, 1977, p. 15.
67. Soerjono, 1968.
68. De Haan, 1952.
69. Faber and Karmono, loc. cit.
70. Kündig-Steiner, 1963, I, p. 1296.
71. Röll, in Imber and Uhlig, 1973, p. 217.
72. Loc. cit., p. 220.
73. Speelman, 1979, pp. 136–7.
74. 'Kali Konto Project', 1979, p. 7.
75. S.M.E.C., 1974; Speelman, 1979, p. 137. — These data are not confined to Jawa. They can be found in many countries. The Archicage reservoir in Colombia, built for the production of hydro-electric power, was closed in 1955. After 10 years, only one-quarter of the original capacity was still available (Dollee, n.d.).
76. Bailey and Bailey, 1960, p. 275.
77. De Haan, 1952.
78. For instance, Bakker and van Wijk, 1951, pp. 56–69.
79. Bakker and van Wijk, loc. cit.
80. Ramsay and Wiersum, 1974, p. 4.
81. Dollee, chapter 7 (n.d.).
82. Edelman, 1947, p. 291.
83. Loc. cit.
84. Vink, 1951, p. 287.
85. Thijsse, 1975-b., p. 3.
86. Uhlig, in Imber and Uhlig, 1973, p. 34.
87. Van Steenis, 1937, p. 641.
88. Ramsay and Wiersum, 1984, p. 9.
89. Loc. cit.
90. Loc. cit., p. 10.

7

ACCEPTING THE CHALLENGE

For many years, deforestation and soil degradation in Jawa have been observed and recorded, but after almost a century of research and countermeasures, the cycle of destruction has not been checked. In the last few decades the Indonesian government has striven to remedy this situation. However they have been hampered by a lack of up-to-date soil maps and other pertinent information. In fact the problem had simply grown too big to be tackled by a government programme, and the interest shown by the local population — so far excluded from the planning process — was only marginal and as a result the situation grew more and more desperate. Officials finally told the peasants concerned to either accept environmental destruction — since they had caused it in the first place — or reverse its effects. The government would support their efforts, but would no longer impose measures on them. The FAO-supported Upper Solo River project in Central Jawa, which between 1973 and 1976 shed much valuable light on soil conservation, recommended in its final report that the programme should be launched only in those villages which required it. The Ministry for Supervision of Development and Environment established in 1978 now stresses the essential dependence in watershed management on community understanding and on the direct involvement of the cultivator himself.[1]

Many of the affected farmers are well aware of the 'great danger that sneakingly draws near', especially since in many cases it is already in full effect. There are innumerable examples where peasants undertook anti-erosive measures on their own, in community cooperation schemes, or within the framework of larger technical projects. But it is obvious that these activities are influenced by a number of factors such as the prevailing land tenure system and the availability of additional labour and of credit to pre-finance such undertakings.[2]

Unfortunately those who cause erosion are not always willing to remedy its effects. Sitting 'up-hill', they do not realise or admit that the damage occurring 'down-hill' is the consequence of their activities. One of the serious problems with anti-erosive measures is that those who cause erosion are very often not the same people who suffer from it. Thus, effective measures can only be achieved if both groups participate.

155

It cannot be said though that Indonesian peasants are normally hostile to innovation. In areas where transport facilities are non-existent or extremely poor, many cropping methods in rain-fed fields have not changed for centuries. But as soon as a road opens up an area, or modern irrigation methods are offered, the peasants become interested in growing cash crops. In the Serayu valley for instance, it was found that the farmers were hardly aware that their cropping methods caused erosion, and certainly no signs of them taking preventative measures have been observed. The changes that had been introduced were minor and directed only towards higher productivity, irrespective of their environmental consequences. Accordingly, all forms of soil erosion can be seen in this watershed.[3]

These examples show that irresponsible activities, though meant to promote progress, may lead to catastrophe sooner or later. If, for example, soil conservation education does not *precede* the introduction of new techniques or the opening-up of new areas, the damage can hardly be remedied afterwards. If precautions are not taken, the population pressure will quickly produce the irredeemable situation of Gunung Kidul (the southern karst mountains) and its eroded areas. Other examples include the island of Madura, which has now lost its forest cover, where the inhabitants survive on fish, cassava and salt production. Timor's erosion problem is beyond control and cassava, a sure sign of poor soils and a nutritionally deficient population, is gaining popularity.[4]

During the Dutch colonial administration, afforestation was mainly confined to the planting of teak trees, a most valuable export wood, but not actually the most suitable type for protective forest. Though the protective function of mixed forests was well known and the need for restoring the depleted stands was obvious, little was done. Exhausted fields were simply given up and abandoned.

The area of *official forest* on Jawa is only about 2% of Indonesia's total area under forest. This explains why, in contrast to the forest policy and wood economy in the Outer Islands, Jawa is not very important as a supplier of government revenues. Politicians as well as foresters have finally realised that their task on Jawa is very much different from that in, say, Kalimantan. The degradation of the environment has reached such a stage that forestry is directed mainly towards protection and restoration of the environmental equilibrium.[5] In the course of time therefore, the forest policy of the Indonesian national government changed. In 1966 the first programmes to counteract the growing deterioration of the Javanese tree-cover were begun. Work has been directed in two broad directions:

Reforestation (herbebossing, reboisasi) is the re-establishment of degraded forests and the replanting of forest areas that have been misused for crop production and other purposes. *Regreening (herbeplanting; penghijauan)* refers to private land usually under cultivation. Here the owners are encouraged to plant multi-useful trees between the crops. Such trees are

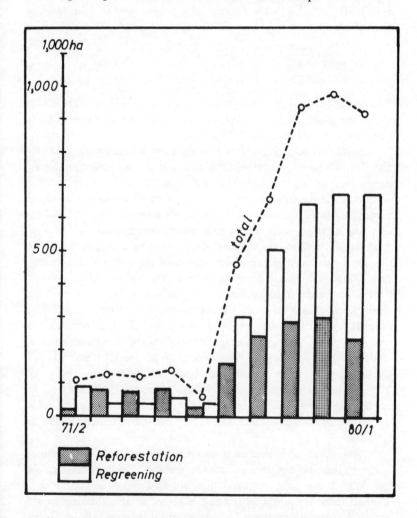

Fig. 19. EFFORTS IN REFORESTATION AND REGREENING IN
INDONESIA, 1971/2–1980/1.

Source: Buku Saku, 1980/1, p. 148.

meant to offer a certain protection against erosion, they may supply timber and firewood, their leaves can be used for fodder or green manure and most trees produce saleable goods such as fruit or nuts.

Quantitatively the programme is well developed. Reforestation as well as regreening now cover growing areas of land in the successive plan periods (REPELITA is the Indonesian acronym for five year development plan):

REPELITA I	(1969/70–1973/4)	510,000 ha.
REPELITA II	(1974/5–1978/9)	2,090,000 ha.
REPELITA III	(1979/80–1983/4)	2,889,000 ha.

In the last decade, the official figures for the whole of Indonesia (of which the bulk refers to Jawa) show an impressive annual increase in replanted land. Whereas in 1971/2, 12,054 ha. came under reforestation and 80,900 ha. under regreening, in 1980/1 the figures were 238,938 ha. and 679,345 ha. respectively.[6] Though at first sight these figures are impressive, they do conceal hidden problems.

First of all, the figures turn out to be planned targets that have not been fully reached in reality. In 1980/1 for instance, only 76% of the reforestation target was met and 85% of the regreening target.[7]

The main problem that makes reforestation an uncertain venture is the fact that deforestation is a consequence of growing population pressure. Reforestation does therefore involve the eviction of poor smallholders — which can be justified morally only if a substitute is offered to them either in the form of other land or of non-agricultural income. Such programmes can best be developed and carried through with the co-operation of the local population concerned. If the government fails to establish such cooperation, the chances of success are slim. Though successful programmes are reported from Central and East Jawa and Yogyakarta, there have been many cases of failure. In West Java for example, between 1962 and 1969, 138,000 ha. of trees were planted, but the deforested surface area increased at the same time from 205,000 ha. to 229,000 ha.[8]

Another handicap is the lack of competent personnel to give technical advice in the field and to motivate the peasants. Often reforestation is not done in the right way and trees are planted on unsuitable soils. Though there are plenty of training facilities in Jawa, lifelong experience and the ability to deal with people can hardly be taught in schools. It is therefore often difficult to explain to the peasants that the long-term goal is to rehabilitate criticial land and preserve watershed systems

'Regreening' involves the planting of useful trees together with the annual crops, possibly along contour lines or even bench terraces.

optimally so that people become strong supporters and defenders of their natural resources. Emil Salim, the minister most involved in the field of environmental defence, has tried hard to convince Indonesia's rural population: 'A failure in putting [*sic*] the people into emotional involvement in regreening and reforestation will accordingly result in the failure in carrying out the programmes sufficiently.'[9]

The second target of the forest policy is the private land under annual crops. These areas generally lack a permanent vegetation cover or other technical measures for soil protection, but are intensively planted with food and cash crops. One of the measures propagated is 'regreening', which involves the planting of useful trees together with annual crops (if possible following contour lines). Such trees are meant to provide soil protection, shade, firewood, fruit, leaves and so forth. Their protective value however, is minor. They may well intercept the raindrops and their roots may contribute to the physical and even chemical stability of the soil, but since it is litter rather than the canopy of a tree which reduces erosion, the planting of scattered trees is not a very promising method.[10] Only in combination with contour ridges can such trees

160   *Land Use and Environment in Indonesia*

develop their protective function. This, however, requires the active participation of the peasants. Pickering, who has looked closely into this matter, has written:

Ridge and tree-planting technology on private lands has been the central technology to date. Unlike 'bench' terracing this does not involve cutting into the slopes of hills to produce tillage areas; rather as its name implies, it means building small parallel ridges along the contours in which fuelwood coppice trees such as *Calliandra* can be closely planted. When trees can be regularly cut back and regrown from coppice shoots an effective barrier against erosion can be established. Moreover, in time rain brings some natural levelling of the tillage area between the ridges which can then be cultivated with reduced soil loss. For the system to be effective however, it is imperative that the farmers, especially in the long rainy season, constantly inspect and maintain the ridges. Since stored water concentrates at the weakest point, a break in the ridge can lead to gully erosion more serious in its effects than the sheet erosion which would occur if there were no ridges. This means that badly constructed or poorly maintained ridges can actually increase soil erosion.

Though evaluation is not yet complete, experience in Java suggests that while farmers can be induced to construct ridges and plant the tree seedlings supplied, they are often less willing to maintain them. In extreme instances seedlings have been pulled up soon after planting, so that the favourable soil of the ridges can be planted with cassava which, as a root-crop, on harvesting tends to increase the soil's detachability under raindrop impact and hence its erodibility.[11]

Firewood plantations offer an interesting and useful combination of protective reforestation or regreening together with the production of a commodity for which there is a constant demand. Firewood has always been the main source of fuel for the rural population. The Dutch colonial administration tried to encourage the peasants to no longer rely on the shrinking natural supply, but rather produce it intentionally in a planned way. Such efforts date back at least to 1910, when 'native firewood cultivations' were established on soils of the Royal Domain in the district of Wonosobo in the upper Serayu valley. However the regulations about the rights and obligations of the population concerned were insufficiently based on enforceable law, and the project failed. Another attempt to establish a cooperative firewood project covering an area of 800 ha. of firewood forests begun in 1921 collapsed after twelve years. These and other examples show that if the population is not persuaded to cooperate, any scheme is bound to fail.[12]

To this very day, new efforts are being made to establish firewood plantations as an integral part of the rural economy. Forest research identified various native shrubs and trees well suited for this purpose (especially *Calliandra, Acacia* and *Leucaena spp.*),[13] and the forest administration is keen to encourage the peasants to use them more widely.[14] Foreign-aided projects often include firewood planting in their activities,[15] but the results remain disappointing. The reasons for this are simple. Food crops and cattle compete with fuel plants for the limited land surface; additional physical work or financial contributions required for the project rouses the peasants' suspicions; and the lack of organisa-tion of the projects in terms of obligations and rights more often than not leads to negligence and finally abandonment. As so often, it is not a technical but rather a social or psychological question. Commercial fire-wood plantations manage to supply power plants perfectly well, but when it comes to providing a village community with fuel from its own land, insurmountable difficulties seem to arise.

Reforestation is a lengthy process even when quick-growing varieties are used. For some time the ground is still exposed to wind and rain as long as the canopy of the young forest remains open. But as we have shown, the erosive impact of rainwater on the topsoil goes on even when the canopy is closed. Therefore as well as reforestation, other protective measures have to be implemented. To support the growing of trees as an anti-erosive technique, a well-recommended method is the sowing of cover crops (also called ground covers) together with the planting of young trees. Recently, much research has gone into plants that are useful as cover crops in tropical and sub-tropical countries.[16]

Cover crops are preferably leguminous but also non-leguminous herbs, shrubs, or convolvuluses that germinate and spread quickly over the ground, covering it with a dense layer of leaves. In this way they provide the protective function afforded by litter in forests or fields, and by grass on pastures. If they are legumes they also supply the soil with nitrogen and this, together with the organic matter provided by the off-fall, improves soil fertility. In fact, cover crops can act as a type of green manure. It is of crucial importance that the plants selected do not compete with the main plant — for instance fruit trees — and that they tolerate sunlight as well as shadow, and that they are well tended, pruned and so forth. Thus the peasants have to both understand the scheme and be sufficiently motivated to take part if it is to be a success.

Up to the present day the cover crops associated with forestry on Jawa have been confined to *Leucaena leucocephala* and, to a lesser extent, *Acacia glauca (villosa)*, whereas in agriculture — particularly under multiannual cultures such as rubber, coffee and tea — a variety of other cover crops have been used. In view of the slow success of reforestation, cover crops could well play an important part in fighting soil erosion, and more species should be selected and tested. Even completely eroded areas can be regreened when the proper pioneer plants are selected. Van der Meulen developed a so-called ecologic-biological method that is completely based on indigenous plant species, and he found that *Mimosa invisa* is best suited as a first soil improver under severe erosion. Using this method, areas of bad and critical erosion can at least be prevented from further destruction and the soils re-established in the long run.[17]

As we have seen earlier, the gradient of a slope tangibly determines soil erosion, and terracing is a technique which can transform a slope into strips of level land suitable for safe cultivation. Even when clear felling of slopes has taken place, terracing reduces the soil erosion to about half that of non-terraced slopes,[18] and in one case where shadow trees over commercial plantations had to be cut due to disease, a leguminous cover crop together with contour ditches provided good protection in lieu of proper bench terraces.[19]

The more the native population engaged in dry-land farming during the colonial period and the more the process of erosion became visible, so the Dutch administration tried its best to introduce terracing. In West Jawa, the Priangan highlands, the area of Cirebon and the 'dying lands', much was tried in this respect. But all these efforts finally failed. It seems that agricultural science endeavoured to solve the problem of soil conservation without properly appreciating the underlying technical, economic and social factors. It is interesting to note that such efforts were rather successful in areas with deep volcanic soils, whereas on flat marly ground the peasants were reluctant to follow technical advice. This is understandable because terracing leads to the destruction of the thin topsoil unless it is removed and replaced at a later date — a lengthy and costly process indeed.

In the colonial period the success of soil conservation and terracing depended on the peasants' confidence in their technical advisers and on their own willingness to change their often desperate situation. There were good examples of cooperation, and where this was the case the administration began to consider watershed schemes — as for instance in the catchment area of the Ci Manuk — in the 1930s and 1940s.[20]

Unfortunately, terraces have to be maintained continuously. Once they are neglected, heavy rainfall causes rills and gullies to form and can lead to severe destruction. If properly maintained, however, they may reduce the soil losses to less than 5% and they are therefore the ideal way

Terraces have to be maintained continuously. Once they are neglected, heavy rainfall causes rills and gullies and leads to severe destruction.

of using sloping land for agricultural purposes. But since the technique is extremely labour-intensive and requires permanent maintenance, it has always been difficult to encourage the local population to participate. To terrace one hectare of land, 1,000–1,500 man-days are required, and in many village communities the peasants are reluctant to engage in such a task. For a long time, these and other technical difficulties prevented the government from carrying through large-scale terracing pro-grammes. Lately however, the peasants have begun to understand their situation, realising that their land, their source of livelihood, will be lost if it is not protected. They are now willing to invest in terracing, and often hire labour from outside their village to do this.[21]

Different types of terracing vary in both their construction and cost. Bench terraces with waterway drop-channels and gully plugs have been developed on dry-land slopes on Jawa with considerable success. The most notable and extensive examples can be seen in the Forest Depart-ment's Upper Solo River Project assisted by the FAO. But on the fertile and porous soils in West Jawa, indigenous systems of level terraces without the costly provision of drop-channels can provide adequate protection against erosion, as studies in the area of Penewangan show.[22]

Another technique which increases the carrying capacity of a certain area of land is irrigation. The Dutch colonial administration — keen to secure high yields — actively encouraged irrigated agriculture.[23] But after colonial supervision ceased, many of the systems fell into disrepair and the water was no longer properly used. The Indonesian government is eager to improve the situation, but as long as water supply remains 'free', the peasants are hardly interested in getting a large economic return from it. Moreover, most of the irrigation schemes depend on rainwater and have no permanent source of supply. This indicates that there is enough water available to increase the carrying capacity of Javanese soils despite the scarcity of arable land and the high percentage of land that is already irrigated.[24]

In discussing soil erosion we have seen the importance of forest cover for ecological stabilisation. We have also noted that the expansion in the area of cultivated land caused by growing population pressure inevitably leads to deforestation with all its unwanted consequences. At first sight it would seem that forest and cultivated land are mutually exclusive and that therefore an ever-growing population would inevitably lead to continuous deforestation and soil destruction, finally rendering region after region uninhabitable. Though this vision may still be valid in the long run — unless population growth is severely curbed — there are a few techniques which can be used to gain time. Apart from the different types of slope stabilisation already mentioned, it is also possible to combine forestry and agriculture on the same plot of land so that both trees and annual crops can be grown. This technique is called the agro-forestry system, but it is better known by the Burmese term of *taungya*. The systems and crops used depend on the local conditions, natural as well as social, but a widely accepted definition is as follows: 'Agro-forestry is a system of permanent land use by which both annual and perennial crops are cultivated simultaneously or in rotation, often in several layers, in such a way that sustained, multiple-purpose production is possible under the beneficial effect of the improved edaphic and micro-climatic conditions provided by simulated forest conditions.'[25]

In 1855, Dietrich Brandis, who taught botany at Bonn University, accepted a post in British-ruled India to manage the teak economy of Burma. After studying the shifting cultivation (slash and burn) of the Karen, he succeeded in persuading them to sow teak seed together with their upland rice and tend the growing saplings for a modest remuneration. In this way, and over the course of time, the stands of *Tectona grandis* spread and the Karen, who formerly destroyed the forest, developed into loyal cooperators with the Forest Department. They gradually abandoned their nomadic life and established permanent rice terraces in the valleys and on soft slopes, together with gardens and fruit tree plantations. This is how the original and simple form of agro-forestry developed in Burma and gave us the term *taungya*.[26]

With the *taungya* or agro-forestry scheme, the farmers plant food crops like maize alongside trees such as *Pinus*.

Outside Burma, it was in Jawa that foresters first introduced the new technique. This is not surprising since teak is an important tree in north-eastern Jawa and there had been an exchange of ideas between the Dutch and Brandis, though the latter never visited the island. *The Official Gazette no. 97* of 1865 for the first time mentions 'agro-forestry'. The *taungya* system was adopted in the teak areas of Pekalongan-Tegal in 1871, and extensively in those of Semarang in 1881.[27]

The amount of land under teak spread more quickly in Java than any other tropical country, mainly under the combined agro-silvicultural system, and now covers about 500,000 hectares. The system is simple. Landless peasants contact the forest office and receive a certain area of forest land on temporary terms ready to be reforested. They are obliged to clear the land and prepare it for cultivation, to collect teak seed and sow it on the plots, interplanting mountain rice, maize, or occasionally cassava. They have to tend the young trees, to mulch the ground and keep it in proper shape. In return, they grow their food and can make use of the brushwood for fuel or even for sale. Economically speaking, the forest administration saves money on nurseries and tending the young plants, whereas the peasant can grow his food and earn some extra income, and his work helps to maintain the forests instead of destroying them. To increase soil fertility, legume trees are planted between the teak, mainly *Leucaena glauca* and *Acacia villosa*, which at the same time supply wood and fodder.[28]

Apart from this classical *taungya* system mainly used to support reforestation — the disadvantage of which is that the peasants have to move after a few years — agro-forestry has developed in various directions. The following systems can now be found in Jawa.[29] A rotational cropping of annual and perennial crops with a natural fallow in between. This in fact is shifting cultivation or *ladang* and has now almost died out in Jawa, but some indigenous rotation systems of tree crops and agricultural crops do exist, for example the so-called *wonosobo* system where *Acacia*, tobacco and vegetables are grown under the *taungya* forest service system, locally called *tumpangsari*. The intercropping of annuals and short-lived perennials in the interspace between perennials, where scattered trees stand on or along agricultural fields, is called *tegal-pekarangan* and is often the result of 'regreening' activities. Mixed cropping in the form of cultivating low and high-growing perennials together has been used in various ways. Where trees and agricultural crops grow together we speak of mixed gardens or *kebun campuran*, but we also find trees and forage crops (for stall feeding), and fodder crops and green manure crops together.

Where multistoreyed cropping is carried on within the housing compound it is known as homegarden or *pekarangan*. Outside the compound it is called either mixed garden, as above, or forest garden (*talun*). Here, long and short duration crops are grown simultaneously. Mixed farming, finally, involves a combination of crop growing and animal husbandry. It can be found on or outside housing compounds, for example when cattle graze in coconut plantations.

All the different agro-forestry systems have certain characteristics in common. Perennial rather than annual crops dominate most of them so that there is a higher nutrient value in the vegetation than in the soil. If undisturbed, the perennials protect the soil from exhaustion, leaching and erosion. If the growth of annuals increases however, the protective layer of litter and herbs decreases and erosion can result. Another characteristic is that two or more vegetation layers occupied by different crops are found in most systems. Optimum results in production and protection depend on the right choice since each crop has other requirements in terms of light, soil, water, space, shade and so forth. Unfortunately, many basic food crops have higher light requirements than vegetables and fruit, making them less suitable for inclusion in most agro-forestry systems. Experience shows that systems which combine forestry and agriculture may have a number of advantages compared with pure forestry or pure agricultural production.

Generally, a large diversity of species together with spatial diversity ensures against the spread of pests and diseases and reduces their occurrence, though some species, particularly newly introduced ones, can be heavily infested. However this diversity of species allows for almost daily yields of food crops, forage, wood and cash crops which are used either for local consumption or for the market.[30] Agro-forestry systems have a good ecological and economical stability provided they are well managed. Under such conditions they give sustained yields with only minimal environmental degradation and act as a safeguard against crop failures, drought and rural poverty.

Agro-forestry can be regarded as a new, artificially created ecosystem, designed to supply man's food and raw material needs. However agro-forestry can, given enough time, adapt to and change its environment, making it more 'natural' and developing its own stability. This process can be speeded-up by man's interference. Provided the necessary technical and organisational changes are made, a new, stable ecosystem may develop which is more 'useful' to man. For example, the relatively low protective value of annual crops on a sloping surface can be increased by levelling or terracing. On the organisational side, land-ownership is considered to be one of the main factors influencing the stability of the system, since it can largely determine the farmer's behaviour towards his soil.[31]

For a better understanding of the benefits extended by a properly managed combination of trees and field crops, we can cite the example of the African state of Malawi. In the south of the country the population pressure, deforestation pattern and cultivation of steep unsuitable soils are comparable with conditions on Jawa.[32] In response to requests from peasant farmers for technical advice on planting trees in or around their villages, a British forester encouraged them to do two things: to grow a common woodlot on village land with quick-growing species (*Eucalyptus*) and to make trials with agro-forestry on their private fields. Two legume trees were introduced (*Acacia albida* and *Leucaena leucocephala*) with the aim of enriching the poor soil with nitrogen and organic matter and reducing soil erosion. Both species have considerable advantages. *Acacia* not only accumulates nitrogen in the earth, its deep roots also draw up vital elements from the soil at a depth beyond the reach of many plants, releasing these through leaf-fall and subsequent decomposition. Since the roots reach far into the underground, the trees do not compete with annual crops for moisture and nutrients. When the rainy season starts, the tree sheds its leaves and so reduces the shade; when the hot season begins, the leaves grow again and offer shade to the under-crops, thus maintaining a favourable micro-climate which tends to even out extremes of soil temperature. *Leucaena* is even more active in nitrogen fixing, but it also produces fuelwood and poles, and the chopped foliage can be used as fodder or green manure. When sown 30 to 40 cm. apart, it can be continuously cut one metre above the ground level. *Leucaena* leaves can be spread as green manure, but the nitrogen activity of the plant means that better yields are obtained under or near trees. Casey sums up the advantages of *Leucaena* thus:

The agro-forestry approach . . . is pointing toward alternative sources of fertility to reduce or replace the costly commercial fertilisers required for growing maize. Research in Nigeria has demonstrated that the application of 10 tonnes per hectare of *Leucaena* prunings resulted in hybrid maize yields equivalent to plots where 100 kg. of nitrogen per hectare had been applied. A farmer keeping livestock is in a position to make an excellent compost which, combined with *Leucaena* prunings could drastically reduce his chemical fertiliser needs. Although this system might increase his work slightly, it could save him a considerable amount in cash outlay. The point is that he should not regard agro-forestry as a separate operation, but as an input of trees into the farming system to be managed along with food crops and livestock . . . Moreover, a greater reliance on organic fertilisers would improve soil structure, increase water-retention capacity and develop resistance to erosion in a manner that is not achieved by chemical fertiliser'.[33]

In Indonesia and particularly in Jawa, the official forestry policy was traditionally directed towards watershed protection, the production of forests for industrial purposes and, finally, wildlife and recreation. In more densely populated areas however, the Forest Department modified its policy and included a form of management which offered more employment opportunities and greater income. On government land where the population density was still comparatively low, agro-forestry was introduced after 1960 with quite good results. Farmers were encouraged to interplant food crops among the rows of young trees, knowing that after two years the land would revert to the Forest Department. By 1978, some 10,000 ha. had come under this system.[34]

The Javanese had had indigenous forms of agro-forestry known as *landbouw-tussenbouw* or *taungya* for a long time. The Forest Department later tried to encourage these methods by offering credit for fertiliser and higher-yielding improved seed.

A more advanced method is the so-called 'prosperity approach'. This encourages the peasants to grow particularly high-yielding species under the forest trees so that a better income may be obtained. One example of this is the combination of forest and pasture with cattle farming, where trees are undergrown with highly productive grass species such as *Pennisetum purpureum*, which can be used for stall-feeding cattle; this has been proposed for the rehabilitation of the upper Solo river area. Undergrowing with fuelwood varieties such as *Calliandra calothyrsus* will produce firewood or the raw material for charcoal production. Other labour-intensive plans which have been developed include: growing and collecting medicinal herbs, growing mulberry trees (*Moracea sp.*) for sericulture, *Melaleuca leucodendron* for the production of *kayu putih* oil and *Schleichera oleosa* to produce shellac. Finally, apiculture is also being introduced.[35] The government of Indonesia is testing and carrying through systems to investigate their technical possibilities, protective value, long and short term production capacity, and the feasability of organising them on a labour-intensive basis aimed at permanent cooperation between the Forest Department and the peasants. The most important point however, is that they be accepted by the local population.

The *taungya* system proper, on government forest soils with a comparatively small population density, generally runs well, but it cannot be denied that the inevitable moving out of peasants after the forest trees have started to thrive led to unfortunate reactions. This is particularly true where the population density is high and the Forest Department wants to reforest soils unsuited for agriculture. In such cases the farmers prefer to stay. For this reason the forestry officer at Patuk (situated in a limestone region) is not optimistic about conspicuous reforestation results: 'In his area over the last 30 years there have been numerous replantings on the stony terraced hillsides. The peasants have a three-year lease of the land under the *taungya* system and interplant with maize, rice and cassava until the newly planted trees are established. The young trees have been surreptitiously felled year after year by the peasants in order to keep this land under crop cultivation for as long as possible.'[36]

Attempts to plant trees on peasants' private land — usually called 'regreening' — have met with even less success. It was found that, in terms of controlling and regulating the farmers' activities, there was a great difference between public and private land programmes, and that the latter called for much greater emphasis on persuasion than was needed for the relatively small population of forest-dwellers on government-owned forest land.[37]

This is the case in spite of the fact that in Indonesia and particularly in Jawa there is a tradition of peasants mixing forest and fruit trees with annual food crops, thus creating 'peasants' forests' (*boerenbossen*), normally found on moderately sloping land. These supply fuel and timber, fruit, vegetables and basic foodstuffs together with animal forage and green manure all the year round, including the dry season. Whereas the stretches covered with peasants' forest were mostly well protected, the steep government land above was usually deforested and eroded.[38]

The government is keen to expand this type of combined land use since it can increase the peasants' income apart from its protective value. After 1973, the Forestry Department developed several initiatives designed to attract the population's interest to such combinations and to include them in their programmes.

To understand the difficulties which such government activities have to face, we must remember that the peasant farmers feel, above all, that some of their minute area of land would be taken away from food production. It needs conviction, training and experience to make them believe that the growing of trees in the fields — done in the right way — would increase their income. 'For the establishment of agro-forestry systems in new areas', writes Wiersum, 'many human organisational aspects have to be considered; the provision of income opportunities during the period when the new systems are not yet providing sufficient yields, training and extension in the necessary new skills for the population. The failure in many of the programmes to establish seemingly stable agro-forestry systems is probably due to insufficient attention being given to these organisational aspects.'[39]

The efforts undertaken during the last few plan periods brought to light two basic considerations which always have to be kept in mind, first: 'If the role of man in ecosystems is to be fully understood, he and his social structures have to be treated as a natural though highly specialised component of the ecosystem, rather than as an unnatural external factor.'[40] This means that man's activities ought to develop an area with, and not against, nature. Secondly, to secure the cooperation of the rural population, we have to switch from the 'top-down' approach so far used to the 'bottom-up' approach. This means that the local people and their local skills have to be part of any programme right from the beginning, particularly because the failures that will inevitably occur can best be solved on the local level.

NOTES

1. Pickering, 1979, p. 61.
2. Eppink and Palte, 1980 (quoted by Wiersum, 1981, p. 40).
3. Struijk (quoted by Faber and Karmono, 1977, pp. 47–9).
4. Bailey and Bailey, 1960, p. 277.
5. Wiersum, 1978a, pp. 303–4.
6. *Buku Saku*, 1980/1, p. 148.
7. *Indonesia Times*, 4 Dec. 1981.
8. Wiersum, loc. cit., p. 311.
9. *Indonesia Times*, 20 Jan. 1982.
10. Wiersum, 1981, p. 16.
11. Pickering, 1979, p. 62.
12. Fokkinga, 1934.
13. National Academy of Sciences, 1980.
14. Thijsse, 1977, p. 445.
15. 'Kali Konto Project', 1979, p. 8.
16. 'Bebossing . . .' 1973, pp. 127–140.
17. Thijsse, 1977, pp. 446–7, refers to van der Meulen (1957) and Cornelius (1978).
18. L. Gonggrijp, 1941, pp. 200–20.
19. Vink, 1951, p. 287.
20. Joosten, 1941, pp. 1063–80.
21. Pickering, 1979, pp. 62–3.
22. Loc. cit.
23. Kündig-Steiner, 1963, I, p. 1297.
24. Blom, 1979, p. 11.
25. Wiersum, 1980a, p. 516.
26. Hesmer, 1970, pp. 94–7.
27. Becking, 1928.
28. Hesmer, 1970, pp. 190–4.
29. Wiersum, 1980a, p. 521.
30. Terra, 1958.
31. The whole complex refers to Wiersum, 1980a, pp. 518–9.
32. For the following see Casey, 1983.
33. Casey, 1983, p. 44. For the characteristics of such trees see National Academy of Sciences, 1980, and *Indonesia Times* which publishes an occasional series on Indonesian flora.
34. Pickering, 1979, p. 61.
35. Wiersum, 1978a, pp. 311–2, and 1980b, p. 7.
36. Bailey and Bailey, 1960, p. 275.

37. Pickering, 1979, p. 61.
38. Wiersum, 1980*b*, p. 7.
39. Wiersum, 1980*a*, p. 519.
40. Jacobs in Dobben *et al.*, 1975, p. 203.

8

POPULATION PRESSURE ON THE OUTER ISLANDS

Jawa provides an impressive example of the ecological consequences of heavy population pressure on soils. However, the 'Jawa model' is not confined to that island; Bali, parts of Lombok and Sumatera are often more 'typically Javanese' than certain parts of Jawa. Though the Outer Islands as a whole are sparsely populated — with densities ranging from 60 to less than 10 people per sq. km. — there are islands or parts of islands where the relations between man and soil are similar to those in Jawa. The carrying capacity of the Outer Islands — like the regions of Jawa — varies from place to place depending on soil type, climate and relief. Except for limited regions, the Outer Islands generally cannot compete with the edaphic advantages that Jawa possesses, with its extended volcanic soils which are periodically rejuvenated following volcanic eruptions. In those limited regions however, we often come close to 'Javanese conditons'.

We have seen earlier that volcanism is important for an area's agricultural carrying capacity. There are, however, differences in volcanic soils according to the eruptive material discharged. In contrast to Jawa, the Outer Islands have little or no volcanic soil. Sumatera comes closest to the fertility of Jawa in its mountainous areas, though some of its volcanoes produce acidic material that weathers to less fertile soils than basic material. A substantial proportion of Sumatera's land area is covered with swamps and swampy soils. In the province of North Sumatera, where a plantation economy has been established since colonial times, and in the province of Lampung in the south of the island where transmigrants have settled since the beginning of this century, population densities of 118 and 139 persons per sq. km. were recorded in 1980 compared to an average of 60 for the whole island.[1]

Kalimantan has no recent volcanic history and is characterised by extensive swampy lowlands and rain forest-clad hills, both with soils of little natural fertility. Its only area of relatively high population density (115 per sq. km.) is in the Hulu Sungai district of South Kalimantan, along the Nagara river, where a small kingdom once flourished; this compares to 12 per sq. km. for Kalimantan itself.[2] The highly mountainous island of Sulawesi has two areas of Quarternary volcanism whose fertile soils support a substantial population. One is in North Sulawesi around Mt Soputan (111 persons per sq. km.), the other in South Sulawesi around Mt Lompobatang (83 persons per sq. km.), compared with 55 for the whole island. In the rest of Sulawesi the soils are mostly laterised and hence not very fertile. Moreover, due to the rugged relief, alluvial areas are few.

The island chain of Nusa Tenggara (Lesser Sundas) stretches eastward from Jawa along various mountain ridges. Edaphically, the islands are very heterogeneous. The northern or inner island arc, comprising Bali, Lombok, Sumbawa, Flores and many smaller islands, is to a large extent volcanic and thus has largely fertile soils. Its productivity, however, is limited by its rugged relief and an annual dry season that becomes longer the farther east one goes. On Bali, fertile volcanic soils and sufficient rainfall have led to an extremely intensive pattern of land use, which supported an insular population density of 444 people per sq. km. in 1980. In some of the southern districts of Bali, however, densities of between 500 and 750 people per sq. km. can be found.[3] In naturally fertile parts of Lombok, densities between 300 and 400 people per sq. km. are not uncommon, and in Sikka, a district of Flores, population densities of over 1,000 have been recorded.[4] The further east one travels in the Lesser Sundas, the population density decreases despite the fertile volcanic soil as the competition for scarce water resources reduces the carrying capacity of the land.[5]

Sumba and Timor are the main islands in the southern or outer island arc which is composed primarily of limestone. Though the soils are not infertile but rather rich in nutrients, they are extremely permeable which, in this area of moderate rainfall, greatly reduces their agricultural productivity, particularly on Timor where the dry season often lasts no less than seven or eight months. The islands are generally rugged, and level areas of alluvial material are scarce. In contrast, some of the tiny islands between Sumba and Timor seem to have a greater carrying capacity despite the unfavourable natural conditions. This is because their inhabitants harvest palms and cultivate small gardens, whereas on the larger islands the people still practise shifting agriculture. Thus, in 1971 population densities of 61 (Roti island), 102 (Savu island) and 232 (Ndao island) were recorded.[6] Compared with this, Sumba, with 16 persons per sq. km. is the most thinly populated.

The island group of Maluku, stretching between Sulawesi and Irian Jaya, has a number of active volcanoes with soils deriving from young volcanism. The land near the volcanoes of Halmahera, Ternate, Tidore and Banda is extremely rich, but due to the lack of vast, flat areas where such material could settle, many of the valuable ejecta are washed into the sea. Soils situated away from the volcanoes are severely laterised. Except for Seram, the administrative centre of the archipelago, the islands are sparsely populated with an average of 19 inhabitants per sq. km., and though nature often favours agricultural production, their isolated situation and trading difficulties have paralysed economic development.

Turning next to Irian Jaya, the western part of New Guinea, we again find an area with extremely high rainfall where luxuriant natural vegetation gives the false impression of soil fertility. Its population density is minimal (3 people per sq. km.), the soil fertility is poor to mediocre, and only on the Kepala Burung (Vogelkop) peninsula has recent volcanic activity occurred. There are pockets of fertile soils in the mountains, but the extended coastal plains are swampy lowlands that have proved a handicap to soil formation. The indigenous Melanesian population has traditionally used the land in the form of shifting cultivation in the mountains or mud-bank farming on the watered plains. They generally apply ecologically protective techniques, but it seems that a growing population slowly leads to environmental destruction, and some areas are now already regarded as 'critical'.[7] The Indonesian government recently began a huge transmigration programme to Irian Jaya, the ecological consequences of which will soon become visible. Among specialists there is agreement that the carrying capacity of Irian Jaya is highly overestimated.[8]

Next to soil qualities, the availability of water determines the agricultural carrying capacity of an area. Here again, non-Javanese Indonesia varies greatly from island to island, though Jawa itself also shows tangible differences between west and east. Whereas the major part of the Greater Sundas enjoys an equatorial rainfall regime, i.e. precipitation well distributed all through the year, the Lesser Sundas only receive monsoonal rainfall and thus have a pronounced dry period. This can mean thirty or more days without rain, but under extreme conditions the drought period has been known to last more than seven months. Parts of the Greater Sundas situated far away from the equator reveal an increasingly monsoonal rainfall pattern, with a pronounced dry period. Such areas are found in the south-east of Sumatera and Kalimantan and in the north and the south of Sulawesi. But even inside the equatorial parts proper, mountains and plains may lead to microclimates with typically monsoonal rainfall.[9] Thus, each of the Outer Islands has developed problem areas where natural constraints keep the carrying capacity low or where, in contrast, natural advantages have attracted too many people and led to environmental problems. We are therefore correct in saying that on the Outer Islands the 'Jawa model' can be found: high population pressure leading to ecological destruction of the living space.

A few examples may help illustrate this phenomenon:

'The common conception of inexhaustible land potential being available on the great Indonesian islands — excepting Jawa — requires correction', writes U. Scholz. 'It is, on the contrary an ascertained fact that considerable parts of Sumatera, which suffer from growing population pressure, are approaching the upper limit of soil-utilisation capacity.'[10]

In addition to natural population growth, certain parts of Sumatera, particularly the province of Lampung on the southern tip of the island, have been the target of many migrants from Jawa who multiplied and encouraged friends, relatives, and former neighbours to follow them to Sumatera. They were attracted by the relatively thinly populated island and the advantages offered by the government transmigration schemes. Van der Kroef, writing thirty years ago observed that: 'Colonies of Javanese emigrants in southern Sumatera do indeed exhibit the same structure of production as that in Jawa itself; the population of the (emigrant) colonies begins with a high degree of density and within a generation it may become so dense that continued colonisation and emigration becomes advisable.'[11] The population density already reached on Lampung (139 per sq. km.) may not seem alarming when compared with Jawa. In 1980, when the average density of Jawa was given as 690 persons per sq. km., that of the three southern provinces of Sumatera was 64, with an estimated 18% of the total area utilised for agriculture. The picture changes though when we consider the natural potential.

Scholz, who has carried out intensive studies in the region in question, concluded that the cultivation of various perennials, preferably rubber, would best meet the ecological requirements. He had however to concede that the produce of such crops has to be marketed whereas food has to be bought. The 'gradually increasing population pressure, steadily accelerated by resettlers from Jawa, will cause a rising demand for foodstuffs though, thus inevitably increasing the necessity for food-crop cultivation.'[12]

Traditionally, the region produced food through shifting cultivation for a few people, but with a rapidly growing population this can no longer serve as an economic method simply in terms of the ecological consequences such as soil erosion, the deterioration of the water balance, and the tremendous waste of highly valuable forestry potential. Thus, it would become necessary to grow wet-rice (after terracing and irrigation have taken place) and annual crops on permanent dry-fields after ploughing and fertilising the soil. *Sawah* cultivation would be the most advantageous technique to use, particularly because after rice has been grown other annuals could be cultivated in rotation on the same terraces. Nevertheless, market gardening on dry-fields, perhaps in the form of *taungya*, should be given a chance in order to save the protective value of the forest.

So far, the resettlement policy has shown that the two hectares of land given to the settlers from Jawa — which was enough to feed a family on the irrigated volcanic soils at home — has proved insufficient given the various soils prevailing in Sumatera. Moreover, the envisaged irrigation facilities did not exist when the settlers arrived, and once they had accepted the new land they were left to their fate. 'In order to survive on a small plot like this, the farmer had no choice but to plant upland food crops as intensively as possible, regardless of the soil fertility. Since he cannot afford fertiliser, the results are rapid decrease of soil fertility and a steady decline of farm income within a few years time. The next problem to be faced then is the rapid ingression of *alang-alang* grass.'[13] Consequently the settlers, or more recently their children, were forced to clear additional land and stick to dry-land farming, to look for an income outside agriculture or even to migrate again. Considering all the natural disadvantages, Scholz feels that so far the biggest handicap has been a lack of capital and that 'with a minimum of capital input, but application of carefully selected crop varieties, and employing certain cultivation techniques under careful consideration of specific ecological conditions of each particular region — a great part of the existing land reserves could be opened and cultivated successfully.'[14] We see that even under not too difficult natural conditions a growing population pressure will inevitably lead to environmental hazards, unless a number of difficult preconditions are met. This is obviously not the case in Lampung, where wet-rice-growing settlers were forced to adapt to dry-field cultivation. 'The remarkable concentration of permanent dryfield cultivation in Lampung', states Scholz, 'is neither the result of an intelligent adjustment to the natural conditions, nor does it derive from the traditional economic way of the resettlers . . . Instead we have to regard it as a reaction forced upon the farmers by a doubtful planning concept that obviously had not at all taken into consideration the natural resources of the planning area and their different aspects.'[15]

Whenever a seminar on environmental problems is held in Indonesia, the case of Sulawesi is discussed. It seems that the local people on this island do not understand or even care about ecological hazards. The topography of Sulawesi is difficult and the mainly mountainous landscape does not offer much economic potential. Any sort of land use leads easily to ecological destruction. The traditional methods of shifting cultivation quicken the erosion of soil and the silting-up of rivers and irrigation channels. Concessional felling of trees has the same effect, especially where chainsaws and tractors accelerate the destruction of forest and soil. As early as 1937, Verhoef reported a case of '*stervend land*' in the Palu valley of Central Sulawesi. With about 500 mm. of rainfall, well distributed over the year, it is the area with the lowest precipitation in Indonesia. Between 1,000 and 2,000 m., the slopes are completely denuded. Higher up the rainfall is somewhat greater and on small patches the growing of trees is possible, but continuous burning renders natural rejuvenation rather difficult. The valley bottom which comprises about 250,000 sq. km. is permanently threatened by *banjirs* of mud and rubble from the slopes above.[16] Here and in other places the denudation process has already severely threatened irrigation projects to such an extent that, for example, the Gumbasa irrigation system had to be closed down after it had dried up: 60,000 cu. m. of sand and sediment had accumulated along an 11 km. stretch of irrigation canal.[17] Attempts to halt the environmental destruction in Sulawesi proved extremely discouraging. An 'old regreening campaigner', in the service of natural conservation since 1962, reported his observations to the then Minister of Environment. Of an area planted with candle-nut trees (*Aleurites moluccana*), cashew nut trees (*Anacardium officinale*) and kapok trees, (*Eriodendron*) only 10% survived. The rest of the land is now covered with sedge grass as before, since nobody cared for the trees. Instead, the people even preferred to work for logging companies outside Sulawesi. The terraces built in the course of the regreening campaign largely went down the slopes since stones were broken out of the retaining walls and sold for road or house construction. Nomadic farming together with expanding numbers of cattle and a lack of understanding on the side of the rural people has thwarted all the conservation efforts undertaken so far.[18]

The westernmost part of the Nusa Tenggara insular chain shares further similarities with eastern Jawa. Bali may in fact have once been part of the larger neighbouring island. The northern division is mountainous and volcanic, being a continuation of Jawa's Central Range. The active volcanoes reach up to 3,142 m. (Mt Agung), 2,276 m. (Mt Bukatan) and 2,151 m. (Mt Abang). The rivers have dug deep ravines into the soft fertile ground and in the course of time have transported the material and built up extended alluvial slopes and plains towards the south. These range in height from 600 m. down to sea-level and are well terraced.

Apart from a barren high plateau in the west, Bali is covered throughout with fertile soil and has reliable rainfall and a balanced climate. This means that a great variety of plants can grow and allows for a high carrying capacity. Whereas in the less favoured north the population density hardly surpasses 100 persons per sq. km., the south, as we mentioned earlier, supports densities of up to 750 per sq. km.[19] It can be seen from this that under such conditions, and provided the necessary care is taken, the carrying capacity of Bali is similar to that of Jawa. The problem of environmental damage due to growing population pressure had already been observed during the colonial period in the 1920s, when the forest cover had shrunk to 13% of its previous total area and was absent mainly from areas where it should have grown for hydrological reasons. In addition, the rivers and rivulets had cut deeply into the loose volcanic soil material and the spread of ravines threatened the neighbouring cultivated land. In 1934, an attempt to halt this destructive process was made using the demarcation of the ravines. In order to protect them from further destruction, the sites within the borderlines belonging to the peasants were bought by the government and given back to them on a loan basis with the proviso that they be cultivated with bamboo, gomuti sugar palms (*Arenga pinnata*) and other perennials. The peasants were permitted to use these thickets for various purposes, but the clearing of these sites was prohibited. The Balinese and their village headmen, intelligent and judicious, accepted these measures and contributed to the rescue of their living space.[20]

This experience shows that even on the best soils the carrying capacity has its limits. Since the disastrous eruption of Mt Agung in 1963, the consequences of overpopulation have become visible again. For centuries the peasants had enjoyed the rewards of their work in the form of high yields without fertilising the soil. In case of volcanic eruption, people moved temporarily to another part of the island and later returned to the land covered with fresh lava to take up the arduous but rewarding task of bringing it under cultivation again. Nowadays, though, empty land for the evacuees is no longer available on Bali, and whereas the average annual emigration rate from Bali amounted to about 3,000 persons between 1950 and 1962, it jumped to 22,598 in 1963. Since then, the annual number of transmigrants varies between 4,000 and 5,000 people.[21] The example of Bali shows that inspite of a well-organised society and a tradition of land use that protects the natural potential to the utmost extent, the point will be reached where any additional population has to look for a new homestead and source of livelihood elsewhere.

It is remarkable to note that the Balinese transmigrants, who mostly left spontaneously, are highly regarded as effective resettlers, and have an excellent reputation, particularly on the neighbouring island of Sulawesi. Their efforts in the area of Parigi, Central Sulawesi, prompted the comment that: 'Wherever they settle, the Balinese have proved to be excellent migrants, and their superior skill in the cultivation of irrigated rice on terraced fields has been a great asset to the settlement area.' And on the south-eastern peninsula of Sulawesi, where shifting cultivators still roam the forested hills, it was noted that, 'If any development is to take place in this province, a further input of manpower is essential along with the knowledge of sedentary agriculture that migrants from Bali and Jawa possess.'[22]

Lombok is an interesting example of an island where the growing population has caused a decline in the people's standard of living from general well-being to starvation. 150 years ago, the rice shipped from Lombok was an important source of income and secured the general well-being of the people. With a high population growth rate (2.39%) and strong resistance to family planning, the island has now reached a density of 483 inhabitants per sq. km. In some areas, densities of between 750 and 1,400 people per sq. km. are found. The consequences are visible everywhere. Forests have generally been replaced by secondary bush and grass formations, and vast areas of exhausted land now abandoned by the farmers are sparsely populated and covered with *alang-alang*. Clearcutting, shifting cultivation and overgrazing have led to serious soil erosion and an increasing shortage of water. Since most of the people depend on agriculture or animal husbandry, the demand for more land has grown steadily. Attempts to relieve the population pressure through transmigration at the beginning of this century were quickly cancelled out by the natural population growth. Between 1967 and 1982, for instance, 37,644 people were encouraged to migrate from Lombok; in the same period the natural increase amounted to 47,000. The peasant families who had formerly cultivated irrigated or rain-fed rice terraces were forced to move up the slopes and engage in shifting cultivation, either as a response to land shortage or because the sedentary dry-land farming did not yield enough to feed a family. Thus, more and more submarginal land was taken under cultivation. Bad harvests were followed by indebtedness among the farmers, illegal wood cutting and squatting on government land caused irredeemable ecological damage, and with the 'critical zones' spreading, periods of hunger became not unusual on Lombok.

It has to be admitted that only part of Lombok enjoys the natural advantages of neighbouring Bali. The north, which has volcanic soil, receives an abundance of rain during the north-west monsoon, but the northern mountain chain that reaches up to more than 3,500 m. prevents the moist air from bringing much rain further south. As a result, the central high plateau and the southern slopes of Lombok are generally dry and the drought often lasts several months. More recently, South Lombok (Lombok Selatan) which is now synonymous with 'critical land', has attracted the interest of the government. Apart from efforts in family planning and inter-insular migration, more attention is being given to the hidden potential of the region. Though the limestone soils are generally poor in nutrients, the main handicap is the lack of water. The construction of water-storage facilities will, it is hoped, help conserve South Lombok's uncertain rainfall (1,000–1,500 mm.) for the production of crops. Emphasis is being placed on the local construction of artificial 'hill lakes' (called *embung* in Indonesia) where rainwater accumulates and can be diverted to the fields, a method that has been used since 1934. Moreover earth dams, neglected till recently, are now being rehabilitated and additional ones built. The Indonesian press has commented favourably on the developments in South Lombok and has hailed it as a new rice-exporting area. They forget that the island's population pressure is many times higher than it was 150 years ago, and that transmigration to the neighbouring island of Sumbawa has not brought a solution. Sumbawa's natural conditions are no better than those of Lombok but the island is less densely settled, yet many a transmigrant family has been disappointed and has abandoned the land. Although it is right to invest in methods which make the fullest use of the limited natural potential, a steadily growing population will certainly outpace all these efforts.[23]

The island of Flores lies in the northern or inner insular arc and though it has deep volcanic soils, unreliable rainfall greatly limits its economic usefulness. Though wet-rice growing is known, usually a sort of shifting cultivation with a fallow and cultivation period of three to four years each was needed to produce the necessary food (using dibble and hoe).

In the *kabupaten* of Sikka, the most densely settled and driest district of the province of East Nusa Tenggara, population pressure forced the inhabitants to cut down the forests on even the steepest volcanic slopes. Consequently the valuable soil went down in enormous streams of mud (*lahars*) into the valleys causing inundations and swamps on the coastal plains. In some places, densities of more than 600 persons per sq. km. are reached, so that dry-field farming became permanent even on rather steep slopes. Inevitably, all this led to a reduction in soil fertility and an increase in erosion.

Here, as in other parts of Indonesia, conservation measures were introduced by the Dutch. They ordered terraces to be constructed, but most local people refused to cooperate. The construction of bamboo anti-erosion fences had little effect. More recently, a new method of 'indirect terracing' changed the picture, partly because the local people realised the precariousness of their situation, and partly because 'model plantations' were introduced to show them what could be achieved under improved conditions. The new technique was particularly successful in densely settled parts of central Sikka where private ownership of land prevailed, in contrast to areas where there was tenant farming or shifting cultivation. In those cases, the personal interest in soil stabilisation was lacking.[24]

'Indirect terracing' simply means the planting of *lamtoro* (*Leucaena leucocephala*) hedges along the contour lines ('*lamtoronisasi*') so that sediment washed down the slope can accumulate behind them. In this way a kind of terrace builds up and the soil can be further enriched by mulching with *lamtoro* cuttings. Since *Leucaena* produces roots that quickly reach substantial depths, it is able to tap aquifers which other plants cannot reach. This allows it to survive drought periods without damage, and since the roots can also penetrate hard soil layers, they improve the soil structure at the same time. Finally, being a legume, it provides an excellent green manure. *Leucaena* leaves also provide a nutritious animal fodder rich in protein, though this has to be supplemented by other grasses and legumes.[25]

In addition, *Leucaena* is able to provide the wood for the production of a high-quality charcoal. Formerly it was shipped to the Lampung iron works on Sumatera, but since a substantial quantity of iron ore has been discovered on Flores, local iron works are now envisaged. The President of the Republic personally gave the order to encourage further planting of *Leucaena*, and the *kabupaten* of Sikka is proud of having changed this once barren land into a green, productive landscape again.[26]

Sumba is an interesting example of how the interaction between natural disfavour and social structures and habits works to the detriment of general development. The island belongs to the outer insular arc of Nusa Tenggara, and has poor to very poor soils without volcanic admixtures, except for some places where tertiary volcanic tuff under the limestone adds to the fertility of the soil. Though the west receives somewhat more rain, the island as a whole belongs to that part of Indonesia with a pronounced dry period.

Historically, Sumba consisted of many minor principalities which tried to annex each other through warfare. Only Dutch interference brought this habit to a halt, but it seems that the present lack of cooperation between groups is a legacy of that period. In addition to this, however, a strong *adat* structure prevails on Sumba. All land belongs to the clan (*kabisu*) and the head, the *tuan tanah*, is the master of that land. He has the sole right of disposal and no land may be used without his consent. In many cases, there is a *raja* above him handling the *adat* who is not necessarily the master of the land. The *adat*, as we know, regulates secular as well as religious matters for both individuals and the community.[27]

Bound into this system, the Sumba peasant has not developed much of an individual consciousness and his agricultural technology has hardly developed; he uses the dibble in the dry-field and in the *sawahs* the buffalo does the puddling. But since the number of buffaloes owned by the peasants is small, only part of the potential *sawah* land can be tilled. On the other hand, a substantial number of horses and cattle are kept on Sumba. Unfortunately these animals are the property of a few large landlords and traders who often require land for pasture that could easily be used to produce food. Thus, we have a typical conflict of interests between the cattle breeder and the farmer. The latter, being in the weaker position, is unable to use the suitable land in the most advantageous way. Only part of it is properly irrigated and cultivated, because the peasants live too far away and do not have enough draught-animals for the field work. In contrast, a few well-to-do animal breeders determine the pattern of land use in vast areas. Training and support of the farming masses together with a change of the social power structure seem to be the prerequisites for better land use on Sumba.[28]

Of all the Outer Islands, Timor has the most unfavourable natural conditions and is one of the driest parts of Indonesia. The average annual rainfall of 1,171 mm. seems reasonable, but it is concentrated within a few months so that there is an extended season of extreme drought. There are no volcanic soils, the relief is rugged, and the poor limestone, marl and clay soils are thin. At the beginning of this century, wars and epidemics kept population and environment in a certain equilibrium. With modern medicine and colonial pacification, the population grew quickly without a corresponding increase in food production.[29] The consequences were a reduction in the fallow time of the prevailing shifting cultivation, further deforestation and the spread of wild grassland (*Imperata* and *Saccharum*) to areas where monsoon forests had exerted some soil protection. Now, in the dry season, wind erosion blows away the topsoil and during the torrential rains fluviatil and sheet erosion.

The cattle-owning people of Timor tended to settle in the mountainous interior for three main reasons. First, this region has a more even rainfall pattern (1,500–2,000 mm.); secondly, there was a good deal of land still to be cleared; and thirdly, though the soils are better in the lowlands, they are heavily infested with malaria, which was to prove a great obstacle to settlement.[30] Thus, on the level coastal plains there are no more than 10 people per sq. km., but in the central highlands that figure reaches 360, together with a great number of cattle kept mainly for prestige and dowry payments.

The environmental consequences are all too visible. The monsoon forest that formerly may have covered vast parts of the island has shrunk to a mere 10% of its original size. The rest was destroyed by extensive shifting cultivation, slash and burn techniques and low-yielding permanent dry-field cultivation. The cultivation of wet-rice is confined to some river plains and terraced slopes, but since the dams are washed away by occasional torrential floods, rice cultivation now depends on rainwater. The water shortage is often so severe that occasionally banana trunks have to be fed to the cattle to supply them with the necessary moisture.

The two colonial administrations — the Dutch in the west and the Portuguese in the east of the island — both tried from an early date to introduce marketable crops in order to offer an income to the peasants. As early as 1815, *Coffea arabica* was planted, and later came coconut palms, rubber trees, sisal agaves and tea bushes — but only coffee flourished as an export crop in Portuguese Timor, often at the expense of the rice fields. Elsewhere, the expected cooperation from the peasants did not materialise, and they remained subsistence-oriented and only grew coconuts and *Aleurites moluccana* (the candle nut tree, used to produce tung oil).

When in 1912 Bali cattle were introduced to Timor on a large scale, a bush, *Lantana camara*, coincidentally also reached the islands and spread over the savannahs. This led to a reduction in cattle-grazing land, particularly because *Lantana* proved unsuitable as a fodder crop. On the other hand, the plants' properties soon came to be appreciated by the shifting cultivators who found that its rapid propagation helped to cover the exhausted soil; five years of fallow were sufficient to clear the recovered ground for another two years of cultivation. But since the pastoralists were in a stronger position, they succeeded in gradually replacing *Lantana* by *Leucaena leucocephala*, known as *lamtoro* in Indonesia. In the area of Amarasi (Western Timor) this process was greatly supported by the *adat* structure on which the power of the local *rajas* was based. They took the initiative, organised the population and forced them to accept the new system. Vast areas were covered with *lamtoro*, which provided the bulk of the fodder needed, and Amarasi became virtually the most important centre of cattle breeding in Indonesia. After only six months, most animals were generally ready for slaughter. The Amarasi experiment was so successful that it became known as the 'Amarasi model'.[31]

The example of Timor provides us with three major lessons: the first is the disastrous consequence of high population pressure on soils that are only marginally suited for food production; secondly it shows that the introduction of a new 'environment', namely '*lamtoronisasi*', based upon a plant that thrives well under the given conditions, can lead to a system of production that does not cause further environmental damage; thirdly it shows that the *adat* power of local leaders (in contrast to what happened on Sumba) may well help to foster development.

We have tried to show that the 'Jawa model', i.e. heavy population pressure leading to environmental destruction, is not confined to the island of Jawa, but can be traced to the Greater and Lesser Sundas as well. Once the population pressure exceeds the limit provided by the carrying capacity, the consequences are ecological damage, declining yields and starvation and a reduction of the carrying capacity. Inputs such as terracing, irrigation, the introduction of new crops and so forth may well raise the limit, but considering the ever-growing population pressure it is safe to say that, as on Jawa, the long-term solution must be birth control and emigration. The question remains however, what will be the environmental damage to the vast virgin areas of the Outer Islands once millions of peasants are moved there?

NOTES

1. *Business News*, 8 July 1981; Zimmermann, 1976, p. 115, reports population densities of 500–700 persons per sq. km. in some of the resettlement areas.
2. Hardjono, 1971, pp. 102–3.
3. Röll, 1979, p. 37.
4. Metzner, 1977a.
5. *Straits Times*, 23 Jan. 1983.
6. Fox, 1977, p. 23..
7. Aditjondro, 1983; *Indonesia Times*, 9 March 1983.
8. Hardjono, 1971, pp. 103–4.
9. Eichelberger, 1924, pp. 103–9.
10. U. Scholz, 1977a, p. 45.
11. Van der Kroef, 1956, p. 744.
12. U. Scholz, 1980, p. 28.
13. Loc. cit., p. 27.
14. Loc. cit., p. 29.
15. U. Scholz, 1977a, p. 50.
16. Verhof, 1937, p. 222.
17. *Indonesia Times*, 16 Dec. 1981.
18. Loc. cit., 13 April 1981 and 16 Dec. 1981.
19. Röll, 1979, p. 37.
20. De Voogd, 1937, pp. 302–10.
21. *Buku Saku*, various years.
22. Hardjono, 1977, pp. 86.
23. Blaut, 1960, p. 196; Hardjono, 1977, p. 11; Röll and Leemann, 1982, pp. 132–45; Hamzah, in *Indonesia Times*, 11 April 1984; Röll and Leemann, 1984.
24. Metzner, 1976a, pp. 224–34.
25. Metzner, loc. cit. and Jones, 1983, pp. 106–10. They observed that the 'Sikka model' is applicable only on steep and volcanic soils, and that the terraces are not continuous, but are often interrupted by patches of shifting cultivation. In such cases, erosive processes can gain a foothold. Cattle may well be kept and fed with *lamtoro*, but for a proper fattening industry, the hedges do not provide enough fodder. In contrast, the 'Amrasi model' in Timor (where the whole area is covered with *Leucaena*), permits a comprehensive cattle fattening programme, and in addition, 'after ten years the micro-climate has normally improved sufficiently to allow gardens to be planted and the cultivation of additional tree crops such as coconuts . . .'
26. *Indonesia Times*, 19 Oct. 1983.
27. For details concerning the 'master of the ground' see F. Scholz, 1962.
28. Dames, 1952; F. Scholz, 1962, pp. 10–29.
29. For further details see Metzner, 1975 and 1977b.
30. Metzner, 1976b.
31. Metzner, 1983. — *Lantana camara*, as a non-indigenous shrub, was looked upon as a useful plant by the forestry service and agricultural experts because of its anti-erosive properties. However, for cattle farming this plant is actually harmful. See Walandouw, 1952, p. 205.

Part III. THE OUTER ISLANDS: LAND USE IN THE FACE OF ECOLOGICAL PROBLEMS

9

THE PRECARIOUS POTENTIAL OF THE OUTER ISLANDS

Uncertain Estimations

From the earliest times, descriptions of what is now the Republic of Indonesia refer to a vast archipelago of lush, yet thinly populated islands, well covered in vegetation. In contrast, the islands of Jawa, Madura and Bali have always been well populated and have given rise to numerous towns and ports. For various reasons described earlier, the population of Jawa and Bali grew rapidly and the people exploited the islands' resources to produce food and export commodities — a process that could not take place without visible damage to the environment. Until recently, the Outer Islands played only a marginal part in this process, and a traveller coming from Europe at the end of the nineteenth century could easily distinguish between the densely populated island of Jawa and the Outer Islands, where the vegetation still looked undisturbed.[1] Though accurate figures are unavailable, we can estimate that at the beginning of the twentieth century roughly 30 million people lived on Jawa, Madura and Bali, while about 10 million may have lived on the Outer Islands.[2] This would mean that about 75% of Indonesia's population lived on Jawa and its neighbouring islands. As a consequence of sound population policies on Jawa and Bali, a net migration from these islands, and a comparatively high natural population growth on the Outer Islands, this relation has to some extent changed. Whereas in 1980 the population of Sumatera still grew at a rate of 3.3%, of Kalimantan by 2.8% and of Sulawesi by 2.2%, the comparable figures for Jawa and Bali were 2.0% and 1.6% respectively. In 1961, Jawa's share of the total population of Indonesia was 66.8%, but had fallen to 63.6% by 1980. At present however, the remaining 36.5% of Indonesia's population lives on 92.8% of the country's total land area.

This population dichotomy has for a long time kept politicians busy

187

considering ways and means of making use of the 'land beyond', the allegedly underpopulated Outer Islands. The two main concerns were to ease the population pressure on Jawa by transferring a substantial number of families to the Outer Islands, and to exploit their economic potential, namely forest products, ores, oil and, last but not least, agricultural production — since practically all the transmigrants would be peasants. Such a policy requires a reliable knowledge of the potential of the Outer Islands. As we know today — and as quite a number of specialists have known for a century or more[3] — the lush vegetation of tropical islands is by no means a guarantee of fertile, arable soils. Quite early on the question was asked, why, over historical periods, did the population concentrate in Jawa and Bali and only in a few selected areas of the Outer Islands? It was obvious that the natural conditions were more favourable in the former islands and accordingly the population settlement was quite dense. This is true at least in regard to a sedentary population who cultivate the same plot of land for generations, occasionally using irrigation. The majority of the indigenous population of the Outer Islands were not sedentary in the strict sense of the word, but practised shifting cultivation.

This made the process of gauging the agricultural potential of the Outer Islands very difficult, especially in the early days of the trans-migration policy when none of the technical survey methods available today could be used. Nevertheless, a few thousand families were resettled successfully in southern Sumatera between 1905 and 1910. Nowadays, however, when the government's fourth Five-Year Plan (1984–9) envisages the transfer of 800,000 families or about 4 million people from the overpopulated islands to the Greater Sundas, to Maluku and even to Irian Jaya, careful preparation of the receiving areas is vital if such schemes are to be feasible.

Climate and soil are the crucial indicators in judging the production potential of the agricultural sector. Stretching a distance of nearly 5,000 km. from east to west, and lying north to south between 6°N and 11°S across the equator, the Outer Islands cannot be expected to have a homogenous climate. And yet the mean temperatures remain between 25°C and 27°C, with a more pronounced variation occurring due to high altitude or the relief. The distribution of rainfall, however, is more heterogeneous. In the northern part of the Outer Islands including the central highlands of Irian Jaya, the annual rainfall generally surpasses 3,000 mm., but it decreases towards the south-east, falling below 1,000 mm. The mean rainfall is lower in summer than in winter, and the changes in the monsoon rains cause seasonal variations between the extremes of the wet and dry seasons and their transitional periods. The seasonal variations are relatively small in the northern and north-western parts, but remain very pronounced in the south-east, particularly in eastern Nusa Tenggara. Furthermore, the relief results in impressive microclimatic differences. Palu in Central Sulawesi has the lowest annual rainfall on record (530 mm.), whereas in Padang, West Sumatera, 4,511 mm. has been measured. Most parts of the Outer Islands, though, receive more than 3,000 mm. of rain per annum. Apart from the Lesser Sundas, characterised by their long dry season, it is true to say that the Outer Islands do have enough water to support agricultural production. However, this rainfall is sometimes periodically and locally so torrential that it endangers cultivated land that is deprived of protective vegetation cover. The heavy rainfall not only waters and erodes the soil, it also leaches it. Valuable mineral substances are dissolved and carried away into deeper levels, and then into springs, rivers and the sea. Finally, lateritic soils are formed on which vegetative growth is nearly impossible, such as the aluminous laterites of Bintan.[4] If we consider the Outer Islands' productive potential as far as water is concerned, they mostly receive enough rain. The problem that remains is how can its destructive force be controlled and the water brought to the fields at the right time and in the right quantity, be it in a natural way (rain) or an artificial way (irrigation). The indispensable counterpart of water for large-scale cultivation of plants is soil. The soils of the Outer Islands show a pronounced heterogeneity as a result of the humidity and the elevation above sea-level. There are three major soil groups on the Outer Islands.

Mountain soils of the humid and sub-humid tropics prevail in the mountain areas covering the western part of Sumatera, the centre of Kalimantan, practically the whole of Sulawesi and the spine of Irian Jaya. The soils situated at a lower altitude and covering substantial areas of Sumatera, Kalimantan and Irian Jaya are classified as lateritic soils of the humid and sub-humid tropics and are covered with either tropical rain forest, monsoon forest or moist savannahs. They are partly bleached and comprise latosols, ferrallitic soils, ferralsols and ferrisols. Mineral hydromorphic soils, such as wet meadow soils, alluvial meadow soils, humic gley soils, gley soils and alluvial soils cover substantial areas in the swamps that spread along the east coast of Sumatera, around the large deltas of Kalimantan, and along the south coast of Irian Jaya.[5]

The podzolic and tropical soils on Sumatera and Kalimantan cover a very wide area and are probably suitable for agriculture. Particularly fertile soils can be found especially in areas of recent volcanic activity where this has led to the development of volcanic ash soils, young lateritic and podzolic soils. This is the case in large areas of the Barisan mountains of Sumatera as well as in parts of North and South Sulawesi. The extensive lowlands of Central Sulawesi are composed of recent alluvium which seems to be fertile, whereas the lowlands of Sumatera, Kalimantan and Irian Jaya consist principally of alluvial and marshy plains, difficult to ameliorate for cultivation, and often containing a high proportion of organic and acid sulphate soils. The soils of Nusa Tenggara and Maluku are usually very shallow and are therefore less suitable for cultivation, except on some of the islands with a recent history of volcanic activity.[6]

The Indonesian Soil Research Institute, Bogor, tried to estimate the extent of the most important soil types of the Outer Islands over an area of 177 million ha. (Table 10), and, at the same time, evaluate their suitability for agricultural use (Table 11).

The most impressive result is that no more than 1% of land is unconditionally suitable for cultivation and another 31% should never be cultivated. The remaining 68%, corresponding to a surface area of roughly 120 million ha., is characterised as 'conditionally suitable'. Baharsjah adds that:

Table 10. ESTIMATED AREA OF SOIL TYPES ON THE
OUTER ISLANDS[1]

Soil type	Area (× 1,000 ha.)
Podzolic complex soil	53,753
Podzolic (ultisol)	47,199
Organosol (histosol)	24,125
Alluvial (entisol)	16,589
Latosol (inceptisol, oxisol)	15,301
Mediterranean (altisol)	6,162
Podzolic (spodosol)	5,012
Andosol (inceptisol)	4,212
Regosol (entisol, inceptisol)	2,261
Rendzina (mollisol)	1,633
Grumosol (vertisol)	319
Others	605
Total	177,171

Source: Soil Research Institute, Bogor, 1975 (quoted by S. Baharsjah, 1978).

[1] excepting Jawa/Madura and Bali.

Table 11. ESTIMATED AREA OF LAND SUITABLE FOR
AGRICULTURE ON THE OUTER ISLANDS[1] (× *1,000 ha.*)

Class	Type	Sumatera	Kalimantan	Sulawesi	Irian Jaya	Nusa Tenggara	Maluku	Total
1–3	suitable	631	—	1,081	—	232	25	1,969
4–6	conditionally suitable	39,293	37,863	7,400	25,588	4,750	5,744	120,638
7–8	unsuitable	7,439	16,137	10,614	16,612	2,056	1,706	54,564
	Total	47,363	54,000	19,095	42,200	7,038	7,475	177,171

Source: Soil Research Institute, Bogor, 1975 (quoted by S. Baharsjah, 1978).

[1] without Jawa/Madura and Bali.

It should be noted that a large part of the potential land resources consists of soil types, particularly the red-yellow podzolic soils (ultisols) and the organosols (histosols), which are categorised as problem soils. These soils are acid and have low fertility. They are considered marginal soils with regard to food production. Careful studies of their properties become necessary in order to formulate suitable cultivation practices, since agricultural expansion in the future will have to be directed to regions with these soil types.[7]

In fact, podzolic soils of various properties cover no less than 60% and ultisols another 14%, so that three-quarters of the 'conditionally suitable' agricultural ground can be cultivated only with care, with specific crops, and by using particular techniques. One important conclusion can be drawn from the list of soil types: the vast hectareage of such soils could easily make them unproductive under poor cultivation methods. Large areas have already deteriorated towards unproductivity, and although there are no exact data on the size of the waste land it is generally agreed that it has reached serious proportions. Shifting cultivation, ecologically adaptable in its classical sense, contributed much to the formation of these waste lands once they came under population pressure. In addition, poor management of the forest stands results in the increase of unproductive waste land, but techniques are now being developed for its reclamation. 'More important, however, are measures which could prevent the increase of the hectareage. It is believed that the problem is more socio-economic in nature than technical.'[8]

With regard to the improvement of transmigration programmes, a number of detailed studies have been made to analyse the potential of the arable soils and to work out programmes for their best possible use without causing their destruction. The researchers do not claim that vast areas of the Outer Islands would better be left untouched. They realise that Jawa, already overflowing with people, needs the space to resettle them. However hardly any study fails to warn of the dangers of unqualified deforestation and cultivation of soils unless such actions are well prepared and guided and correspond with the specific ecological requirements of the area in question.

East Kalimantan was, for a couple of years, not only the subject of serious research but also the destination of a large influx of settlers.[9] Information taken from satellite surveys showed the land to be susceptible to severe erosive processes due to its geological and pedological structure and rainfall pattern. Even without human interference and despite the natural vegetation cover, the erosion is so heavy that outside intervention would only lead to uncontrollable devastation. This is the tenor of the whole report. Detailed investigations into more than 20 soil types showed many of them to be completely unsuitable for cultivation. Others would require either detailed ground surveys to identify the limited areas suitable for cultivation, or massive investments in drainage and fertiliser: 'The aforementioned facts demonstrate that the mapped region as a whole is unsuitable for any land use. It should rather be preserved under natural conditions. Even forestry activities except selective cutting of trees could disturb the ecological balance severely and cause strong destruction, as can be seen from shifting cultivation . . . Any damages in the natural soil conditions occur very fast and are irreversible, even expensive soil conservation measures may be in vain.'[10]

A study of the province of Bengkulu in Sumatera concluded that at present only 7% of its 2 million ha. are used for agriculture and that 83.1% remains under forest and swamp forest. Wide areas of the province, including coastal plains as well as hills and mountains, can be described as unsettled, and people are widely scattered over various natural regions. Because of steep slopes and poor soil properties, the agricultural potential of these regions is often strictly limited. Thus the land considered as 'reserve' should be reduced to areas which are both unsettled and cultivable. According to one expert, the amount of land which meets these criteria does not exceed 382,000 ha., corresponding to only 21% of the unused surface area including swamps.[11]

Moreover, analysis of the potential of specific areas to accomodate additional families (transmigrants) generally produces pessimistic conclusions. A study of the situation in a part of West Kalimantan bore this fact out. The area of Sambas has relatively little land suitable for the expansion of agriculture. Where it is suitable, permanent cultivation exists already. At present, there is rarely any artificial irrigation, so production in such areas could possibly be intensified in the future. Rain forest grows generally in the lower reaches which are swampy, making drainage difficult. The land on higher areas is not swampy, but except for some ferrisols, the soils lack nutrients. Areas suitable for wet-rice cultivation are already farmed. The potential for tree crops is greater and *Hevea* is the ideal plant, but without the input of fertiliser, amelioration of the soil is impossible. The opportunities for immigrants are very restricted. 'Virgin soils', required for transmigrants, are rare, and their potential is limited. Wherever there are good soils, such as ferrisols, they are usually densely settled, and in the areas with shifting cultivation the carrying capacity has long been reached and it becomes necessary to change to permanent dry-field cultivation. This is only possible with fertilisation. Transmigration is therefore not recommended. The expert, comparing Jawa and Kalimantan, concludes with a statement that could well apply to most of the Outer Islands: 'The moderate population density of the area . . . should not lead to the conclusion that these surfaces were "empty" and therefore were reception areas for the surplus population of Jawa. Such considerations fail to understand that only something can be "empty" that also can be filled, and that the unequal population densities are not coincidental, but strongly reflect the unequal potential of the two islands.'[12]

This does not mean that the idea of transmigration to the Outer Islands has been shelved. It does however point to the need to prepare the transmigration programmes properly and to refrain from over-optimistic expectations. Recently, even the normally cautious World Bank became involved in a transmigration project. With the help of aerial photographs, topographical maps and soil analyses, suitable locations are to be identified for the resettlement of some 100,000 families, i.e. roughly 500,000 people a year. The sites will be properly prepared to receive the transmigrants, who will also be trained in soil conservation. Finally, technical assistance in the fields of development and environment will be given to the ministry concerned in order to improve the ecological compatibility of the transmigration projects.[13]

Judging the carrying or absorptive capacity of a certain area of land — dealt with earlier — depends very much on both the criteria used and on the techniques applied. Usually, we envisage the *physical* absorptive capacity, namely the ability of an agricultural settlement under the given soil and other conditions to support its inhabitants on a subsistence basis. The *economic* absorptive capacity goes one step further. It comprises the kinds of crops that can be grown and marketed and their cost/benefit ratio; both are usually considered when a transmigration project is envisaged and prepared. Since in general most receiving areas are not completely empty, but are inhabited by a minority of indigenous people, two more points must be kept in mind.

The *cultural* absorptive capacity determines to what extent newcomers will be accepted by the local population. For instance, tensions may arise between sedentary *sawah* peasants (newcomers) and *ladang* peasants (local people). The *religious-based* absorptive capacity is strongly determined by peoples' cultural, economic and ecological behaviour. For example, the pig-rearing Hindu Balinese were disliked by the Muslims of Lampung and repatriated Javanese from Surinam, modern in outlook and partly Christianised, were driven away when they tried to settle in the Minangkabau area of Sumatera. Swasono goes even further: 'Without trying to exaggerate, the author believes the political element to be a greater limiting factor than any of those mentioned above.'[14] This is certainly worth considering, since the transfer of millions of Javanese to the Outer Islands is bound to intensify the feeling among the local people of being 'Javanised', in the sense that they feel that they are being colonised. This problem will certainly reach dangerous dimensions on Irian Jaya where the modern world (Javanese, Malays) comes in contact with a minority of 'Stone-Age' people (Melanesians).[15]

Since the physical absorptive capacity depends on the environmental conditions as well as on the techniques of soil use applied by the new settlers, it is important to properly train and watch over the transmigrants so that they can quickly adapt to their new environment. This has not always been the case. As early as 1957, Kampto Utomo warned of the physical limitations of the soil, as he described the complete failure of some existing agricultural settlements in South Sumatera due to the condition of the land: 'The settlers move after the first or second harvest to find other land, burning down another available forest, and expecting better fertility while leaving behind them their wasted lands covered by the wild grass.'[16]

The phenomenon of recently cultivated land being abandoned because of rapidly declining soil fertility is, unfortunately, not unusual, nor is it confined to a few early transmigration areas where experience was lacking: much earlier settlements than these became overcrowded in the course of time. The population of a transmigration province such as Lampung in southern Sumatera at present grows at an annual rate of 5.23%, 2.35% of which is due to natural growth and 2.88% to people coming in from other regions. North Lampung already has a density of

53, Central Lampung 140, and South Lampung 209 inhabitants per sq. km. The municipality of Tanjung Karang has 433 persons per sq. km. Because of the increasing natural deterioration of this province, caused mostly by illegal wood-cutting and farming in protected forest areas, efforts have been made to relocate settlers from the most densely occupied regions into new intra-provincial transmigration projects. Such a project was started in 1980 when the population of Lampung had already reached more than 4.6 million and the environment was visibly deteriorating. The project was planned with the dual aims of conserving natural fauna and flora and maintaining the area's hydrological potential for river-basin irrigation.[17] Obtaining a reliable estimate of the Outer Islands' potential for transmigration is extremely difficult. Even if — by using unlimited funds and technical expertise[18] — the soils could be classified in minute detail and suitable areas identified so that the 'physical absorptive capacity' could be calculated, it remains uncertain what the settlers would do with their soil. As we have seen, even sensitive soils can be used economically if the proper techniques are applied. In many cases however, such techniques are either unknown to the transmigrants or require too high an investment to be feasible. Thus, any estimate has to allow for a substantial reduction due to soil devastation caused by inappropriate methods.

Even if soil fertility, irrigation, and relief favour the transmigrants' agriculture, enabling them to produce a surplus, they may be hampered by the area's low 'economic absorptive capacity' caused largely by lack of transport facilities and the area's remoteness from markets. Some of the Maluku islands provide a good example of this phenomenon.

Official statements and press reports usually refrain from giving estimates which deal with more than just a specific project area or a certain period of time. In most cases, the reports are so vague that it remains unclear where the figures actually refer to and they do not distinguish transmigrants from their families. Moreover, studies undertaken by independent observers came to different conclusions. Land reserves on the Outer Islands vary, according to them, from between 25 to 35 million ha.[19] If we take into consideration a remark in a speech made by President Suharto in 1983 that 'as a whole, this transmigration programme would involve the moving of around 30 million farming families',[20] and assuming that a sedentary farmer family could make a living on one hectare of irrigated land — then these figures seem roughly contiguous. They are however extremely vague, and nobody can expect that 'the land beyond' will be offered to the transmigrants as irrigated land

'Javanese style'. In fact, such figures are no more accurate than the many remarks made in a great number of studies. The official authorities, and above all the Ministry of Transmigration, are not ready to commit themselves by releasing figures.[21]

It is, therefore, more realistic to look at areas of the Outer Islands which are not so densely settled: the forests, the swamps and the grasslands as possible settlement areas for Indonesia's future generations.

Forests

Indonesia is one of the most heavily forested countries in the world with about 122 million ha. of the land surface of 192 million ha. being administered by the Directorate General of Forestry. If we deduct the areas of denuded and bare land where the forest has been destroyed, we come to an actual forest cover of 113 million ha. or 59% of the whole. The forests of Jawa, Madura and Bali count for no more than 2.2% and are of only minor importance in Indonesian forest policy. The principal Indonesian forests are those of the Outer Islands, of which 32.4% are in Kalimantan, 25.5% in Irian Jaya, 22.5% in Sumatera and 10.0% in Sulawesi.

Responsibility for these forest areas lies with the Directorate General, but it seems that even this official authority has difficulty obtaining reliable statistical information. If we compare the data published in the *Statistical Yearbook of Indonesia, 1982*, with those in the *Statistical Pocketbook of Indonesia, 1980/1* (given in brackets) the anomaly becomes clear. The protected forest area covers 27 (11)%, nature conservation forest is given as 27 (8)%, production forest at 16 (57)% and the remaining 30 (24)% is called reserve forest and has not yet been designated for a specific purpose.[22]

Various types of forest and other vegetation reflect a region's climate, its soil properties and its altitude. (Fig. 20 conveys an impression of the distribution of the main forest types of Indonesia.) It shows clearly the tropical evergreen rain forest which covers no less than 73% of the area and stretches along the hills of Sumatera, covers large parts of Kalimantan, Sulawesi and Irian Jaya, but decreases towards the southeast where the climate becomes drier and monsoon forest and raingreen forest take over. The mountainous backbones of the islands are clearly reflected by upland rain forest which, on Irian Jaya, is overtopped by tundra vegetation. The east of Sumatera and the south of Kalimantan and Irian Jaya hardly reach above mean sea-level, and provide suitable

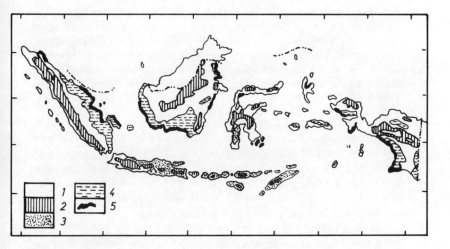

Fig. 20. MAIN FOREST TYPES OF INDONESIA

1 = Tropical evergreen rain forest
2 = Tropical upland rain forest
3 = Tropical raingreen (monsoon) forest
4 = Tropical floodplain and swamp forest
5 = Mangrove swamp forest

Base map: Meyers Grosser Weltatlas, 1979, pp. 240–1.

areas for various types of swamp and floodplain vegetation. Swamp forest covers 10.7, mangrove and coastal forest 0.8, and peat forest 1.2% in the Outer Islands.

As the population of the Outer Islands grows in spite of government family planning programmes,[23] and as the number of migrants begins to surpass the natural population increase, so the forest area of Indonesia has become the target of politically planned and spontaneous individual initiatives. As the director of forest planning stated:

The forest policy is aimed at the utilisation of forest lands for the benefit and prosperity of people, based on sustained and maximum yields. For gaining the goal mentioned above, the existing forest lands will be kept as permanent forest, except for about 25 million ha. reserved with the possibility for conversion into other land uses. The presently available standing stock will still be utilised either for log export or for industrial purposes, while the non-productive forests and bare forest land will be reforested for economic and other objectives. Nature conservation is another responsibility of forestry. A National Park system will be one of the models to be applied in nature conservation. The management of protection forest mainly to maintain the hydro-orological balance, will be and is starting to follow the model of watershed management. In connection with community development, the management of forests in densely populated areas follows the pattern of the multi-crop management. All of these policies are mentioned in the Forestry National Plan covering the period 1975–2000.[24]

This brief quotation shows clearly that the forest authorities of Indonesia are well aware of the methods that should be followed to maintain a reasonable stand of trees. Due to the regulation that any excision of forest land for other purposes must be replaced by at least the same hectareage in newly-planted trees, a net loss of designated forest land has in recent times hardly occurred in Jawa, in contrast to what has happened on the Outer Islands. Here, half a million hectares a year will be converted from forest to agricultural use.[25] The distance of the Outer Islands from the central authorities in Jakarta has for many years enabled the illegal cutting of trees to take place. In 1937 Groeneveldt remarked that, for twenty years, a process of forest destruction has taken place in the Outer Islands without anything being done by the central authorities.

A combination of various factors resulted in a rapid penetration of primary forests in order to exploit the allegedly fertile virgin soil and to maximise profits from the growing export economy. The spreading network of tracks and roads opened up the forested country to an unprecedented extent. Without considering the consequences, man entered the jungle and extracted wood, cleared the forest and cultivated perennial as well as annual crops according to the market situation. The development of modern transportation by road and rail made it possible to market all types of products in large quantities.

Once the soil was exhausted or the prices dropped, the land was abandoned and often left without any protective cover so that *Imperata* grass took over. Where a few million jungle dwellers had produced their food through shifting cultivation perhaps for a period of over a thousand years without perceptibly destroying their natural environment, a change in land use took place and every year hundreds of thousands of hectares were felled or burnt down. In view of the environmental consequences, Groeneveldt warned explicitly 'that a transfer of Javanese colonists on a large scale to the Outer Islands would be irresponsible without a preceding minute analysis of the arable soil of the area in question.'[26] Although at that time the interdependency between rain forest, soil fertility, shifting cultivation, burning and the advance of *alang-alang* was the subject of lively discussions and was viewed quite differently from today, a number of researchers and experts showed great concern with what was going on. Moreover they deplored the indifference shown not only by the authorities but also by some foresters and other observers to the forest destruction in the Outer Islands.[27]

Forests have always acted as a land reserve for a growing population. Vast areas of Europe and North America once covered with densely wooded land have been cleared for agricultural and industrial purposes. Occasionally, catastrophic soil destruction caused by water and wind has indicated that the protective function of forests has often been ignored.[28] In the temperate zones however, much of the damage could be healed thanks to a favourable climate and fertile soils, provided the right measures were taken in time.[29] Yet the vast, continuous boreal forests have largely given way to cultivated land and now cover only the marginal areas of the northern hemisphere. The tropical forests now represent one of the last remaining reserves of unused land in the world and they are becoming one of the main targets of 'development'. The growing demand for tropical timber in the rich countries leads to the extraction of the most valuable species and so often impoverishes the rain forest in terms of species composition; but in the poor countries, competition for food-producing land leads to wholesale forest destruction. For the peasant seeking land and food, forest is an obstacle that has to be removed.

The future of the (remaining) tropical rain forest as a biotope and the future of the ever-growing hungry population depends very much on increasing our understanding of the biological and pedological interdependency between plant and soil in tropical rain forest conditions. For several years, European and American scientists and technicians have struggled to find theoretical unanimity on these questions, but unfortunately they remain largely divided over what measures are needed. Moreover scientists do tend to work in a sort of 'vacuum', whereas the technicians in the field who are confronted with the tasks of settling and feeding people are bound to be more optimistic. One of the basic questions to be answered is how the primary tropical rain forest manages to be the most productive ecosystem in the world, and yet when annual crops are grown on the same soil, they show rapidly decreasing yields after only one or two harvests. The conclusive question is: are there techniques that could be applied to increase and stabilise soil fertility for agriculture?

The fact is that the enormous green biomass of the tropical lowland rain forest grows on virtually sterile soils, under excessive rainfall and a superabundance of pathogens and pests. Photosynthesis is rapid under the propitious combination of light, moisture and temperature which commonly prevails, while an elevated carbon dioxide concentration speeds the process in the shade beneath the canopy. Therefore, nutrients are the only other factor which can limit photosynthesis, and indeed it is through the study of nutrient dynamics that the jungle is revealing some extraordinary mechanisms. The rapid decrease in crop yields after cleancutting led to the suggestion that the forest functioned quite independently of the soils, and that soil samples correlate less with vegetation in the lowland tropics than in temperate regions. Goodland and Irvin, who undertook their research in the Amazon jungle, gave an excellent summary of this unique system:

It was suggested that the tightly-closed nutrient cycle of the vegetation permitted the evolution of the soil to proceed rapidly and independently in these high rainfall areas. Soil supporting the forest and water in streams draining the forest are both singularly deficient in nutrients — further evidence of a tight and independent cycle. The postulated tight and rapid nutrient cycle has now been demonstrated by radioactive tracers and it has also recently been shown that at least one nutrient cycle is quite independent of soils. The 'direct nutrient cycling' hypothesis states that 'nutrients are obtained from dead organic litter on the forest floor by mycorrhizal fungi which pass nutrients to living tree roots . . .' It therefore can be extrapolated that any structure or process delaying the downward movement of substances containing nutrients contributes to direct cycling. The jungle is replete with such 'delays': structural configurations such as axils, crotches and spines, all trap litter or intercept its through-fall. Much decay occurs above ground, thereby enriching the through-fall which sustains the dense surficial, absorptive rootmat. The immense number of species, the diversity of physiognomic life-forms, and the wide age-spread of individual plants all contribute to reduce nutrient competition.[30]

Another important reason for low outputs on cultivated land in comparison with that of the rain forest is that herbivores can thrive all year round in tropical climates. The ecosystem of the rainforest, however, has evolved mechanisms effectively defending its components against such pests:

Firstly, the extremely low population density of individual tropical species protects such species from host-specific pests. In other words, the rarity of any particular species, or the distance between hosts, extenuates damage by pests. Secondly, repellent or toxic chemicals are notoriously characteristic of tropical plants and serve as a major defence against pests and pathogens. Much metabolic energy is diverted from growth and reproduction in the manufacture of these compounds, which indicates how essential they must be to the plant. Many of these substances have been found to possess valuable therapeutic . . . or other uses . . . for humans. Lastly this ecosystem comprises an entire spectrum of predators and parasites. This big component of the ecosystem acts as a control, dampening population increase of the pests once they have located a host. For example, rodents are controlled by other vertebrates, including birds. Birds, bats and carnivorous insects control insect pests, while some insects may even reduce pathogenic fungi.[31]

The authors conclude with the following statement:

The understandable yet dangerously erroneous illusion of the richness of tropical soils has long since been dispelled in scientific circles. This optimistic illusion persists undimmed, however, in the political and governmental ambit. The historical growth of the illusion is easy to comprehend. Astonished by the stupendous size of the trees and by the luxuriance of the vegetation, early explorers naturally assumed a prodigiously fertile soil: biomass was equated with fertility. The realisation that this conclusion is not so slow in coming. Repeated experience of the sharp slump in yield in the second and third harvests from any given clearing provided the first evidence. Indeed, some clearings failed to provide a second harvest.[32]

To date, scientists and technicians do not completely agree upon the uselessness of the cleared jungle soil, but at least there is no longer the unjustifiable enthusiasm about its fertility.[33] Most of those who still regard the tropical rain forest as a valuable land reserve for agricultural production refer to extremely complicated and difficult techniques — often tested only in the laboratory — and probably too expensive to be ever applicable to the large-scale production of food.[34] Others, more realistically, recommend methods of enriching the cleared ground with compost and fertiliser (e.g. through mulching), to help support the development of new humus and to gradually supply the necessary nutrients to change the physical structure of the soils. Such methods are practicable in a small-scale, horticultural type of land use and require much technical and educational support. Many peasant families do the same with their house gardens, but it is doubtful whether they could apply such methods to large-scale projects.[35]

In Indonesia, both the government and the timber business hardly cared for ecology for many years. Recently, environmental consciousness has spread not only among the authorities but also among the public thanks to the indefatigable information work of Dr. Emil Salim, formerly Minister of Development Supervision and Environment and now Minister of State for Population and Environment.

On the government side many measures have been taken to stimulate the development of the country in general and silviculture in particular, but this requires capital and the government soon found out that this could be provided by the large-scale exploitation of the forests. This of course led to grave contradictions in the policy, since drawing a maximum of revenue from the forests is not compatible with their rational, careful exploitation. The results of this can be seen mainly in the islands of Sumatera and Kalimantan. Here, there seems to be more interest in boosting wood production for export, establishing wood processing industries, and replacing foreign technicians by Indonesians than in controlling logging, natural rejuvenation and the protective function of the forests. All this is well-known to the authorities, and occasional actions prove it, but it seems that at the decision-making level there is still insufficient insight for a drastic change of the situation.[36]

On the research side, much remains to be done. Though the National Biological Institute (LBN), BIOTROP and the Man-and-Biosphere Programme (particularly in East Kalimantan) do their best to shed light on the environmental complex, and some of the universities contribute to the work, the lack of funds and highly-qualified manpower is a severe handicap.[37] Besides, the so-called botanical explorations or surveys conducted by many forest officers are too superficial in their approach and biased because all their activities are directed only to timber producing species with a view to forest exploitation and export purposes.[38]

Results of various types of research show, as far as the tropical lowland forest is concerned, a rather grim outlook. Whereas deforestation in the Asian-Pacific region continues at the alarming rate of some 5,000 ha. per day, and forest management and afforestation programmes have so far made little impact, Indonesia has the sad distinction of being the country with the highest deforestation rate of the region, with 500,000 ha. a year, followed by Thailand with 333,000 ha., according to UNEP, the Environmental Programme of the United Nations.[39] The causes of this problem lie in the often inconsiderate logging practices, in industrialisation and mining which takes place at the expense of the forest cover and, last but not least, in the clearing of forest for the settlement of transmigrants.

Among the provinces of Indonesia, writes Kartawinata, East Kali-

Indonesia's deforestation rate is the highest in South-east Asia (500,000 ha. lost every year).

mantan stands out because of the extent of its forest reserves. Because of developments in both the oil and the forestry sector, the province has the highest rate of economic growth in Indonesia.[40] Its population of only 733,536 in 1971 is being considerably augmented by migrants who come to share in the economic boom. When a law conducive to foreign investment was passed, the logging of productive forest developed quickly after 1967, and in 1973 East Kalimantan was responsible for about 50% of Indonesia's total wood exports. 'Because of the rapid rate of exploitation, the extent of disturbed natural forest has increased proportionately. According to some predictions, it may not be long before hardly any lowland forest remains. Moreover, logging may lead to the extinction of many species that might otherwise be of economic importance in the future.'[41]

Investigations in specific logging areas showed that it is not only the loss of wood that changes the biotope; the remaining vegetation and the soil are also violated.[42] At Beloro in East Kalimantan it was found that after mechanised logging in the lowland forest — where an average of twenty-five trees per ha. were extracted — only half of the residual trees were left behind undamaged, and 30% of the ground was damaged.[43] Mechanical logging generally leads to the compaction of soil, increases the soil erodibility and adds to the loss of soil fertility.[44] Certainly, a forest damaged in this way may recover, but it is a very slow process and can take more than forty years.[45]

Hopefully these examples will suffice to illustrate the situation in the forest areas in the Outer Islands of Indonesia. This is a renewable resource, but it is a fragile one that can easily be partly or completely destroyed. 'It may be possible', writes Manning:

to justify the dispensation of large forest areas to small producers, contractors, and non-business organisations on the grounds of their present contribution to production and revenue, but it is incorrect to represent these developments as part of any planned process. In a society where there are numerous centres of political and administrative power, individuals both within and outside the government have played an important role in determining the pace and nature of the timber boom. Combined with the interest of the provincial governments in rapid proliferation of concessions, these have placed severe constraints on the operation of central government policy. Given the social need to preserve forest resources and develop processing, such developments serve neither long-term economic and social interests of the region nor that of the nation . . . The ultimate success of future policy will depend, however, on much more than the tightening of legal provisions regulating forest exploitation. Above all, it hinges on the government's capacity to implement new laws, regulations and contract clauses. Here, sufficient political power given to those entrusted with enforcing regulations, and increased administrative efficiency are crucial for the effectiveness of the government program.[46]

Swamplands

Various types of swampland cover a substantial part of Indonesia. There are an estimated 35 million ha. of inland and coastal swamps, more or less equally divided between the islands of Sumatera, Kalimantan and Irian Jaya, but smaller swamp areas can be found on practically all the islands. It is estimated that of this total swamp area about 6 million ha. of coastal swamps have an agricultural potential, especially for the cultivation of wet-rice, though this does have its problems. One million hectares are located on both Sumatera and Kalimantan, whereas another 4 million ha. is the estimate for Irian Jaya.

Historically, the early states and empires on Sumatera and Kalimantan grew up at or near the mouths of many of the larger rivers which provided the only practicable routes to the jungle-clad hinterland. The alluvial lowlands, in contrast, were hindered in their development by the width of a swamp belt covering the coastal margins. Although significantly narrower in early historic times, the swamp belt must have been of quite considerable extent even then.[47] Over the centuries, well-adapted planting techniques and schemes for avoiding damage caused by floodwater or salination have been evolved to grow wet-rice and coconuts by the various peoples who went in search of new land. The Banjarese, natives of Kalimantan, migrated from the uplands to selected coastal areas in the south of their island, reclaimed and developed the land, and grew wet-rice and coconuts. The Buginese people of Sulawesi, highly-skilled boat builders and navigators, were faced with land scarcity on their own island. Some crossed the Java Sea and became settler-residents in the swamps of Riau, Jambi and South Sumatera, where they reclaimed and developed the land along the major rivers and later even dug Dutch navigation canals by hand.[48]

Physically, the coastal swamps occupy broad, flat and poorly drained zones between the mountains or peneplains and the sea. These swamp areas and their forests are the result of a land building process where material eroded from the uplands is deposited at the coast. Sand, silt and clay are colonised by vegetation, and successive floods bring more material. Since anaerobic conditions prevail, organic matter is rapidly accumulated and builds up the surface of the land. As the land extends seawards, the older swampland becomes less saline and less frequently flooded.[49] Though wide estuaries were formed by the drowning of river mouths, they were later at least partially filled with huge quantities of detritus deposited by the truncated rivers flowing into the shallow epicontinental seas. The very process of drowning led to an increase of the rate of sedimentation, and along many of the shallower and less disturbed shores of the archipelago, particularly on Sumatera, Kalimantan and Irian Jaya, vast belts of swamps were built up.[50]

The land-building process goes on as long as soil transportation from the upland continues. This is visible everywhere. In eastern Sumatera for instance, evidence of the greatly accelerated erosion in the uplands of the interior can be seen. The colour of the streams betrays erosion and so does the silting-up of the rivers and the outward growth of the mangrove belt. Ports which in 1942 could be reached by small coastal steamers have lost their function as shipping points along the east coast of Sumatera,[51] and the Trisakti Oceanport of Banjarmasin, South Kalimantan, which serves two provinces, suffers from so much sedimentation that it is now uneconomical to keep it open. About 3.6 million cu.m. of mud a day accumulates in the 14.3 km. long waterway since routine dredging was halted.[52] The silting process can be found on all the Indonesian islands to some extent and is particularly impressive where it forms deltas in the huge jungle streams. Extended swamp areas stretch behind beach ridges and the natural levees along the rivers built up by sedimentation. Even in the interior, extended freshwater swamps and lakes have developed behind mountain ridges in closed depressions with inward-directed drainage. One of the largest and most important is the Kutai Basin (a typical *banjir*) which is located about 100 km. upstream of the Mahakam river. Such levees are often heavily settled, but people also live in lake-dwellings supported by piles or rafts.[53]

The main characteristics of soils prevailing in swampland have been summarised by Soewardi *et. al.*:[54]

Soils in tidal swamplands are much more complex compared to those in upland areas because of the additional soil building factors associated with tidal influences and saline water conditions. There are three major soil groups found in swamps: entisols, inceptisols, and histosols. The chemical composition and fertility of these soils vary. Entisols are fertile soils but, due to high sodium content and unripe organic matter, they are not suitable for agriculture. Inceptisols (mineral soils) are generally found next to entisols. They are also fertile and have a lower sodium content and more ripe organic materials than entisols. These soils therefore have a higher agricultural potential. Histosols (organic soils) are mostly found between rivers in saturated or inundated conditions. These soils generally have the lowest concentrations of sodium of the three soils and their fertility depends heavily upon the state of development and depth of their organic materials.

Apart from salinity, swamp soils have various other disadvantages. Intercalated iron sulfide layers are common; on drying and oxidizing, these become toxic to crops. Whenever such catclay underlays organosols their usefulness is severely reduced, unless the water content can be continuously controlled. Water control is an important factor. Alluvial soils, usually rich in nutrients, depend on it in depressions and valleys. Periodic inundations from river flow or tidal currents, salt intrusions from the sea, particularly in the dry season, and the lower layers rich in sulfides make the swampy zones problematic unless water control is secured. If it is secured, rice production is possible. However in general, the soils of swampy zones belong to soil class 5 which means that they have a rather low fertility.[55]

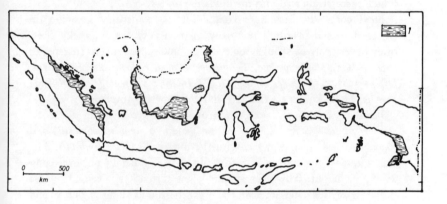

Fig. 21. DISTRIBUTION OF SWAMPLANDS IN INDONESIA

1 = swampland including mangrove forests.
Base map: Kötter *et al.*, 1979, pp. 56–7.

If we take East Kalimantan as an example, the vegetation of the swamp areas between the foothills and the sea can be divided into four forest types: the freshwater-swamp forest, the brackish-water forest, the littoral forest and the tidal mangrove forest.[56]

Freshwater-swamp forests are extremely widespread over alluvial soils that are flooded for long periods with fresh water. Their occurrence coincides with huge river basins, such as that of the middle Mahakam river and its tributaries, or the basins of the Sesayap, the Sembakung or the Sebuku rivers in the north of the province. These forests are similar to the lowland rain forests, but their structure is less layered, diameters and heights of the trees remain smaller and there are less species. The most important trees prevailing in this type of forest are *Canosperma, Alstonia, Eugenia, Canarium, Koompassia, Calophyllum* and *Melanorrhoea.*

The enormous importance of these areas lies in their function as water reservoirs and sedimentation basins for the big rivers. Therefore, any large-scale deforestation would undoubtedly have serious effects on the water supply to the lower river basins and would lead to climatic change.

Between the freshwater-swamp forest and the littoral forest, an area of transition is formed by a brackish water forest zone. The topography is slightly above mean sea-level and is mainly caused by depositional landforms along the coast, river estuaries and deltas. The prevailing species growing here are *Bruguiera, Xylocarpus, Aegiceras* and *Lumnitzera*, which successively lead into the lowland rain forest.

Next come the littoral or coastal forests, sometimes growing on emerged coastal plains. They grow on non-salty, sandy, rocky sites, often in very narrow strips along the coast, and support major tree species such as *Barringtonia speciosa, Terminalia catappa, Calophyllum inophyllum, Hibiscus tiliaceus, Thespesia populnea, Casuarina equisetifolia* and *Pisonia grandis*. These forests are rich in epiphytes, especially orchids.

Finally, the tidal or tidal mangrove forest grows along the coast proper where we find species of the genera *Avicennia, Sonneratia* and *Rhizophora*, often intermixed with the nipa palm (*Nipa fruticans*). The tidal forest is one-storeyed, the tops of the trees generally not exceeding 25 m. in height. This very specific plant formation is caused by the influence of tides and sea water and can be found along the coast and along river mouths, especially the big deltas of rivers such as the Mahakam.

As in other countries of South-east Asia, the continued existence of the mangrove areas in Indonesia is under threat. To fully understand this danger we have to study the function of this ecosystem. Wherever the land surface is being extended seawards by the deposition of silt, the shoreline is characteristically fringed with a belt of mangroves, which, thriving in half consolidated mud, help to consolidate the latter more fully and thus extend the land still further out to sea. This process continues, when undisturbed, along all the more sheltered stretches of the gently shelving shorelines and particularly near the mouths of many of the large rivers. In these mangrove belts — the width of which varies from 100 metres to a couple of kilometres — can be found as many as thirty species of mangroves, all of which vary in their adaptation to marine inundation. Economically, mangrove swamps are very useful, providing firewood, charcoal, building material and tannin obtained from the bark.[57] But apart from its function as a protector of the coastline and as a source of raw materials, the mangrove belt has an important ecological function. In 1981, the United Nations Economic and Social Commission for Asia and the Pacific (ESCAP) began to show great concern about the question of the rapid disappearance of mangrove forests in Indonesia. This process threatens fish and shrimp production, since the tropical trees serve as spawning and nursing grounds for many marine species.[58] This has already been recorded in other South-east Asian countries such as Thailand.[59] Fortunately, Indonesia is well aware of the dangers posed by mangrove deforestation. At the Gulf of Kau on Halmahera (Maluku), the National Institute of Oceanology (LON) in cooperation with the Indonesian Institute of Sciences (LIPI) has established a mangrove research base, and further work is being done on Seram. The research workers are particularly interested in the ecological consequences of exploiting mangrove stands. The results obtained so far have already led to the cancellation of concessions to cut mangroves and export the wood to Japan.[60] Similar efforts are being made elsewhere. In West Jawa for instance, a mangrove reforestation programme is being run in conjunction with a fish and lobster breeding programme. This 'aqua-forestry'[61] serves to both protect the coast and offer a source of income for the local inhabitants. However the success of these measures is by no means certain. Apart from the fact that townspeople want to convert the mangrove coast into fishponds for quick financial gain at the expense of ecology, any effort to counteract such plans with well-balanced alternatives suffers from a lack of precise knowledge. The ecology of the swamp — like the dependence of its productivity on nutrient flux from the adjacent lowland systems, and its response limits to nutrient inputs from urban and industrial wastes — is little known and documented. Swamps, especially inundated swamps, dominate the freshwater landscape of the tropics because of the high precipitation. For the most part the productivity of inundated swamps, especially when forested, appears comparable to the lowland rain forests. Although the ecosystem of the swamp is structured more simply, the litter production is similar, though its decomposition is slower in inundated swamps. They conserve water, minerals and energy in decomposing organic matter and may therefore be called 'nutrient and energy traps'. From these swamps, about 40% of the nutrients are supplied to the neighbouring riparian systems. When these swamps are transformed into rice fields the ecosystem is seriously damaged and the ecological consequences of this should be studied.[62] From a broader environmental perspective though, successful swampland development involves elements other than rice production. Ecologically destructive activities are also initiated by forest cutting and

The sago palm (*Metroxylon sagu*), an excellent carbohydrate source which can be grown in swampland.

canal construction. If the degradation process proceeds at a rate which exceeds the combined beneficial effects of natural regenerative capacity and local management capabilities, the area is likely to be abandoned. This constitutes a major resource and environmental problem.[63]

Before examining the attempts to settle and exploit the swamplands, we have to deal briefly with another most important though neglected swamp product, the sago palm (*Metroxylon sagu*). This plant, whose trunk supplies a substantial amount of starch, is characteristic of some of the coastal parts of Indonesia where the basic carbohydrate intake of the people is derived from it. During the Dutch colonial period, sago plantations were established and the total area under sago palms was estimated at 7 million ha., with Maluku and Irian Jaya as the main producing provinces. Today, there are only about 1 million ha. of sago forest. Some areas have been transformed into rice fields, but most have simply been abandoned. The sago palm can be cultivated on marginal soils such as peat, and it is capable of maintaining a full canopy throughout the year to get maximum benefit from the sunlight. The more light the plant receives, the more starch it produces in its trunk. In addition, it forms a blanket vegetation cover which provides a humus-maintenance and weed-control system of great efficiency. Thus, there is no doubt that the sago palm is a useful plant for growing in swamps, especially because it does not break any ecological cycle. And since Indonesia has plenty of such swamp areas, experts believe that this crop has an enormous potential.

As a food, sago starch contains about 80% carbohydrate and could well serve as a substitute for rice. Though it has a lower vegetable protein content than rice and maize and a lower calorific value, its production is very much easier. In order to promote sago as a basic food, after being fortified with proteins and vitamines, sago cakes are fed to the armed forces in Irian Jaya. The starch is also valued as a raw material in the production of ethanol, glue for the plywood factories and for other purposes.[64] In the province of Riau (Sumatera) there is a daily production of 2,000 tonnes of sago; in 1982, they produced nearly 22,500 tonnes of processed sago from about 15,000 ha. of sago plantations and shipped it to Jawa and Singapore.[65]

Serious government interest and the first limited soil surveys in swamplands only began in the 1920s, and the evolution of a tidal rice cultivation programme (*pasang surut*) shows how these wetlands attracted the interest of Dutch as well as Indonesian engineers. Though the native population demonstrated the possibility of growing rice on swamplands, the engineers struggled to increase the efficiency of the traditional techniques of land development.[66] And yet there was little government-sponsored work in the field of swampland reclamation before the Second World War. It only took off in the late 1960s when the transmigration programmes began to receive a high priority in the development policy. Both government-sponsored and spontaneous activities (the latter mainly undertaken by the Buginese and Banjarese) with the aim of settling farming families and in particular transmigrants date back to 1939 and 1959, when such schemes were tried in South Kalimantan and later on in Sumatera — but the results were often disappointing. As experience in planning and executing such schemes grows, Indonesia is able to reclaim increasing areas of swampland.

During REPELITA I (1969/70–1973/4), 32,000 ha. of swampland were reclaimed and during REPELITA II (1974/5–1978/9), the figure rose to 221,000 ha. The target of REPELITA III (1979/80–1984/5) was no less than 400,000 ha. But whereas at the end of the 1970s there were large areas, particularly in Riau and Jambi (Sumatera), still unsettled, by 1983 most of the reclaimed land had been brought under cultivation.

The planned target for REPELITA IV (1984/5–1988/9) foresees the reclamation of 500,000 ha. of swampland and the settling of 140,000 families or no less than 20% of all the transmigrants in the planned period.[67]

In practical terms, the reclamation, settling and ecologic use of swamp areas has to be approached from different social, techno-economical and ecological aspects: 'With good management, careful site selection, reasonable levels of funding, and adequate attention to the needs and the rights of existing settler groups, transmigration into marginal environments could lead to stabilised agricultural communities',[68] whereas neglecting only one of these criteria will sooner or later result in failure.

Even in swamp areas there are indigenous inhabitants whose rights have to be considered when the land is reclaimed and settlers from outside are brought in. Otherwise, the lack of cooperation which often occurs between old and new settlers will turn into open rivalry.[69] Decisions about the forest, the land and water resources were traditionally made by a local leader or a local council, and in thinly populated regions they often decided to expand the amount of land they cultivated. Once the modern reclamation projects were begun, however, representatives of many government departments or foreign aid agencies with varying levels of interest appeared on the scene, often with very controversial ideas of what should be done. Reports on such projects often reveal that they were hastily conceived, and that there was a lack of proper research into the prevailing social and non-social conditions in the field. It has to be kept in mind that such projects are not meant to exacerbate social tensions but to ease them. EUROCONSULT, a Dutch consulting firm intimately engaged with such projects, stresses 'that an early identification and selection of the beneficiaries of the project, as well as the exclusion of claimants from outside the project area, are of urgent importance. An effective control to secure the presently still vacant areas from squatters and speculators has to be established. Usufructuary rights allowed to people who do not till their own land should be declared invalid.'[70]

Such rules are by no means far-fetched, since both legal and illegal wood-felling concessionaires, have, at the end of their concession, sold the çlearings to interested settlers.[71] All over the world people try to exploit development projects not actually meant for them.

Technically, the reclamation of swamplands can be achieved in three different ways: empoldering, water control by tidal effect and low-cost technology and, finally, water control by tidal effect in combination

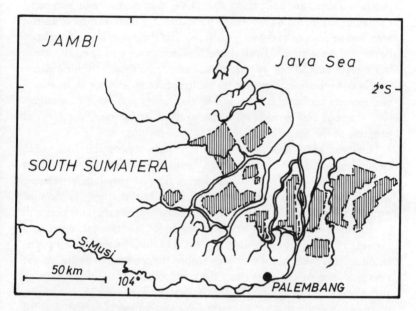

Fig. 22. MODEL AREA FOR SWAMP RECLAMATION IN SOUTH SUMATERA

Note: The hatched areas include two IBRD projects and government funded reclamation areas.

Source: IBRD, 1984–c.

with pumps. Apart from the planned but very limited implementation of a gigantic Kalimantan polder system between 1948 and 1953,[72] and a few small areas geared more to urbanisation than to land reclamation, empoldering does not play an important part in swamp reclamation. And given the costs and technical problems of pumping in the pilot projects, water control by tidal effect was the chief method used. There was undoubted success in settling families and clearing the land, but problems and technical drawbacks led to disappointing results in some of the projects: there were the problems of canal maintenance, severe weed competition, poor water control, low productivity, lack of farm power, salt-intrusion, the low standard of cultivation practices, pests, diseases, and probably soil fertility problems.[73]

In the first instance, the reclamation of swamp areas begins with the construction of the necessary infrastructure. To protect against floods, protection dams are built along the rivers, then canals are dug to carry mountain-fed irrigation water. Other canals are then used to drain the area, but this has to be done carefully, so that the cultivable land will not be dried out too much. Bridges and farmhouses have to be constructed, but people should not be settled before a year-round drinking-water supply and other social services are secured. Swamp settlement, in principle, requires the same caution and level of activity as other transmigration projects, and in addition it has to consider the specific ecological situation of the site.

However ideas about extending the potential of the Indonesian swamplands and the techniques to be used in order to make them productive vary greatly, although there is now a general understanding among scientists that swamp ecosystems are just as fragile as those of primary forests. All these resource systems have an innate value that will be destroyed when they are radically altered for resettlement. Altering a system can also upset the biological control of the animal and plant population so that pests and weeds may reduce the economic viability of the new land use. At the same time, the development of such systems may alter the conditions of land outside the development areas. For example, removing the forest surrounding a river system can alter the seasonal flow of water, which can make the control of floods and the supply of irrigation water more difficult and expensive. Therefore all these side-effects have to be carefully controlled.[74]

Some of the project areas are no longer covered with primary forest since, being situated close to rivers, they had been intensively logged even before the project began and the remaining trees are of limited economic value. Nearly half of the area of the IBRD-sponsored Second Swamp Reclamation Project near Palembang in South Sumatera is now covered with bush and small diameter cajeput trees (*Melaleuca leucodendron*). The rest is overgrown with secondary swamp forest. The trees (*Macaranga, Loropetalum, Dyera* etc.) have a density of 550 per hectare and an average diameter of 50 cm.[75] Such a reduction of the vegetation cover cannot take place without causing severe hydrological problems.

In the Rawa Sragi Swamp Development Project (Lampung, Sumatera), extensive clearcutting of the forest led to a reduction in the water-retention capability of the Sekampung river's watershed. Consequently, as the run-off peak data grew, the average downflow in comparable periods increased from 460 to 662 cu.m. per second. There was a

corresponding increase in the erosion of the fertile soil and the inundation of cultivated land. During the dry season, in contrast, the reduced
flow of fresh water in the river allows salt water from the sea to intrude
further upstream, which can increase the salinity of the soils in the
swamps and reduce their agricultural potential. Thus the growing need
for flood protection makes the project more and more expensive.[76]

Similarly, the areas above the swamp project can be negatively influenced by changes in the ecology. The penetration of the mountainous
forests by new settlers together with the diminution of the shifting
farmers' cultivation cycle has reduced or at least thinned the vegetation
cover. The situation deteriorates further if the regions in question have a
very pronounced relief. In the planning area of southern Sumatera, for
example, more than one-fifth of the land slopes at a gradient of more
than 70%, and where this coincides with deforestation it can lead to
catastrophic flooding, which is occasionally reported in the Indonesian
press.[77] Pelzer observed other changes consequent upon human interference with the hydrological balance that once supported a specific type
of land use.

Extensive sections along the coast once producing large quantities of copra are
now being taken over by swamp vegetation. The coconut palms have been
killed because of the drastic changes in the drainage. Rubber trees are dying off
because the estates can no longer maintain adequate drainage, since the land
downstream has been occupied by squatters whose main aim as wet-rice subsistence farmers is to slow down drainage [and he concludes], unless there is
understanding and cooperation between neighbours in the same drainage basin,
the action of one may harm the economic interest of the other.[78]

This suggests the need for regional or watershed planning, a method
which, as we have shown, can best integrate economic and ecological
interests. It would be unilateral to view the Indonesian swamplands only
as future wet-rice production areas and forget the ecological function of
swamps and coastal vegetation. Hanson has written:

In some marginal ecosystems, good income and nutritional levels will be met
only through exploitation of a wide range of resources and multiple cropping.
Thus, an intensive and integrated philosophy of area development program-
ming is required. Fisheries, forestry, erosion control, optimal crop scheduling,
and careful monitoring of all essential factors important in land abandonment
require attention . . . The need to find ecologically satisfactory solutions to
marginal land use will become increasingly important in Indonesia as highway
networks expand and deforestation continues . . .[79]

Grassland

Next to rain forests and swamp areas, tropical grassland is an important form of vegetation cover on the Outer Islands. The exact surface area is not known, but estimates vary between 16 million and 20 million ha., with an annual increase of 100,000 to 200,000 ha.[80] On Sumatera and Sulawesi it covers substantial areas, but it is now also spreading on Kalimantan, where it was first observed in the upper Mahakam watershed around 1870.[81] On the islands of Maluku and Nusa Tenggara, forests have largely been replaced by grassland, especially as the extended dry season increases the number of forest fires — a great hazard to natural reforestation. On some of the Lesser Sundas, the grassland is also associated with stock raising, for instance on Sumba and Flores and in the savannah vegetation of Timor.[82] Fig. 23 shows the distribution of tropical grassland over the Indonesian islands.

The existence of extended grassland is usually considered to be the result of human action and particularly shifting cultivation, which exhausts soil fertility and thwarts the natural rejuvenation of the forest. Improper logging techniques may also encourage grasses to advance. Finally, when agricultural land is exhausted and abandoned it is very susceptible to destruction by fire, and *alang-alang* invades the site and makes the land unsuitable for agricultural use. Reports dealing with the problem of *alang-alang* eradication date back to before the Second World War.[83] To understand the disadvantages of this process, it is necessary to know the botanical characteristics of the noxious and combustible tropical grasses. Cogongrass (*Imperata cylindrica*), in Indonesian called *alang-alang* or *lalang*, is generally regarded as the most important component of the Indonesian grassland, though some scholars doubt whether it actually dominates the other species.[84] However all the studies dealing with the Indonesian grassland issued recently refer exclusively to *Imperata cylindrica*. The main characteristics of this grass are that it thrives on particularly poor soils, produces a strong system of underground rhizomes, and is very easily ignited. The roots also produce an exudate that inhibits the growth of other grass crops such as rice and maize. The flat-bladed coarse grass reaches up to two metres in height, but the leaves are only palatable for cattle for a few weeks after sprouting. After this period, the plant becomes virtually useless except for its ecological function. The rhizomes protect the soil against erosion but, at the same time, speed the surface run-off of rainwater, thus decreasing its percolation into the ground.

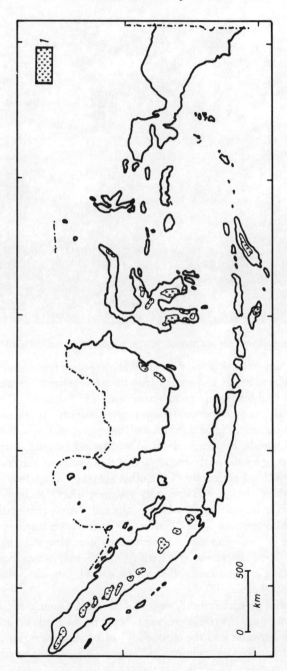

Fig. 23. DISTRIBUTION OF TROPICAL GRASSLAND IN INDONESIA

1 = *alang-alang* and other grassland

Base map: Kötter *et al*., 1979, pp. 56-7.

Alang-alang engulfing a pepper plantation abandoned due to soil exhaustion.

The most important ecological function of *alang-alang* is in connection with fire. *Alang-alang* is known as a 'fire-climax vegetation' because it burns easily without being totally destroyed, and because fire helps maintain its dominance over any competing vegetation. At the end of the dry season when the grass is dehydrated, lightning, spontaneous combustion, or a fire lit by farmers leads to widespread burning. Farmers wait expectantly for the fresh sprouts so that they can graze their cattle, but the fire will kill practically all the other seedlings which may have developed within the grass. It does not, however, affect the *alang-alang* rhizomes which lie well protected under the soil. 'After two or three days the weed is growing again. Within months it reaches maturity and within a year or so the fire sequence is repeated. Since *alang-alang* has no leaf-drop, it does not renew the fertility of the soil, which rapidly becomes dry, yellow and hard. Nitrogen taken up by the *alang-alang* is volatised during burning.'[85]

The spread of grassland in the tropics reflects the behaviour of different social groups and the consequences of their land use methods. In the course of his intrusion into the untouched natural vegetation, man has brought with him certain traditional techniques such as cattle breeding,

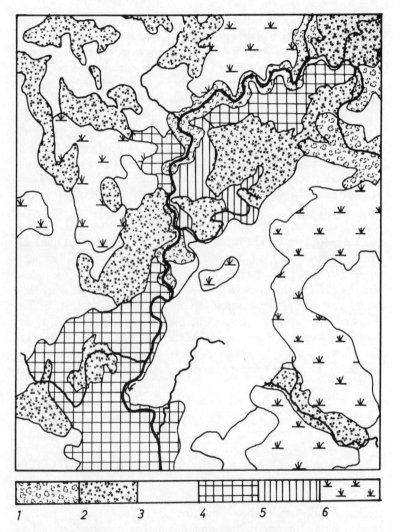

Fig. 24. DENUDED FOREST AREA IN EAST KALIMANTAN

1 = Primary forest
2 = Secondary forest, woods and brushwood
3 = Shifting cultivation, partly abandoned
4 = Cultivated areas
5 = Abandoned cultivated areas
6 = Tropical grass, partly burnt.

Source: Voss, 1979 (simplified).

shifting cultivation, or sedentary agriculture. As a legacy of his activities he leaves behind secondary ecological associations such as *belukar* (secondary forest) or grassland. Since fire prevents the regeneration of forests, the timber line is pushed back and the extension of permanent grassland continues. It is now clear that man did not follow existing grassland with his cattle as was once thought, but rather, that grassland followed in the wake of man and his cattle[86] (see Fig. 24).[87] In Indonesia it can be shown that specific social groups with their own social organisation and land use methods have created quite different forms of secondary ecological development. Keleny compares two ethnic groups on Sumatera. The Minangkabau in central Sumatera, who farm intensely irrigated rice fields and produce high yields, apply forest compost and use transplantation. The whole area has the appearance of forest, often even tropical rain forest, with cash crops such as rubber, cloves, cinnamon and nutmeg being grown. The Batak area in North Sumatera looks quite different. The tree cover has disappeared except in some steep gullies and is replaced by open grassland both on the mountainsides and on the upland plateaux. This is the case even where there is high rainfall, where one would expect tropical rain forest and intensive rice cultivation. In fact, the Bataks are a pastoral people who own large numbers of cattle and water-buffaloes which are responsible for the extension of grassland at the expense of forest, and who use fire as a permanent feature of their husbandry. Efforts at reforestation in this area with *Pinus merkusii* have failed.[88]

Given the great demand for arable land in Indonesia, and the substantial areas covered with *alang-alang*, the question that arises is why is this useless grass not eradicated and the ground underneath used productively to grow food and plant forests?

The reason normally given is that the local farmers, who have fairly simple tools, are not physically capable of eradicating the grass. The underground rhizome system of *alang-alang* is so strong that the labour force of a whole family is barely sufficient to keep one hectare of land free of grass during the year, as has been found in transmigration areas where *alang-alang* land was given to the settlers.[89] Some peasants try to help themselves by burning the grass so that their cattle can feed on the fresh sprouts, but this, in the long run, strengthens the hold of the grass.

Sherman, however, found that the Bataks, who live near Lake Toba on Sumatera, were quite capable of making practical use of the grassland. Apart from using the young sprouts to feed cattle and the older grass as thatch or mulch, they eradicate *alang-alang* when they want to cultivate the land: 'The field may or may not be burned over if the vegetation is dry enough. The ground is then broken up and turned, a task usually done by four or five men who drive their mattocks into the sod, section by section, in unison, and then heave and pull the section of sod over. On average they can turn a three foot long, one foot wide, eight inch deep pallet of sod at a time. They move in rows back and forth across the field. It requires some 35 man-days to open an acre (0.4 ha.), and the result actually resembles a field which has been ploughed by means of a mole-board plough.'[90] The rhizomes are then dried in the sun and finally raked out and burnt. It is extremely hard work, and though such techniques are applied in South and East Kalimantan, Lampung, and South Sumatera,[91] they are not used very intensively. Most, if not all of these cases can be called 'incipient grassland farming'; first, because the grass is not eradicated from the plot under cultivation, and secondly, because it will unavoidably resprout and send out additional rhizomes during the growing season, which seriously limits crop yields.[92]

Since grassland farming in one form or another is as widely spread as grassland itself, it is interesting to compare the cycle of nutrients in the forest with that in grassland. As we have seen, the nutrients of a rain forest are mainly stored in the biomass, which has to be burned or mulched in order to release them. In the grassland on the other hand, nutrients build up in the sod matrix inspite of periodic burning. The decay of the rhizomes provides a gradual build-up of humus, and burning stimulates the production of additional rhizomes which in turn decay and raise the level of humus. It is therefore possible to practise a sort of shifting cultivation on grassland as is done in the forest, by clearing the grass, cultivating the soil, and abandoning it for a fallow period. Such 'integral grassland farming', however, seems to be rare. The Bataks employ it along with other forms of land use including irrigated rice, and the Batainabura in the central highlands of New Guinea, living in an area which is almost entirely grassland including *Imperata*, successfully establish 'gardens'. 'They cut and burn, [the grass] they turn the soil and work it into mounds, a process that may be repeated consecutively for several years before the plot is abandoned to a fallow period of five to ten years under grass, but even they have a few gardens in the forest and thus do not farm grassland alone. Generally it can be said that some tribes do successfully eradicate *Imperata* when they set out to do so in a given plot. They have at least "integral" subsystems of grassland farming.'[93]

The central issue however, is not whether a few indigenous inhabitants succeed in cultivating a few hectares of *alang-alang*. The question is rather, can these enormous areas be turned into productive farmland where transmigrants may be settled and where they can grow crops for themselves and for sale? Since transmigrants — normally used to *sawah* cultivation — cannot be expected to switch to grassland cultivation (with all the hardship and the low yields that it entails), ways and means have to be found to prepare at least part of the plot for them prior to their arrival. This means that a technique for the eradication of *alang-alang* has to be developed. Efforts in this direction can be traced back several decades. Trials in 1951 using chemical weedkillers to eradicate *alang-alang* as well as *glagah* (*Imperata cylindrica* and *Saccharum spontaneum*) produced rather unsatisfactory results and were prohibitively expensive. In contrast, the use of ploughing harrows for shallow mechanical cultivation, combined with fertilising or manuring and the planting of leguminous plants, led to the eradication of the grasses.[94] The most recent results of experiments undertaken at Baturaja, South Sumatera, and elsewhere, suggest that vegetative elimination might be possible. The technique involves however not only the eradication of *alang-alang*, but its replacement by a plant which can successfully compete with its re-invasion and which could restore fertility to the leached soil. The technique is neither simple nor cheap, but if the *alang-alang* could be eradicated and the soil cultivated again, the present area of land under agricultural use could easily be doubled, which would certainly justify the effort.

The process begins with careful burning, butting and dragging of the *Imperata* grass; then follows the spreading of phosphate fertiliser and finally the broadcast sowing of a legume seed. The most suitable is *Stylosanthes guianensis*, a perennial forage legume, which can also be propagated from cuttings. This legume is a fast-growing cover crop which can prevent the *Imperata* from regrowing out of its underground rhizomes. Once the *Stylosanthes* is well-established, it can be cut periodically and the cuttings mulched for later cultivation with food crops. Depending on marketing conditions, the settlers can then feed cattle on *Stylosanthes* grass and plant commercially valuable trees.[95]

Though the experiments on Sumatera showed that *lamtoro* (*Leucaena leucocephala*) did not do well on the Baturaja soil even when it was carefully cleared of *alang-alang* before planting, this legume was very successful in eradicating the grass in Nusa Tenggara. It replaced the useless weed by a fodder legume and halted its encroachment into a coconut plantation when planted in hedges between the palms.[96] This example shows the importance of continued experimentation.

Thus it seems that the Indonesian grassland under *alang-alang* can be used as a reserve for new settlers providing the land is properly prepared prior to their arrival. In contrast to primary forest and swampland where the soil does have a certain agricultural value, *alang-alang* land is only of marginal use to man. Agricultural land and forests are both capable of indefinitely producing returns, provided the resources are properly managed. Primary forest and swamp forest have major commercial tree species capable of sustained wood production which would be lost if the forests were cleared. Moreover the relationship between swamp forests and fisheries, and primary forest and soil conservation is very important. However resettling transmigrants in forest areas necessarily involves the destruction of a self-renewing resource. The most beneficial scheme therefore is agricultural production on cleared land, minus the opportunity costs, i.e. the value of production from the same land before it was cleared. Using such criteria, *alang-alang* grassland will produce the largest net benefit because the opportunity costs are virtually nil. Tidal swamp would have the second largest net benefit and upland forest the third. The quicker forest land is cleared for transmigration, the lower the rate of timber recovery will be, consequently making the use of grassland much more attractive. The development of effective techniques for the vegetative elimination of *alang-alang* would certainly help to speed-up this process.[97] In other words, where we deal with *alang-alang* grassland, we have a system that has already been changed from forest to grassland and is maintained by fire to prevent natural reforestation. Man only gets a few benefits from the new system such as soil stabilisation, a habitat for certain species of wildlife and a marginal income from agriculture. Changing the system so that its agricultural potential is increased does not imply the loss of a major productive resource.[98] The basic question is whether or not the soil man intends to liberate from *alang-alang* is suitable for agricultural purposes.

'It has not yet been proved', admits Daroesman,

that sustainable dry land annual crop agriculture is possible in the red-yellow podzolic soils which predominate in a large part of the Outer Islands, now covered by *alang-alang*. Several segments of a sustainable farming system are known, but these have not yet been successfully combined over any long period. It is sometimes assumed that enough chemical fertiliser, if applied to these soils, can overcome all problems. This proposition is questionable even for medium quality soils; it is almost certainly incorrect for difficult ones. Even for the period during which it appears to work physically, it may not pay in economic terms.[99]

Suryatna and MacIntosh are less pessimistic and 'consider the red-yellow podzolic soils . . . as potential rather than problem soils.' They refer instead to the fact that given enough rainfall, cassava and other drought-resistant crops can thrive, and that the soils respond quite well to fertiliser: 'These soils are responsive to fertiliser and do not possess unusual "fixing capacities" or nutrient element deficiencies. Consequently it seems reasonable that we could increase the fertility of the soil by judicious use of fertiliser and maintain it by returning as much of the residues [as straw . . . or as stable manure] as possible'.[100]

Whatever pattern of land use is adopted, it must be remembered that the grassland was once forest soil.

A sensible agricultural system attempting permanent cropping will, therefore, as far as possible duplicate the function of the forest cover. Permanent tree crop agriculture is of course the closest approximation to the forest and its protective and nutrition holding function. With annual crops, the more the cropping pattern maintains crop growing all year round, the more available nutrients are saved and tied up in the crop to be released again when the crop residues are returned to the soil. The more the agricultural system keeps the soil covered by leaves and by mulch from crop residues or other vegetative material brought from outside, the less physical damage is done to the soil.[101]

This indicates that the farmer on former grassland has to establish and maintain soil fertility by using techniques that he may not necessarily be used to. Extension work is therefore indispensable.

The conversion of *alang-alang* land into pastures of palatable and digestible grasses and legumes would greatly contribute to the improvement of livestock production in Indonesia. Plants suitable for use as fodder crops include *Panicum maximum, Stylosanthes guianensis, Centrosema pubescens, Pueraria phaseolodes* and *Desmodium ovalifolium*, and to enable them to compete against the aggressive regrowth of the native grass, phosphorus fertilisation after burning is recommended.[102] In fact, stock breeding has never played an important part compared with plant production, and apart from draught-animals there were hardly any cattle kept by the indigenous population.[103] Similarly, research in grasses was exclusively botanical and rarely took place outside Jawa. Little knowledge has yet been acquired about the value of wild growing grass varieties. Much is known about the botany of grasses, but little about how better grass in greater quantities can be produced: 'Whenever we talk to agronomists, veterinary surgeons or cattle farmers in Indonesia about grass, it becomes obvious that we only think of grass like *Bengaals, Olifants,* or *Mexicaans,* grass that is cultivated in specially established gardens. This way of thinking reveals that we practically do not know ordinary grass as it is grown in other countries, and mainly in the temperate climate.'[104] In other words, cattle breeders for a long time thought in terms of imported grass varieties on specially fenced-in pastures, instead of looking for native grasses and improving their growth.

Considering the lack of animal protein in the diet of the Indonesian masses, the propagation of 'mixed farming', i.e. the inclusion of animal husbandry into farm production, is very desirable, particularly when transmigrants are settled in new areas and have to adapt their techniques to the new environment of the Outer Islands anyway.[105] Since 1972 a certain amount of good quality but dry grassland has been taken over for cattle-ranching on the eastern islands. On Sumatera and Sulawesi, about 500,000 hectares of poor quality watered grassland, namely *alang-alang,* is used for this purpose. Next to fencing and the provision of drinking water, state and private ranchers did carry out some supplementary planting of legumes in dryer areas, and of *Panicum maximum, Stylosanthes, Centrosema* and *Brachiaria* in wetter areas to replace *alang-alang* and balance the pasture. Their success was so encouraging that this industry may well become an important pioneer in developing those large areas of deteriorating land. Accordingly it may be judged that, 'although livestock, particularly buffalo, will probably play an important part in the development of littoral swamp and floodplain regions, the most suitable areas for livestock will be the large areas of secondary forest and *alang-alang* grassland.'[106]

In summarising the facts and considerations mentioned above, we come to the conclusion that the enormous and growing areas covered with *alang-alang* constitute a considerable land reserve on the Outer Islands. Although the eradication of the noxious grass is difficult and somewhat expensive, once done, the ground could be converted into cultivated land, pasture, tree cultures, or forest as conditions permit. Before more lowland forest and swamp forest is cleared away for ever to obtain land for settlers, it would be advisable to direct greater efforts into research and experimentation and the practical conversion of the tropical grassland.

NOTES

1. Kievits, 1891, pp. 710–12.
2. Keyfitz, 1953, p. 64.
3. In his *History of Sumatra*, first published in 1783, William Marsden wrote: 'Notwithstanding the received opinion of the fertility of what are called the Malay Islands, countenanced by the authority of M. Poivre and other celebrated writers, and still more by the extraordinary produce of grain . . . I cannot help saying that I think the soil of the western coast of Sumatra is in general rather sterile than rich. It is for the most part a stiff, red clay, burned nearly to the state of a brick, where it is exposed to the influence of the sun. The small portion of the whole that is cultivated, is either ground from which old woods have been recently cleared, whose leaves had formed a bed of vegetable earth some inches deep, or else ravines into which the scanty mould of the adjoining hills has been washed by the annual torrents of rain. It is true that in many parts of the coast there are, between the cliffs and the sea beach, plains varying in breadth and extent, of a sandy soil, probably left by the sea, and more or less mixed with earth in proportion to the time they have remained uncovered by the waters; and such are found to prove the most favourable spots for raising the productions of other parts of the world. But these are partial and insufficient proofs of fertility.

   Every person who has attempted to make a garden of any kind . . . must well know how ineffectual a labour it would prove, to turn up with the spade a piece of ground adopted at random. It becomes necessary for this purpose to form an artificial soil of dung, ashes, rubbish and such other materials as can be procured. From these alone he can except to raise the smallest supply of vegetables for the table. I have seen many extensive plantations of coconut, *pinang*, lime, and coffee-trees, laid out at a considerable expense by different gentlemen, and not one do I recollect to have succeeded; owing, as it would seem, to the barrenness of the soil, although covered with long grass. These disappointments have induced the Europeans almost entirely to neglect agriculture' (W. Marsden, *History of Sumatra*, 3rd edn, 1811, pp. 78–9.) Sceptical views about the limited soil fertility of the Outer Islands may also be found in works by Sir Thomas Stamford Raffles, 1817 and 1830.
4. Khan, 1973, pp. 222–3.
5. Ganssen and Hädrich, 1965, pp. 76–7.
6. Woelke, 1978, p. 3.
7. Baharsjah, 1978a, p. 81.
8. Loc. cit.
9. Prior to 1971, the province of East Kalimantan had less than 700,000 inhabitants. The census figure of 1980 was 1.2 million. This means there was an annual increase of 5.6% from natural growth and transmigration. In 1981, about 12,000 families settled in the province. (*Statistical Yearbook of Indonesia*, 1982).
10. Voss, 1979, p. 178.
11. Rieser, 1975a, Ch. 11, p. 1.
12. Löffler, 1982, pp. 130–1.
13. *Südostasien aktuell*, May 1983, p. 196.
14. Swasono, 1969, p. 114.
15. Callick, *Indonesia Times*, 18 Nov. 1983; also see Lagerberg, 1979, and current

reports in the Indonesian press.

16. Utomo, 1957 (quoted by Swasono, 1969, p. 109).
17. Soepriyo, *Indonesia Times*, 21 April 1984.
18. Muljadi and Dent, 1979, present a quick, efficient and inexpensive method for the evaluation of Indonesian soil and land resources; see also Mock, 1973, for water resources appraisal.
19. See, for example, Dequin, 1975; Woelker, 1978.
20. *Indonesia Times*, 30 July 1983.
21. Personal communication through the good offices of the West German Embassy, Jakarta, 1984.
22. *Statistical Yearbook of Indonesia*, 1982, table v. p. 33; *Buku Saku* 1980/1, table iv.3.2.
23. Chernichovsky and Meesook, 1981, p. 47.
24. Daryadi, 1978, p. 118.
25. Loc. cit., p. 119.
26. Groeneveldt, 1937, p. 323.
27. J.W. Gonggrijp, 1937, pp. 793–9.
28. Vogt, 1948.
29. Richter, 1976.
30. Goodland and Irvin, 1975, pp. 28–9; in West Germany, Weischet (1977 and 1978) has presented this complex in detail to the reader; see also White, 1983, pp. 3–46.
31. Goodland and Irvin, 1975, pp. 28–9.
32. Loc. cit., p. 29.
33. See for instance U. Scholz, 1980, p. 29.
34. Revelle, 1976, pp. 165–78, who still wants to feed 40,000 million people on the earth's surface.
35. Welte, 1978, pp. 634–8; Hartge and Wiebe, 1978, p. 636.
36. W.S. Reksodihardjo, 1971; Wiersum, 1978*a*.
37. See Kartawinata, in Kartawinata *et al.*, 1974, pp. 169–75; Kahirman and Mardjuki, 1978, pp. 125–7.
38. Rifai in Kartawinata, *et al.*, 1974, p. 167.
39. *Indonesia Times*, 23 March 1982.
40. See Daroesman, 1979, pp. 43–82, with the comments by Kartawinata, 1980, pp. 120–1.
41. Kartawinata *et al.*, 1978, p. 8.
42. Sudiono and Daryadi, 1978.
43. Tinal and Palinewen, 1974.
44. Inansothy, 1975.
45. Meijer, 1970; Soedjarwo, 1975.
46. Manning, 1971, pp. 59–60.
47. Fisher, 1964, p. 106.
48. Hanson, 1981, pp. 221–2; IBRD, 1984*c*, p. 3.
49. Soewardi *et al.*, 1980, p. 58.
50. Fisher, 1964, p. 19.
51. Pelzer, 1968, p. 277.
52. *Indonesia Times*, 22 Oct. 1981.
53. Uhlig in Imber and Uhlig, 1973, p. 35.

54. Soewardi *et al.*, 1980, p. 60.
55. Rieser, 1975*b*, p. 77; IBRD, 1984*c*, p. 11.
56. For the following see Voss, 1979, pp. 94–102, and, for more detail, Bohlander, 1978.
57. Fisher, 1964, p. 47.
58. *Bangkok Post*, 1 June 1981.
59. Donner, 1984, pp. 60–1.
60. *Indonesia Times*, 2 Feb. 1982.
61. Hamzah, *Indonesia Times*, 7 March 1983.
62. Furtado in Furtado (ed.), 1980, pp. 797–8.
63. Hanson, 1981, p. 226.
64. Tobler and Ulbricht, 1945, pp. 23–4; Simanjuntak, *Indonesia Times*, 2 Feb. 1984.
65. *Indonesia Times*, 16 July 1983.
66. Hanson, 1981, pp. 221–2.
67. Soewardi *et al.*, 1980, p. 58; IBRD, 1984*c*, p. 3.
68. Hanson, 1981, p. 231.
69. Frauendorfer in Brendl, 1982, p. 683.
70. EUROCONSULT, 1980, p. 14 (quoted by Frauendorfer in Brendl, 1982, p. 686).
71. Frauendorfer in Brendl, 1982, p. 675.
72. Hanson, 1981, p. 222.
73. Blom, 1977, p. 14.
74. For this complex see Blom, 1977, p. 14; Soewardi *et al.*, 1980, p. 57; Hanson, 1981, p. 232.
75. IBRD, 1984*c*, pp. 12–13.
76. Soewardi *el at.*, 1980, p. 58; Frauendorfer in Brendl, 1982, p. 676.
77. Rieser, 1975*b*, p. 72. In mid-1982, short but heavy rainfall in the districts of Ogan Komering Ulu and Muara Enim (South Sumatera) caused flooding which left 155 dead, 38 missing and 5,340 homeless. Schools, houses, administrative buildings, cars, cattle, bridges and irrigation works were damaged. (*Indonesia Times*, 16 June 1982).
78. Pelzer, 1968, pp. 277–8.
79. Hanson, 1981, pp. 232, 235.
80. Soewardi *et al.*, 1980, p. 59; Uhlig, in Imber and Uhlig, 1973, p. 34, estimate about 8 million ha.; Gourou, 1953, p. 288, on the other hand, estimates as much as 30% of Indonesia or 57 million ha. (certainly an exaggeration); Suryatna and MacIntosh, 1976, calculate the area under grass, mainly *alang-alang*, to be 64.5 million ha. of which, however, only 20 million is suitable for cultivation.
81. Eichelberger, 1924, p. 110.
82. Loc. cit., pp. 109–13; Keleny, 1960, p. 129.
83. Rudin, 1935, pp. 346–8; Gerlach, 1938, pp. 446–50.
84. Eichelberger, 1924, p. 110, identifies the high grass varieties on the Indonesian islands as *alang* (*Imperata arundinacea*) and *glagah* (*Saccharum spontaneum*) which grow mainly on marl, lime, sandstone and tuff soils as well as on laterite. Van Steenis, 1937, p. 641, maintains that *Imperata* grass plays only a minor part in the composition of the grasslands where *Themeda gigantea* together with other varieties prevail. Walandouw (1952), in his article about the grasslands in Indonesia, conveys the impression that *Imperata* is only one among more than 300

grasses, leguminoses, shrubs etc. on the wild pastures of Indonesia. It is not even mentioned in the text, whereas the role of *Lantana camara* as a destroyer of grassland is widely considered.

85. Daroesman, 1981, pp. 83–4.
86. Keleny, 1960, p. 128. — Grassland undoubtedly impedes natural reforestation. Hagreis (1931, p. 607) refers to a number of plants that could penetrate the grass and serve as pioneer vegetation: lantana (*Lantana camara*), pine tree (*Pinus merkusii*), gordonia (*Gordonia wallichii*), the Malacca tree (*Emblica officinalis*) and others.
87. The sketch is simplified and based upon the findings of Voss, 1979, pp. 127–8, who evaluated aerial photographs.
88. Hagreis (1931, pp. 603–4) clearly refers to the *grasladangbouw* of the Toba and Karo Bataks. See also Keleny, 1960, pp. 128–9. It may also be added that Terra (1952–3) studies the connection between patrilineal and matrilineal groups and the ecological impact of their land use.
89. Daroesman, 1981, p. 87.
90. Sherman, 1980, pp. 116–7.
91. Gerlach, 1938, pp. 446–50.
92. Sherman, 1980, p. 133.
93. See various references in Sherman, 1980.
94. Van Alphen de Veer and Vink, 1952, pp. 97–116.
95. See Daroesman, 1981.
96. Metzner, 1976*a*, pp. 231–2.
97. Burbridge *et al.*, 1981, pp. 108–13.
98. Soewardi *et al.*, 1980, p. 64.
99. Daroesman, 1981, p. 85.
100. Suryatna and MacIntosh, 1980, pp. 2–3, 6.
101. Daroesman, 1981, p. 86.
102. Teuscher *et al.*, 1978, p. 63.
103. Kündig-Steiner, 1963, I, p. 1300. Recently some efforts have been made to improve animal husbandry — which is still distributed rather unevenly. Whereas in Bali and Nusa Tenggara there are 125.4 livestock units per 100 ha., and on Jawa 101.9, the figures for Kalimantan and Sumatera are only 20.5 and 35.8 respectively. Notwithstanding a statistical increase, one can conclude that animal husbandry is still an underdeveloped branch of Indonesian agriculture which has only recently received proper attention. Woelke, 1978, pp. 180–94.
104. Schoorl, 1953, p. 6. — The author refers to *Panicum maximum, Danthonia elephantina* and *Euchleana mexicana*.
105. Schoorl, 1953, p. 6; Zimmermann, 1975*b*.
106. Leake, 1980, p. 74.

10

THE ENVIRONMENTAL IMPACT OF DEVELOPMENT

Although areas of extensive *sawah*-type cultivation exist outside Jawa, namely around Lake Toba and in the western highlands of Sumatera or in the south-western peninsula of Sulawesi, only about 4% of the Outer Islands land area is cultivated — this was the case at least some twenty years ago. Whereas on Jawa *swidden* agriculture, usually called shifting cultivation, has practically disappeared, on the Outer Islands 90% of the land is still farmed in this way. In principle therefore, it can be said that Indonesia is divided into two ecosystems, *swidden* and *sawah*, each with its own population density, modes of land use and agricultural productivity.[1]

Increases in population, be they caused by natural growth or migration, resulted in certain changes in the patterns of land use. Once all the suitable land was irrigated, the peasants generally turned to dry-land cultivation, relying on a generally not too unfavourable rainfall regime. In this way, they often managed to penetrate into the *swidden* area. The shifting cultivators, for their part, first reacted to the population increase by expanding the areas where they shifted their fields, but having met their limits, they reduced the cultivation cycle. This led to a serious increase in the environmental impact of *swidden* agriculture.

As we have seen, shifting cultivation involves clearing virgin forest, cutting and burning the original tree and bush vegetation, and thus releasing the nutrients stored in the biomass. They are added to the exposed soil, and food crops such as millet, mountain rice and maize are planted. Rapidly declining yields will soon announce the beginning of soil exhaustion after only a very few harvests, sometimes only one or two, depending on the underlying soil, the mulching effect, ash fertilising and so forth. Then the *swidden* farmer abandons his plot and shifts to another piece of virgin forest to start the clearing work again.

The abandoned plot is soon covered with some type of forest, and the longer the fallow period lasts, the better the regeneration will be. Usually, complete regeneration does not occur, and this means that there are fewer and therefore proportionately more fire-resisting species, shorter trees and fewer storeys. It is, however, maintained that this does not have a substantial effect on the yields of future crops. Likewise, the soil structure under a deflected climax forest would not be substantially less favourable or the soil nutrient status much lower: 'And, in the tropics, weathering processes quickly repair the damage resulting from erosion in all but extreme combinations of resistant rock, high slope, and heavy rainfall.'[2] Blaut's surprisingly optimistic statement may be true in regard to volcanic soils, but it is less applicable to most of the areas under shifting cultivation on the Outer Islands. In the highlands of Irian Jaya, where the indigenous population practises a very advanced form of shifting cultivation, it was found that the farmers always prefer to open up virgin forest rather than secondary vegetation. This is for the simple reason that the former area always gives better yields, while the amount of labour needed is not much greater. Moreover the shifting cultivators use a very considerate technique: they cut down the biggest trees, lop the branches off the others and burn or leave them rotting so that part of the trees survive and regrowth is therefore much quicker.[3]

Secondary and tertiary forest can be distinguished by the composition of their vegetation. This is determined not only by the effects of cutting, lopping and burning, but also by climatic factors such as soil insolation and heavier rainfall, and by the introduction of new plants from outside which have an influence on what will grow after the peasant ceases cultivation. It thus depends very much on what he is actually doing to the soil. If he retreats in time, i.e. before the fertility has declined too much, a *belukar* (secondary forest) will start to grow, the seedlings of which can oppress wild grasses. All authors stress that the secondary succession of primary forests is of a very complex nature and differs considerably from place to place. Various factors determine the floristic composition of *belukar*. An important factor is the extent of the area where the original vegetation has been destroyed. Minor plots will simply disappear again in the surrounding vegetation, but extended clearings depend very much on the floristic composition of their immediate surroundings and on the season, in conjunction with the seed-bearing period of the neighbouring vegetation. It is often claimed that a secondary growth, if left to itself, would ultimately create primary vegetation again. Kostermans, however, stresses that there is one restriction: megaspores should be available within a reasonable distance. This prerequisite is no longer present in numerous areas of Indonesia, which means that the original vegetation will never reestablish itself and only deflected climaxes result.[4]

There are numerous other factors determining the characteristics of *belukar* such as the soil prevailing on the plot in question; the way the original forest has been cleared; the duration of man's interference; the climate, especially the direction and strength of winds which disperse the seed and the coincidence of rain needed for a favourable germination; the influence of seed-dispersing animals and finally the character of the plant species available in the vicinity to invade the denuded area. It is evident that shifting cultivation does result in a slow but sure destruction of the primary forests and that a great number of factors are responsible for what follows this destruction. But 'secondary forest in Indonesia is without doubt poorer in species than primary forest except if compared with the poorer kind of primary vegetation.'[5]

However, since the number of *swidden* farmers is more often than not on the increase the cleared plot is used for too long to allow natural reforestation to take place. The soil is extremely depleted of nutrients so that only tropical grasses invade. Also, such plots are used as pasture for cattle once the tilling comes to an end, and the ground is periodically burnt over to encourage the grass to sprout. Then, of course, photophil (light-loving) grasses such as *Imperata cylindrica*, *Saccharum spontaneum*, *Melastoma spp.* and ferns take over and give rise to the extended areas of grass dealt with in detail earlier.[6]

To sum up we can make the following observations. Shifting cultivation is an extremely extensive form of agricultural land use that is justified under certain conditions. However, once the rotation cycle becomes too short, for whatever reasons (e.g. because of population growth or the curtailment of the area which had been at the group's disposal) shifting cultivation will lead to the devastation of the forest cover and consequently the soil. The interests of *swidden* farmers on the one side and forest economy and hydrology on the other often provokes conflicts. In fact, the loss of forests and soil seems to bother relatively few people, and the arguments for conservation are usually not very strong. They comprise aesthetic, scientific and economic viewpoints which are not always compatible. It is thus important that clear boundaries are established to protect watersheds and their protective forests from encroachment by shifting cultivators. To achieve this, the cooperation of various authorities will secure quick success.

Obviously, *swidden* agriculture cannot be eradicated overnight. Without outside guidance and support, and acceptance and cooperation from the groups concerned, little change can be expected — unless a 'clearance sale' of nature prevents any more 'shifting'. Once the forest is cleared from a certain area, the grassland has to be attacked. After this takes place, normal dryland farming is the only form of land use that remains possible. It would therefore be advisable to anticipate this development and take the necessary precautions. The ideal solution would be to integrate shifting cultivators into regional or watershed planning and successively transform their modes of land use according to the natural conditions. *Sawahs* should be introduced into these plans as much as possible and dryland farming encouraged for annual and perennial (tree) crops where appropriate. Under upland conditions, depending on soil and climate, an ecology-bound type of polyculture or intercropping of upland rice, maize, peanuts, beans, chillies, etc., would make the best use of the soil. Provided a careful treatment of soil fertility through mulching, composting and fertilising is kept up, the productivity could well be maintained, unless its reduction compels the peasants to switch to the monocropping of tubers like cassava. Over most of the surface, some sort of forest has to be re-established or protected.[7]

It is safe to say that shifting cultivation is not only a form of land use but also a complete way of life. The *swidden* farmer is heir to a very long tradition and tries his best to maintain the ecological equilibrium, because only this can secure his future survival. He depends completely on his environment and if everything goes well, he is self-sufficient and independent. He is generally suspicious of innovation. 'It is almost unthinkable', writes Blaut,

that a shifting agriculturist will abandon his traditional system entirely and convert immediately to pure commercial production of crops he cannot if necessary eat himself. If land is abundant, it is a rather simple matter for the farmer to employ both systems side by side . . . the conversion process becomes a kind of game, with interesting rewards and no sacrifices. If, however, there is a land shortage . . . the situation is quite different . . . shortage of land leads to efforts to compensate with materials at hand, namely, labour, and this absorbs any surplus needed for the permanent cash crops; and, of course, any diversion of land to the cash crop makes the land shortage for food crops much more severe.[8]

In practice however, this is not the case. The commonly held view that *swidden* systems produce very little of commercial value is not wholly accurate. The economic history of Indonesia and particularly that of the Outer Islands proves that a substantial proportion of export crops such as pepper, coffee, benzoin (aromatic gum), copra and rubber were produced by shifting cultivators. Pepper (*Piper nigrum*), probably brought to the archipelago by the Hindus after AD 600, was perhaps the

first perennial grown by shifting cultivators to enter world trade. The
Sumatran pepper grower, for example, clears a good portion of old forest
and plants cuttings of the coral tree (*Erythrina variegata*) together with
upland rice. The tree is needed to support the pepper vines which reach
their prime after seven years, but the pepper garden is usually exhausted
after fifteen years and then reverts to *belukar*. There are very early reports
(including one by the great fourteenth-century traveller, Ibn Battuta) of
Batak *swidden* cultivators planting seeds of the benzoin or benjamin tree
(*Styrax benzoides*) together with upland rice. After seven years, the tree
can be tapped and produces a balsamic resin, benzoin, which is used to
make incense, medicine and perfume, for a period of about eight years.
Coffee (*Coffea robusta*) came to Sumatra in the early nineteenth century,
was soon being grown by shifting cultivators, and by 1829 was exported
from the port of Padang. In southern Sumatera, the *swidden* farmers
planted the coffee seedlings together with upland rice and maize in their
jungle clearings, and during the second year intercropped sweet potatoes.
In the third year the trees became productive, but after three to four
years *belukar* took over the land for eight to twenty years. The interest in
coconut production started when the demand for copra began to grow in
the later part of the nineteenth century. In northern Sulawesi, Sangrinese
and Manadonese shifting cultivators soon started to plant coco seedlings
together with upland rice and maize, even in the second and third year of
the production cycle. They also interplanted *Derris malaccensis*, a plant
which naturally repels insects, peanuts, soya beans, pineapples, tobacco
and cotton. In Sumatera, the *swidden* cultivators interplanted with
robusta coffee, bananas and even cocoa. Up to 95% of Indonesian copra
is produced by smallholders, and a very high percentage of the export
crop comes from the *swidden* regions of Sulawesi.

Rubber (*Hevea brasiliensis*) has only been planted systematically since
about 1900 when world demand grew rapidly. Though the colonial
administration did not regard it as a smallholders' crop, the Chinese
traders, who bought forest products from the shifting cultivators, supplied them with seed and the smallholders started converting *swidden*
into rubber gardens on Sumatera as well as on Kalimantan. For two or
even three years they planted a catch crop (third harvest) and thus
delayed the development of *belukar* or the invasion of *alang-alang*. Ecologically speaking, a *swidden* rubber forest is closer to the secondary
forest than a rubber plantation. The smallholders avoided the planters'
costly mistake and did not practise clean weeding — with its concomitant soil erosion. Intuitively they chose the correct method
which the European rubber planters only learned slowly.[9] This is

achieved simply by the use of a cover crop, a function that is taken over unintentionally by a food or cash crop. We have tried to show that *swidden* agriculture did contribute successfully to the economy over and above the farmers' subsistence, a contribution usually only attributed to the plantations. In fact, tens of thousands of Indonesians in Sumatera, Borneo and in other parts of Indonesia followed the example of the plantations and planted commercial crops, above all rubber, on their old *ladangs*, thereby converting hundreds of thousands of hectares into permanent gardens. On the whole, they too selected the most fertile areas for this cultivation and thus reduced the potential areas for colonisation.[10]

When, towards the end of the nineteenth century, more and more land was brought under plantation economy, particularly on Sumatera, an ecological impact of this sort of development had to be accepted. Reading through old reports or more recent analyses, one finds a continual debate over whether the native shifting cultivators or the foreign planters did most of the damage to the ecological equilibrium on the Outer Islands. Gonggrijp for instance states: 'According to my opinion, the shifting cultivation causes much more damage than the European plantations';[11] Kostermans, however, differentiates: 'The destruction of the vegetation is not solely attributable to primitive man living at subsistence level, but is partly due to modern man, who wants to transfer woodland into agricultural land; this, however, is carried out on the base of conservation';[12] whereas Pelzer writes: 'No Sumatran shifting cultivator has done as much damage to the land as was done by the tobacco planters.'[13]

At present, however, official opinion once more turns against the shifting cultivator. What was formerly the plantation economy, namely the producer of public and private revenues, is now the forest economy, which receives more sympathy and understanding, though neither can justly claim to be a great protector of the environment.[14]

Pelzer observed the destructive land use practices of the tobacco plantations in northern Sumatera over a period of more than twenty-five years. During the colonial period, powerful Dutch tobacco companies obtained concessions from local rulers, such as the sultans of Deli, Langkat and Serdang on Sumatera, and by 1940 had an area of 255,000 ha. at their disposal. The indigenous peoples, living in overcrowded villages along the coast or on the lower, navigable rivers, were hemmed in by the plantations and given no room for expansion. The population pressure was further increased by the great numbers of immigrant Javanese plantation workers who settled down in the area after the termination of their contracts. Since the Sumatran tobacco planters used the shifting cultivation system — with a bush fallow time of seven to eight years — only a minor part of the land controlled by them was under cultivation. These areas of secondary growth were favoured by the peasants who encroached on plantation territory and established *swidden* or smallholdings. The process became particularly catastrophic after independence, when the Dutch plantations were nationalised and the sultans lost their power. Pelzer, who visited Medan again in 1955, found 'an almost completely deforested plain. The town itself had expanded greatly at the expense of the surrounding plantations. Furthermore, every depression and all the low-lying land along rivers had been converted into wet-rice fields. Gone was the forested landscape of the pre-war days . . .' Finally, in 1967, he saw the enormous areas of former tobacco plantation being squatted by thousands of peasant families trying to eke a living from the soil.

The foreign planters laid the blame for the destruction of the planta-tion economy on the political changes in Indonesia. They 'ignored or even denied the fact that the top-soil of the upland tobacco plantations had already been lost decades ago as a result of deforestation for tobacco cultivation. This is why the upland plantations were abandoned in the 1920s. We have ample photographic evidence of this development on the hillsides of Deli, Langkat, and Serdang', writes Pelzer.[15]

Next to *swidden* and the plantation economy, the timber indus-try has also produced a tangible ecological impact. In 1980, the govern-ment of Indonesia introduced a new forest policy with the aim of banning the export of logs until 1985 to help build up a national wood-processing industry. Henceforth export-permits would only be granted for processed wood (see Fig. 25). This scheme follows a reason-able principle of development policy, though it was greeted with dismay in Japan. If we ignore the cases of log smuggling,[16] this policy was suc-cessful in helping to establish many plywood factories on the Outer Islands.[17] By the end of 1982, 60 plywood plants with a total capacity of 3.6 million cu. m. were operating, another 48 plants were under con-struction and 45 more in the planning stage. By the end of 1985, the Indonesian government envisaged having 150 plywood plants with a total capacity of 9 million cu. m. These steps resulted in the closing down of plywood factories in Japan and South Korea, and Indonesian plywood began to make an impact in the world market with 600,000 cu. m. being exported in 1981 and 1.9 million cu. m. in 1983.[18] And although the new sector could not initially meet American demands for quality and quantity,[19] there is a good chance that in the near future Indonesia will soon become the largest plywood exporter in Asia. Similarly, national pulp and paper production is on the increase. The establishment of Centres for Wood Economy in Jawa, Sumatera and

Kalimantan, equipped with research, processing and shipping facilities, will further this trend.[20] This development cannot continue however without causing a tangible impact on logwood production and, by extension, on the forest environment. Whereas the export of logwood was once the second most important foreign exchange earner after raw petroleum, the production of logs dropped after their export had been curtailed. However, the demand of the growing national wood-processing industry for logwood is increasing, and the forest administration is now planning tree planting programmes to help meet the demand for timber in the next century.

Before the growth of the wood-processing industry, the forest administration's efforts were largely devoted to logging. However the large amounts of public revenue and foreign exchange earned by forestry led the government to encourage logging through the granting of concessions, preferably to foreign companies who, from the beginning, used heavy equipment. Between 1967 and early 1970, forest concessions of about 12 million ha. were given to foreign and national timber companies. But this area soon doubled, as the powerful companies (whose activities on the Outer Islands could not be easily controlled) sought to maximise their profits.

Emil Salim, the foremost environmentalist in Indonesia, did not hide his uneasiness about what was going on in the extended forests. He stated that environmental preservation was being almost ignored in the face of forest exploitation while economic interests seemed to have become the primary objective. In a statement before the House Commission for Research he said that field studies had shown that holders of forest exploitation licences (HPH) were only concerned with reaching economic targets and were causing forest destruction and endangering the environment.[21] There are of course certain regulations which have to be followed, the most important one being the PTI, or Indonesian Selective Tree Felling Regulation. It forbids the cutting of trees below 60 cm. in diameter so as to maintain further growth of younger trees and avoid the devastation of the forest. When the Minister of Agriculture spoke publicly about the subject in 1981, he expressed great concern about unselective tree felling in tropical forests.[22]

Only a few months later, however, 'wholesale felling' became a matter of passionate debate among both specialists and the public. Wholesale felling involves clear felling of the whole concession lot

The four pictures *overleaf* show features of soil erosion after the destruction of the tropical forest in East Kalimantan. Once the forest cover and undergrowth are destroyed, the destruction of the soil cannot be checked.

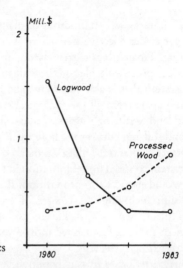

Fig. 25. EXPORT
REVENUES FROM
LOGS AND PRO-
CESSED WOOD,
1980–83

Source: Based on
Indonesian government statistics

with the intention of replanting it immediately with selected tree species. Because of bad experiences with the recent exploitation of productive forests by concession holders, officials and government agencies
sought ways and means of avoiding such problems in the future. The
idea of wholesale felling seemed to be attractive. This method ensured
that there could be some control over whether denuded portions of the
forest were replanted and the best quality seedlings could be chosen.
Such an artificial forest would yield better products, and in greater
quantities than before. There was an option to select certain types of
trees for replanting according to specific purposes such as wood supply
for pulp and paper mills.[23]

At first sight these arguments seem convincing, particularly since
the system has also been used in the temperate zones — though its
utility is now under question. Under tropical conditions however, and
considering the weak control-system on the Outer Islands of Indonesia,
this technique does involve tremendous risks. Once the concession plot
has been denuded immediate replanting cannot be guaranteed. This
means that, for a considerable time, the bare soil will be exposed to rain,
sun and wind — long enough, at any rate, to initiate the process of
pedological and hydrological deterioration dealt with at length earlier.
Even a quick replanting leaves it without enough protection for many
years, by which time the damage has been done to the ground that will
never be healed again. But quite apart from disturbing the balance
between soil and vegetation, wholesale felling leads to the destruction of

fauna and flora as well. Hundreds of plant and animal species that can still survive under selective tree felling will inevitably be eradicated after clear felling. Finally, it is extremely doubtful whether the replanting of specific tree species suitable for the industries of Indonesia will create a new ecosystem that is able to survive and flourish under tropical conditions. And apart from all the ecological considerations, the opening-up of forest land with logging tracks has already attracted irresponsible settlers and shifting cultivators; how many more settlers would clear-cut areas attract? 'Total tree felling', remarked a leading forest developer, 'is a good system to speed up reforestation in the country' and as a businessman he would support this proposition. But because of the consequences for the soil, including the danger of erosion, for ecological balance between plant and animal life in the forest, he would urge the government to make further and more elaborate studies about the effects of the clear felling system.'[24]

Various participants in this nationwide discussion, particularly local authorities but also representatives of the wood industry, suggested an alternative; namely that innumerable hectares of barren land and *alang-alang* grassland across the country could be planted with artificial forests.[25] In fact, the country's 'critical land' surface is increasing. In the tiny province of South Kalimantan (3.7 million ha.) for instance, no less than 26% was classified as 'critical land' at the beginning of 1982 with an annual increase of some 180,000 ha. in the last few years, caused only by illegal logging. The provincial administration suggested a regreening and reforestation project based on cooperation with the local people, using coconut, clove and cashew-nut trees, but also legumes, *Acacia* and *Albizzia*. If the plan could be maintained up to the year 2000, no critical land would remain.[26] By 1980, roughly 7 million ha. were officially classified as 'critical land', half of it inside and half outside the area administered by the forest authorities.[27] Thus it is obvious that Indonesia does not need clear felling to plant selected species in artificial forests.

So far, the idea of general clear cutting for logging purposes has not caught on. After the creation of the new Ministry of Forestry in early 1983, forest policy up to the year 2000, including the next three Five-Year Plans (1984–99), was formulated. It obviously lays emphasis on the expansion of the production forest, on the rehabilitation of damaged stands, on the reforestation and regreening of deforested land and on watershed protection. And though the projections are ambitious and quantitatively hardly realistic, the policy is moving in the right direction.

Within the next fifteen years it is envisaged that 5–6 million ha. of production forest will be reforested, providing a yield of some 90 million

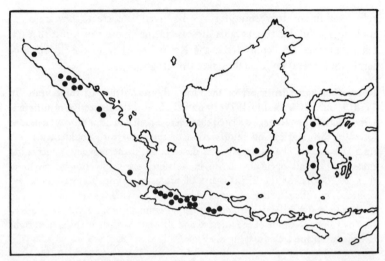

Fig. 26. WATERSHEDS SUBJECT TO REHABILITATION BY THE
MINISTRY OF FORESTRY (1984/5–1988/9)

Source: Indonesia Times, 13 April 1984.

cu. m. of timber a year. Selective felling will be continued on 58 million
ha., producing 60 million cu. m. of timber annually. Thus, in the year
2000, the forests of Indonesia are expected to supply no less than 150
million cu.m. of timber per year, enough to feed the national wood
industry. As to reforestation, the first priority under the current plan
(1984–9) is being given to very arid land. Finally in terms of land conservation, twenty-two selected watersheds will be rehabilitated under the
current plan to improve the condition of the land, prevent routine floods,
erosion and water shortages, and thus increase the farmers' income. The
twenty-two watersheds comprise river systems in Jawa, Sumatera,
Sulawesi and Kalimantan (see Fig. 26), most of which are in a critical
condition and suffer from high sediment transport.[28]

Efforts in reforestation and regreening often result in impressive
figures in official statistics. However when compared, they are obscure
and highly contradictory. The Indonesian press is full of reports about
the destruction not only of natural forest land, but also of freshly
afforested ground. Apart from cases where village communities succeeded in establishing firewood plantations on barren land (e.g. with
Calliandra calothyrsus) or undertake regreening schemes with fruit trees
such as cloves, coconut, cashew-nut, mango and the like, many efforts
ended in failure as a result of inconsiderate farming. Thus, while the

reforestation and regreening figures may well be correct, they are often no longer valid due to a parallel increase in the amount of forest that has been destroyed. A report from the Kerinci regency in the province of Jambi (Sumatera) helps to illustrate the situation:

Regreening has fallen short of the target figures in the last three years. In 1978/9, only 14%, and in 1979/80 only 37.5% of the respective plan figures were reached. The Regent declared that there is a total of 50,000 ha. of denuded forest in the regency and reforestation of the wasteland would take much time and funds. The environmental destruction in the regency is increasing since the peasants open new farms in restricted areas; covering so far more than 20,700 ha. of the 275,000 ha. of protected forest. Reforestation will continue, but the efforts are hampered by illegal activities of local farmers who do not understand the importance of the government programmes to preserve the environment. Some of the forested and afforested areas have been used right to exhaustion and are now lying idle.[29]

Willem Meijer, an American university teacher and experienced forest specialist who travelled around Indonesia in the mid-1970s, is rather sceptical about the long-term success of government activities. When logging and shifting cultivation in the Indonesian forests causes more and more devastation, he observed, so press releases about government schemes for the replanting of trees also increase. His verdict on official efforts is that they are ridiculous. Within seven years, the government wants to replant an area about half as large as that planned for deforestation to settle farmers, and one tenth of the total area to be logged. And he describes the efforts as utopian because the forest authorities are so weak that they cannot protect the law in the field; staff numbers are too small, they are not sufficiently trained or supplied with funds, and whereas the size of the forest service remains static, concessions are continually being given away. Similarly, he feels that the research work is worthless and a waste of money.

Meijer does not envisage much of a future for the government forestry service, and therefore suggests that private enterprise and private interests should be given more responsibility: 'Let the timber companies expand their activities in the field of forest surveys, regeneration studies, staff training, plantations, and even agricultural extension work inside their concession areas. They would be able to put up solid training programmes, rigid work schedules, and good pay scales.' This would of course lead to increased costs and would require tax reductions in return. And he concludes:

Whatever occurs, in case the country does not want to sell out its timber resources to private enterprises and to a capitalistic, monopolistic, élite or ruling class of (ex)military managers and technocrats, there still remains a task for the government to enforce some of the rules of the game and to collect some of the revenue for general purposes.

In case the idea prevails that all timber land will be loggable and suitable for crop production, this can be tried. But the final result will be less ideal than the Indonesian Forest Law and the skillfully written Masterplan foresees. The mismanagement in the provinces of South Sumatera and Lampung with their extensive degraded *alang* fields should be a warning. Clear felling under the promise of plantation schemes which may never materialise, would be an ecological and economic disaster in the long run.[30]

Entering and 'developing' the natural forest cover of the Outer Islands does unfortunately cause more damage to the ecological equilibrium than logging and shifting cultivation already do.

When describing the geographical setting earlier in this work, some mention was made of the remarkable zones of wildlife found in Indonesia. In the western zone comprising Sumatera, Kalimantan, Jawa and Bali, we mainly find the Asian element, namely the great mammals such as elephants, buffaloes, tapirs, rhinoceroses, apes and monkeys, wild boars and, except on Kalimantan, the tiger. In the eastern zone comprising the provinces of Maluku and Irian Jaya, the Australian element predominates. There are no great mammals, but there are wild boars and many small marsupials, rodents and the ant-bear, in addition to many birds, including the bird of paradise. The third zone comprising Sulawesi and the Lesser Sundas is a transitional one where both elements mix. We must however admit that human interference has tangibly reduced and often nearly eradicated one or other of the species.[31]

The coexistence of humans, animals and plants on the islands of the archipelago has gone on for centuries if not for millenia, but with the growing population and increased economic importance of the soil, forest plants and animals, this coexistence came to an end. As man penetrated the jungle, cut and burnt his plot and started to cultivate his crops, so the habitat of the wild animals and many plants began to shrink. And since some of the large mammals need substantial areas of land in which to feed and mate, their eventual extinction became only a matter of time.[32]

Nowadays, international organisations like the World Wildlife Fund, and national institutions like the Natural Environment Protection and Conservation Service (PPA) are trying to protect what is left of the once multifarious wildlife of the Indonesian islands. However their task is a difficult one. First, and most importantly, the conservationists are hopelessly understaffed, funds are scarce, and there are powerful economic interests ranged against them. The peasants, particularly the transmigrants, who often have to face wild animals destroying their crops, killing their goats and sometimes attacking humans, demand protection. The logging concessionaires simply cannot be considerate towards wild animals: once an area is given over to logging, we might as well forget the animals that once lived there. The loggers maintain that the number of wild animals is still higher than usually reported and that there is no reason to fear for their extinction. The conservationists, on the other hand, do admit that — thanks to their work — the situation is now not so bad.[33]

The official conservation policy of Indonesia has created, as a visible sign of its intentions, an impressive number of so-called 'Natural Preserve' areas. By 1971 there were 158 such areas scattered all over the country, covering a total area of 3.5 million ha., but for the most part there is little known about the conditions prevailing there. It does seem however that the conditions on Jawa are quite satisfactory. In the Meru-Beriti Preserve, for instance, the Java tiger can now be found again. On the Outer Islands, particularly in two areas — Gunung Leuser Wildlife and Natural Preserve in Aceh, North Sumatera (416,500 ha.), and Kutai Preserve in East Kalimantan (200,000 ha.) — the World Wildlife Fund, Man-and-Biosphere and BIOTROP (University of Bogor Programme for the Study of Biological Subjects in the Tropics) undertake research work. An inventory of the Natural Preserves in North Sumatera, South Sumatera, Lampung and East and Central Kalimantan showed a generally satisfactory situation. Rhinoceroses, tapirs, elephants, wild buffaloes, gibbons, orang-utans, proboscis monkeys and wild oxen are found there. But it must be remembered that timber companies are continually extending their activities into these forest and wildlife reserves.[34]

The Indonesian press openly discusses the violation of protective laws by timber companies and the difficulties of settlers who are confronted with wild animals. Universities send teams of scientists to study the problems on the spot, the minister responsible for protecting nature takes an active interest in the matter, and 'Ganesha groups' from the armed forces succeed in driving endangered elephants into new reserve areas.[35] Other positive examples of conservation in action include the establishment of orang-utan rehabilitation centres where tame young animals (often the offspring of a mother that has been killed) are prepared for eventual release into the jungles of Sumatera and Kalimantan;[36] the marking-out of wildlife sanctuaries on Flores and Timor (East Nusa Tenggara) to protect the local fauna and encourage deer previously endangered by hunting to breed;[37] and lastly, government legislation forbidding trade in protected animals and curbing controlled hunting has been passed. Customs officers often catch people trying to smuggle abroad protected animals, both live and stuffed, especially rhino horns and skins. Though more than eighty nations have signed the Convention on International Trade in Endangered Species of Wild Fauna and Flora (CITES), Singapore, the major trading centre in South-east Asia, has not done so, thus encouraging poachers all over the region.[38] Hunting, once a means of controlling rapid increases in the animal population, is now unfortunately undertaken for profit and pleasure.[39]

A lack of coordination between the different authorities and the prevailing system of doing a favour to a good friend, often called corruption, always works to the detriment of nature. When the Aceh provincial government issued permits to establish sawmills in the Gunung Leuser Natural Preserve, the protective agency fought a losing battle. Increasing numbers of loggers penetrated the Preserve and extracted wood, albeit illegally. 'Some years ago', a representative lamented, 'the forest was fresh and the air clean. Now the hills become bald, and the climate is hot day and night. The road along the Preserve is dotted with sawmills.'[40] The crucial fact is that hardly anybody cares. Willem Meijer observed 'that people living close to the protected forests and wildlife preserves in Sumatera do not care to obey official prohibitions meant to protect the forests and the preserves. The plank boards erected to display prohibitions against hunting, tree-felling and farming in the protected areas were not at all respected by the people.'[41]

Little can be hoped for the future of nature where 'progress' is on the march and input/output thinking dominates decisions, unless we can prove that protection will pay in the short run. However, it is not only human influence and man's economic rapacity that slowly destroys the ecological equilibrium and the original image of nature. We know that soil erosion is going on all over the globe, even in totally uninhabited areas. Similarly, self-ignition is a natural phenomenon that destroys tropical and sub-tropical forests without human interference. In particular cases however, the cause is not always clear. The Indonesian press periodically reports the outbreak of forest fires in the Outer Islands, affecting areas ranging from a few hundred to thousands of hectares.[42] Such fires destroy primeval jungles as well as newly afforested areas, and usually the tree cultures of smallholders are not spared. As the authorities report, arson cannot be ruled out, and the fire-fighting teams prove generally helpless in the face of such catastrophes. Unfortunately, the population is in most cases indifferent towards reforestation and the damage done by forest fires unless they suffer a personal loss. In many cases one suspects that such fires, if not started intentionally, are welcomed because they open up more jungle land for cultivation.

Certainly, prolonged periods of drought facilitate self-ignition, but arson and negligence on the part of peasants as well as foresters may have been responsible for many a conflagration. Nobody has so far discovered the cause of Indonesia's worst forest fire this century that ravaged an estimated 35,100 sq.km. of tropical forest between February and June 1983 in East Kalimantan. The International Union for Conservation of Nature and Natural Resources (IUCN) said that 'the area, normally one of the dampest on earth, had experienced two years of severe drought as the result of the behaviour of the ocean current nicknamed El Niño. This, together with the region's underlying deposits of soft coal and peat and the presence of easily ignited undergrowth caused by selective logging operations are blamed for the conflagration. In these spots, where the earth literally caught fire, the forest had no chance of survival. . . . The fire destroyed the equivalent of two-thirds of the annual loss of forest cover worldwide.'[43]

Though — after the normalisation of the climate and the return of the rains — plant regeneration begins, the calamity is by no means over. Apart from the destruction of hundreds of thousands of giant mahogany trees and other commercially valuable trees, representing a loss of saleable timber of about US$5,600 million, countless numbers of birds, insects and animals such as leopards, bears, deer, pigs, civet cats, and rodents were killed, leading to the extinction of many species. But 'it was also an important historical scientific event, because it occurred in an ecosystem that we, perhaps naively, thought was relatively stable', said one renowned ecologist. 'It makes us realise how fragile a natural system can be, and the capriciousness of the environment in this case has caused us to look at all kinds of ecosystems in a new light.'[44]

We should come to realise that tropical nature is sensitive and finely balanced, it does not need human interference to damage it. But if, under population pressure, millions of human beings are forced to live in it or on it, man becomes a main agent for destruction. It will be impossible to exclude the tropical forest areas from man's economic domain in the future, but his first obligation should be to avoid destructive practices and aim at a coexistence advantageous to both parties.

NOTES

1. Geertz, 1963, p. 13.
2. Blaut, 1960, p. 191.
3. Kostermans, 1960, p. 334. Spencer, 1966, p. 39, however comments: 'Virgin forest, in the sense that it has never been cleared by human or natural agency, may actually exist in numerous areas. It is likely, however, that most of the mature forests of the Orient today are not virgin forests in the proper sense, but merely old forests that have reached a fairly stable equilibrium of ecological succession after some earlier clearing by human or natural means.'
4. Kostermans, 1960, p. 333. According to Henderson's *Dictionary of Biological Terms*, 1979, p. 251, a megaspore is a larger-sized spore of dimorphic forms in reproduction by spore formation; the larger spore of heterosporous plants which give rise to the female gametophyte. In seed plants it is also called an embryo sac.
5. Kostermans, 1960, p. 335.
6. Hagreis, 1931, p. 606.
7. Hagreis, 1931, pp. 620-1; Kostermans, 1960, p. 332; Blom, 1979, pp. 14-15.
8. Blaut, 1960, p. 195.
9. Pelzer, 1971, pp. 263-70.
10. Pelzer, 1946, p. 139.
11. J.W. Gonggrijp, 1937, p. 793.
12. Kostermans, 1960, p. 332.
13. Pelzer, 1968, p. 278.
14. In a discussion, the chief of a South Kalimantan office of the Ministry of Agriculture denied the allegation that the holders of forest concessions were responsible for the denudation of the areas at the foot of Mt Meratus. It was the shift-and-burn farming that has denuded the area, he asserted. *Indonesia Times*, 18 Nov. 1981.
15. Pelzer, 1968, pp. 277-9.
16. *Indonesia Times*, 13 May 1981.
17. Satoto, *Indonesia Times*, 19 Jan. 1982.
18. *Indonesia Times*, 2 April 1983.
19. *Sinar Harapan*, 7 April 1983.
20. Ostasiatischer Verein, *Ländernachrichten*, no.5/82, p. 8.
21. *Indonesia Times*, 25 Sept. 1981.
22. Loc. cit., 9 Nov. 1981.
23. Loc. cit., 8 May 1982.
24. Loc. cit., 1 May 1982.
25. In 1975, the official forest statistics for the whole of Indonesia reported 16 million ha. of bare land, *alang-alang*, and other neglected land, and 23 million ha. of damaged forest and brushwood (Woelke, 1978, p. 211).
26. *Indonesia Times*, 30 Dec. 1981, and 4 Jan. 1982.
27. *Buku Saku*, 1980/1, p. 148.
28. *Indonesia Times*, 23 March 1984 and 13 April 1984.
29. Loc. cit., 5 Nov. 1981.
30. Meijer, 1975, pp. 94-6.
31. U. Scholz in Kötter *et al.*, 1979, pp. 60-1.
32. *Indonesia Times*, 2 April 1984. — Periodically, there are reports in the local press about such encounters between humans and wild animals.
33. Sommer in *Die Zeit*, 26 Oct. 1984.

34. Reksodihardjo, 1971, pp. 83–4.
35. *Indonesia Times*, 2 Sept. 1981; 16 March 1982 and 10 Feb. 1983, dealing with the Air Sugihan resettlement project and the transfer of 50 to 70 elephants into a new habitat.
36. Sommer, loc. cit.
37. *Indonesia Times*, 24 June 1982.
38. Shelton, 1983; *Indonesia Times*, 10 Feb. 1983 and 4 March 1983.
39. *Indonesia Times*, 5 Nov. 1981.
40. Loc. cit., 14 Feb. 1983.
41. Loc. cit., 11 Aug. 1981.
42. In 1983, nearly 1,000 ha. of wild orchid forest belonging to the Kersik Luway nature reserve in East Kalimantan was destroyed by fire. It took four days for the fire to be put out, with heavy equipment by digging a moat around it. *Indonesia Times*, 6 May 1983).
43. *Indonesia Times*, 5 April 1984. — Some scientists noted that the government of Indonesia was reluctant to release news of such a disaster. They suggested that the officials felt the reaction to news of the fire might lead to the curtailment of logging concessions. Though these make forests more vulnerable to fire, they provide substantial financial contributions to the nation's economy. *Australian Financial Review*, 4 May 1984.
44. *Australian Financial Review*, loc. cit.

11

ASPECTS OF THE TRANSMIGRATION POLICY

We have dealt in detail with the problem of a growing population facing limited natural resources. Now that the island of Jawa has been cultivated to the limit of its physical resources, the transmigration policy seems to be the only safety-valve, but the more people are settled on the Outer Islands, the more will environmental problems also become visible there. Everybody is now aware that the lush tropical rain forest with its enormous production of biomass is by no means indicative of fertile soils on which high-yielding annual food crops could be grown. The Indonesian agriculture and forestry authorities are well aware that the potential of the Outer Islands for the accommodation of transmigrants is limited. The weaknesses of the system have been exposed: many transmigration settlements were abandoned because the settlers were not able to make a decent living under the given environmental conditions. A degraded landscape was often left behind — eroded, invaded by *alang-alang*, or at best covered with *belukar*.

For a population and transmigration policy to succeed it is important to have at least a rough idea of the potential which the Outer Islands can offer to migrants — in other words, how many people could Indonesia possibly accommodate with the prospect of living space and a source of income? We have shown earlier that such data are still rather scarce and need constant revision. This is quite normal since it is not easy to estimate the potential of areas which are little known to science. With continued research work, though, such figures will become more reliable. Unfortunately the transmigration policy is proceeding more quickly than the research work that should actually provide the basic knowledge for it.

For the period of REPELITA III (1979/80–1983/4), the Indonesian authorities reached an estimate of the potential of the Outer Islands to accommodate transmigrants in that five year plan period. They came to a gross potential of roughly 10 million people, but since some of the provinces will surpass their carrying capacity with their own expected natural population increase, they should transfer inhabitants to other islands rather than accept more transmigrants from outside.[1] Using these criteria, the net potential for settling transmigrants would come to about 8.3 million persons. Table 12 gives an idea of the more recent estimates,

249

Table 12. POTENTIAL OF THE OUTER ISLANDS TO
ACCOMMODATE TRANSMIGRANTS DURING REPELITA III

	Carrying capacity	Population increase	Transmigration potential
Aceh	327,200	193,856	133,344
North Sumatera	270,000	657,312	(– 387,312)
West Sumatera	129,810	158,237	(– 28,427)
Jambi	407,700	124,277	282,723
Riau	822,600	160,527	662,073
Bengkulu	108,300	86,479	21,821
South Sumatera	708,300	465,001	243,299
Lampung	157,300	672,927	(– 515,627)
West Kalimantan	2,373,700	207,732	2,165,968
South Kalimantan	154,100	531,543	(– 377,443)
East Kalimantan	1,108,600	211,294	897,306
Central Kalimantan	1,571,300	111,813	1,459,487
South Sulawesi	150,156	347,729	(– 197,537)
Central Sulawesi	642,500	109,306	533,194
South-east Sulawesi	184,900	47,082	137,818
North Sulawesi	13,760	137,658	(– 123,898)
Maluku	897,700	127,578	770,122
Irian Jaya	2,729,300	94,651	2,634,649
Total	12,756,526		9,941,804
Overpopulation			(– 1,630,290)
Net potential for migrants			8,311,514

Source: *Business News*, Jakarta, 27 May 1981, according to the Director of Urban and Provincial
Planning, corrected after consultation with Indonesian authorities.

according to which Irian Jaya, West and Central Kalimantan, but also
parts of Sulawesi, Sumatera and Maluku seem to still have a substantial
capacity to absorb transmigrants.

Such estimates are, of course, of limited value. We do not know the
premises under which they were made, nor whether they only include
peasants to be settled or, in addition, those to be employed in the non-
agricultural sector. Though the resettlement of 8.5 million persons or
1.7 million families within a period of five years is technically out of the
question anyway, this potential by no means covers the future prospects
for transmigration. If we add — with all necessary reservations — the
areas regarded as suitable for being converted into agricultural land,
namely 25 million ha. of swampland and 20 million ha. of grassland, we
come close to reserves of 50 million ha. Since irrigation will be available

in the beginning to only a tiny minority of the transmigrants, we may as well calculate on the basis of 5 ha. holdings for a family of five persons. Such holdings could be cultivated with annual food crops, used partly as pasture and partly planted with trees. Under such conditions, some 50 million people or 10 million families could be transferred to and settled in the Outer Islands. Given that the present annual population increase in Jawa is 2.02% or 1.8 million people, and presuming that this increase could migrate simultaneously, Jawa would exhaust the soil reserves of the Outer Islands within twenty-seven years. We have, however, to allow for the population growth on the Outer Islands themselves, which at 2.7% per annum totals almost 1.5 million people. Taken together, Jawa and the Outer Islands' population increase would occupy the present land reserves within no more than fifteen years.

Many other factors could modify this picture. The success of family planning could reduce the population growth, and more intensive land use, for instance through irrigation, could increase the carrying capacity of the 50 million ha. But at the same time, much cultivable land would be lost due to erosion, leaching and so forth. Therefore, as far as settling peasant families is concerned, it seems that calculations based on 50 million ha. and 10 million families are sufficiently cautious for the time being. And this means that within the next 15–20 years, the Indonesian population increase would have to come to a complete standstill to avoid an environmental catastrophe in the Outer Islands. It is not the aim of this study to make estimates of its own, but rather to point to the environmental consequences of all these possible activities. Since we shall deal later on with activities in other sectors, we shall restrict ourselves now to the most important environmental aspects of transmigration in the agricultural sector.

Agriculture remains the predominant type of land use in the field of transmigration for various reasons. First, practically all transmigrants are peasants and want to remain so, though under better conditions. To achieve this they are prepared to migrate.[2] Secondly, historically or traditionally speaking, transmigration has always been a movement of peasants to new land, to their own holdings or plantations or both; and it is only recently that other forms of occupation and income have been considered. Thirdly, since Indonesia is still short of food — rice production, though generally on the increase, does not necessarily supply the internal demand — the government welcomes all efforts to expand the food producing surface of the Outer Islands. Thus it is understandable that transmigration has so far been primarily regarded as an activity of the agricultural sector. Only recently, when the handicaps and limitations became visible, have the politicians considered not only establishing transmigrants outside the agricultural sector, but also switching from a land settlement strategy to a policy of development planning, regional planning and watershed development outside Jawa.[3]

We have to remember that natural vegetation and animals form a symbiotic system of co-existence (a biotope) which, particularly in the tropics, consists of an unimaginable biological diversity covering thousands of different species. They all depend upon one another, thus forming an extremely complicated ecological system or equilibrium. As long as it remains undisturbed, none of the species will get the upper hand, the water economy and the humus and nutrient levels will remain stable and erosion will largely be prevented. Lowland tropical forests have survived under such systems for millennia, unless a natural catastrophe destroyed them and a new system had to develop.

The intrusion of man — biologically part of the ecosystem but thanks to his abilities in a position to subdue it to his own interests — threatened the natural equilibrium. As long as his activities were limited, the biotopes he entered changed only marginally. But as his numbers increased, so he started to eradicate species that were not directly useful to him (such as weeds) and began to rely more on his ingenuity than on the wisdom of nature. Consequently he had to develop costly techniques to replace the services that nature had provided. By destroying the habitat of certain animals whose principal source of food was insects he was obliged to use insecticides; by introducing monoculture he encouraged the spread of crop-specific pests, making the application of pesticides unavoidable; by clean cutting forests, doing away with fallow and cleaning the fields of residuals, the introduction of chemical fertiliser became necessary. These techniques often resulted in the destruction of the remaining systems, thus requiring the application of more and more techniques, more complicated chemistry, and so forth. In the industrialised countries we have now reached the point of reconsideration and are slowly trying to reestablish the natural systems neglected for so long.

When transmigrants enter the tropical rain forest — and the resettlement schemes mainly consist of such intrusions — the change to the ecosystem is enforced within a few months. We have dealt with the environmental consequences in detail earlier. Considering the fact that most of the soils in the Outer Islands are poor in nutrients, we can expect low and decreasing yields when we convert lowland rain forest into fields with annual crops. The sad experience of many transmigration projects shows that the choice of the site and its preparation before the arrival of the settlers left much to be desired. Generally it can be stated that the transmigration process gains its own momentum, so that the preparatory work remains insufficient. It is understandable that the growing population pressure requires quick decisions, but the conversion of tropical jungle, grass and swamp land into cultivated areas simply requires time and should not be done in a hurry.

The officials who are responsible for the proper implementation of the transmigration programmes do not play down the difficulties and failures. Quantitatively, it seems that the programmes have been realised, but qualitatively much remains unsatisfactory. The press does not evade discussion nor do the authorities; a healthy attitude that may help to improve the situation. Looking back to the achievements of REPELITA III (1979/80–1983/4), the government officials admit that they are not fully satisfied though 527,000 families have been resettled, which surpasses the planned target of 500,000 families. Over 2,000 families left the new settlements and returned to their original homes for various reasons, voluntary or otherwise, mainly because the land allocated to them was unsuitable for cultivation. Twenty-seven officials who abused their authority — resulting in a 2,000 million rupiah loss to the ministry — had measures taken against them. In carrying out the fieldwork, 193 contractors were involved in opening up new land and 560 companies involved in housing projects.[4]

More recently, even the President of the Republic warned the officials responsible not to ship transmigrants to their new homes before the sites were properly prepared to receive them. This is certainly a step ahead, but the environmental viewpoint still plays a marginal role. It is also wise to make use of the transmigrants' labour to clear the settlement areas where the use of heavy equipment is not unavoidable, and to train them before leaving for the new sites. Extension workers who are familiar with local conditions should be stationed in each transmigration area. However, environmental problems start with the selection of the site.[5] To facilitate this decision-making the Indonesian Department for the Environment recommended that:

The national system of reserves should hold lands representative of the major ecosystem types used in transmigration areas on each island. In addition, there is a need for recognition of special areas in transmigration planning and project management. These should include archaeological, religious and other culturally important sites, wildlife and fisheries, protected areas, and botanically important areas. Transmigration planning provides an opportunity for protection of such areas but also presents the hazard of unintentional damage, unless there is full appreciation of management needs for land and water reserves.[6]

Naturally, the conditions vary from site to site. If we take the resettlement in East Kalimantan, for example, the authorities have had the advantage of access to various research work already done by foreign experts over the last few years.[7] A few of the findings and recommendations may illustrate the environmental dilemmas which transmigration projects generally face.[8]

The preparation of 1.5 million ha. of land and the settlement of 500,000 families gives some idea of the scale of the project and of the management responsibilities involved in such an enterprise. Quite apart from the destruction of the ecological and meteorological equilibrium, it is possible that up to 20% of the settlers may abandon their fields after the exhaustion of soil fertility and leave it to *alang-alang*. Thus, quite a large area of primary forest would have been felled in vain. But since the felling of primary forest is officially no longer tolerated — which does not, however, prevent spontaneous settlers from encroaching on logged forests — secondary forests, devastated areas, fallow land and grassland should be prepared for the settlers, especially as it is estimated that these soils are better than others still covered by primary forest.

A number of rules must be followed when choosing sites and clearing land.[9] These include not clearing sloping areas above a gradient of 25°, leaving areas with infertile soils and hill tops under forest cover, and keeping at least 50% of the project area under protected forest to give a chance of survival to plant and animal species. Usually, the upper layer of soil or A horizon is, under tropical forest conditions, extremely thin and covered with leaf litter that decomposes quickly due to humidity and warmth. Once exposed to rain and sun, this moderate fertile stratum is rapidly destroyed unless it is properly protected. Therefore, experts recommend caution when cutting and removing the logs so as to reduce damage of the topsoil to a minimum. They also recommend that all the residual matter like branches, tops, bark and stumps be chipped and incorporated with the foliage into the soil where it will rapidly decompose and increase soil fertility. Similarly, large single trees and groups of trees should be left standing on the cleared areas to give shade, reduce wind damage, add litter to the ground and make the site more attractive.[10] Burning is strictly to be avoided, so as to help protect organic matter, and preserve the nitrogen and sulphur content of the soil.

It has to be admitted that there are certain disadvantages, the most severe being the fact that neither in Kalimantan nor anywhere else in the

During the initial years, tropical forest soil may produce good but decreasing yields if the site is well selected and the soil properly managed.

tropical rain forest biome has there been any large-scale experience with a non-burning technique — since the traditional procedure is slash-and-burn (shifting cultivation). The costs are much higher than those of the destructive methods involving heavy machinery and burning. Another potential problem is the regrowth of spontaneous vegetation which will certainly occur quicker than after burning, and has to be suppressed by cover crops and additional weeding. Moreover the tree stumps and wood piles may be a source of pests and diseases as well as a home for termites requiring further control. Yet all this has to be compared with the increased economic returns in the form of wood, improved soil fertility and reduced soil erosion. Extended tests should be made and, if successful, the method should become obligatory in settlement areas.[11]

Clearing the forest means entering virgin territory in more senses than one. In many cases we intrude into land which has never, or not for many years been used for agricultural production, and we try to develop a method or methods of doing this without creating environmental damage. Under extremely low soil fertility, low cation exchange capacity and low soil pH, the permanent cultivation of annual food crops may, if at all, be feasible only with the use of sophisticated farming techniques. This so-called eco-farming aims at the establishment of an ecological system as close as possible to the original, e.g. the rain forest. Research carried out on podzolic soils in South Sumatera showed that the cultivation of only one hectare requires an amount of labour far beyond the possibilities of a transmigrant family. Unfortunately, the majority of settlers are not familiar with the conditions under which they have to practise sustained dryland farming, and extension services are provided only occasionally. Therefore only simple techniques can be used, and the area under annual food crops has to be kept to a minimum.

In order to make use of the ground placed at the disposal of each transmigrant family (ideally 3.5 ha.), it is recommended that emphasis be given to tree crops which, in three to six years time, will supply a regular income to the farmer without too much field work.[12] The plantation of coffee, clove and rubber trees, as well as oil palms and an area reserved for firewood that can later be planted with fast-growing varieties such as *Calliandra, Albizzia* etc., would reduce not only the peasants' risk but also the ecological risk, because it creates an environment similar to the original one, protects the circulation of water and nutrients, and smoothes climatic fluctuations as it reduces soil erosion.

Areas developed as pastures when planted with elephant grass (*Pennisetum purpureum*) or leguminosae may well be included in transmigrants' land use programmes, though large ruminants have not been part of transmigration land preparation schemes until very recently. Land development could be much more successful if clearing procedures made better use of nutrients stored in the natural vegetation. Even if clearing cannot be achieved without burning the residues, the remaining litter and ashes could be better used as a medium for growing legumes which can be fed to cattle. This would result in a greater availability of animals for farm traction, a vital consideration where each settler family is given 2 to 5 ha. of land which they are unable to cultivate with family labour alone.[13]

Among the useful trees to be planted, certain palms may be recommended which provide either the raw material for handicrafts, like the rattan palm (*Calamus rotang*), or foodstuffs like the sugar palm (*Arenga*

The Javanese transmigrants, mostly experienced in growing wet-rice, were promised irrigated land in the new settlements (*above*), but found only dryland (*below*) and had to change to rain-fed cultivation.

saccharifera). In the extended swampy regions of Indonesia the sago palm
(*Metroxylon sagu*) also plays an important part in providing basic food.
Only recently, the government of Indonesia showed fresh interest in the
sago palm, a plant which may have been used in South-east Asia since
prehistoric times.[14] We dealt earlier with the characteristics of this plant,
but the sago palm offers other possibilities within the framework of
transmigration and land use. Certainly, sago growing areas are not easily
accessible to modern equipment, but for the local population this does
not present a problem. Moreover, the fact that the palm can be harvested
after only four to six years does not mean it is at a serious disadvantage
when compared to other tree crops. Undoubtedly the advantages out-
weigh any drawbacks. It grows well in swampy areas that can be
developed for other crops only at high cost; as a perennial, it is better
suited to humid tropical lowlands where permanent cropping of annuals
without irrigation remains a problem; and it also surpasses all other
crops in the production of food energy.

There is no doubt that the cultivation of sago should be encouraged,
perhaps as an additional crop, especially in areas where transmigrants'
plots include swampy land. Its cultivation as an additional crop should also
be promoted. M. Flach, of the Agricultural University at Wageningen in
the Netherlands, who has been conducting research into the sago palm,
sees many possibilities in the field of food and raw material supply from
this tree. However he warns that:

> The exploitation of natural stands of sago palms on any sizeable scale will have an
> effect on the natural vegetation. The natural swamp is one of the very few natural
> environments that has remained virtually untouched by man. In most such areas
> the local population lives, at a low level, in a unique harmony with this
> environment. If exploitation on a sizeable scale is undertaken, it should be
> accompanied and preferably preceded by research aimed at preventing the detri-
> mental effects of such exploitation.[15]

The majority of land allocated to new settlers will, however, consist
of rain-fed upland. Only a few of the transmigrants are familiar with the
cultivation of such land since they are either *sawah* farmers used to
irrigation facilities, or dryland farmers of comparatively fertile volcanic
soils. Without irrigation, low yields have to be expected, and many a
transmigration project has failed because the promised irrigation
facilities were never provided. Obviously every effort must be made to
make sure that in future the provision of irrigation facilities is given a
high priority when transmigration areas are selected and technically
prepared.

Irrigation water is usually supplied from surface resources or from groundwater. On Jawa, surface water is usually diverted from rivers and sometimes reservoirs are built. However the shifting cultivators on the Outer Islands did not use artificial irrigation. Thus, in most of the transmigration areas the peasants depend on rain. In some cases, they try to establish *sawah* cultivation within the framework of an integrated irrigation project which includes watershed protection. The World Bank-assisted Toraut Project in the province of North Sulawesi is one rare example, but it shows the difficulties arising when the indigenous population continues burning and cutting the forest cover.[16]

In fact a much easier technique could be used, namely the construction, on a local scale, of so-called 'hill lakes', whereby simple earth dams are used to bar a natural depression, thus storing rainwater in the wet season that would otherwise be lost. Such techniques — already implemented on Lombok — should be promoted wherever possible. In addition to providing irrigation, such lakes serve the ecological equilibrium in many other respects.[17]

Tube-well irrigation using groundwater is far more expensive, but the government has supported this technique since the end of the 1970s by setting a target of sinking 2,580 deep tube-wells in order to irrigate 169,000 ha. of land. Apart from Jawa and Bali, the drought-ridden islands of West and East Nusa Tenggara gain most from this policy. The two pre-conditions are, of course, that there is enough water in the ground and that the peasants are ready to cooperate. It seems that the people welcome the chance to irrigate also outside the rainy season and grow rice, tobacco, water-melons, soya beans and maize (often with several harvests). In these tube-well irrigation projects, the peasants are organised in water-user organisations (HIPPA) which act on their behalf in requesting extension services and credit. The peasants are at any rate happy to now have an abundance of drinking water. Lifting groundwater in substantial quantities may, however, also have environmental consequences. Therefore, control drillings have to be sunk in order to undertake consecutive groundwater balance studies to avoid any over-exploitation of the subterraneous reserves.[18]

The modernisation of agriculture includes a better use of potentials as well as protective measures, both in the interest of the individual peasant and the national economy as a whole. In many cases however, conflicts arose between long-established peasants and those who introduced new techniques,[19] particularly when these innovations required substantial alterations in people's surroundings (such as the construction of a water reservoir).[20] In such cases, next to the environmental impact, social conflicts also have to be observed and, where necessary, ameliorated.

After what we have seen in the field of land use and its environmental consequences in the Outer Islands, the opinion expressed by Soewardi *et al.* receive this author's support: 'There are only two choices in utilising marginal soils as are found in the Outer Islands — *swidden* cultivation or intensive farming.'[21] Shifting cultivation, though, becomes more and more unrealistic as the population density per square kilometre grows. Conversely, extended fields or plantations are not the best way to cultivate the difficult soils of these islands except in some particularly favourable areas. In order to make the best and most protective use of the environment and to resettle the maximum number of families, the intensely cultivated 'garden' with a large share of tree crops will be the most promising model. This model does not differ much from agroforestry, and this means it is as close as possible to the dense tropical vegetation that previously covered most of the land.

To implement all the sound recommendations which refer to the environment of the transmigration projects for the benefit of both farmers and the nation as a whole, the following recommendations should be kept in mind:[22]

1. More importance should be given to the selection of favourable transmigration areas. In particular, the regional ecology, soil fertility, and acceptable development possibilities should be carefully considered.
2. The selection of transmigrants is of paramount importance. It is not the quantity, but the quality of transmigrants that decides the success or failure of the programmes, above all their readiness to accept the new living conditions on the Outer Islands. Those who are willing should be carefully trained and prepared mentally as well as technically for the coming task.
3. Since it is now well known that deforestation for the sake of gaining arable land is accompanied by a great number of disadvantages in the field of climate, hydrology and soil, conservation methods should be elaborated and implemented, beginning with the clearing process and ending with cultivation practices. The settlers should be well educated and trained in these methods, and supported as well as controlled during the initial years of the scheme.
4. The economic success of a transmigration scheme largely depends on the necessary infrastructure such as roads and marketing, but also the provision of housing, drinking water, hygiene, health and education. As far as possible, such facilities should be provided before the settlers arrive;

much, however, can also be done with their participation.

5. A laborious life at subsistence level for the first few years often causes a certain quota of migrants to return to their previous homes. To reduce this rate to a minimum, transmigrants should not only be properly advised and prepared before they leave (see no. 2.), but supervision in successful agriculture, and long-term advice and control of agricultural activities in the field are indispensable. To this end, 'nucleus estates' should be established in areas of similar environmental type to carry through experiments, acquire experience, and advise settlers through an extension service. Establishing non-agricultural employment would greatly reduce re-migration.

6. An extremely high birth rate among the transmigrants on the Outer Islands as a whole might result in a 'slum effect'. Family planning has been slower here and the logistics of reaching a widely-scattered population will present quite a challenge. There is no evidence that the programme would face religious or socio-cultural obstacles on the Outer Islands, but dealing with transmigrant families requires a special empathy. It seems clear, however, that the focus of the struggle to hold down the size of Indonesia's population demanded an extension of the programme to those areas situated far away from Jakarta.[23] It is quite possible that the isolation of these areas causes small families to increase their numbers, and therefore everything should be done to supplement the usual family planning activities by additional social events in the field of culture, politics and sport.

We cannot conclude this chapter without stressing again that, next to a reduction of the population increase, the protection of the environment is the highest priority. Without preserving the natural potential on which they live, the people face a bleak future. We have seen the consequences of this in overpopulated Jawa which has, comparatively speaking, highly fertile soils. How much more should we worry about the future of the marginal soils of the Outer Islands now they are exposed to the redundance of immigrants? 'If soil conservation is not 100% effective from the moment of the opening-up of new lands, soil fertility will be lost, erosion will advance and food supplies will become inadequate in quality and quantity. Malnutrition will be the final harvest.'[24]

So far we have dealt with the transmigration programme and its environmental impact mainly from the viewpoint of agricultural land use. Obviously the Outer Islands offer other possibilities, for instance in the field of handicraft, industry, mining, services and the aquatic sector. Since, however, this study is not intended to criticise the transmigration

programme in general, we shall deal briefly with the non-agricultural
sector under a more general heading which includes Jawa with Indonesia
as a whole.

It should be reiterated that the carrying capacity of an area depends
greatly on incomes drawn from the non-agricultural sector once the land
is used to a great extent. Jawa is a splendid example of this. Had not a
substantial part of the largely rural population, namely people with little
or no land of their own, earned their living from non-agricultural activities, the present population density would have long ago led to a social
catastrophe. Unfortunately, recent trends in development give cause for
concern, as Horstmann points out:

Non-farming activities have a long history in rural Jawa. They had become an
indispensible means of accomodating surplus population where agricultural
carrying capacity could not be expanded any more, or where out-migration
could not offer sufficient relief. At present, conditions of labour-intensive rural
manufacturing have become critical. Its viability is endangered by the industrialisation process concentrating manufacturing in big labour-saving factories
at central places. The decline of handweaving in the Majalayah area south-east
of Bandung is one example of this. What will happen to the carrying capacity
and social conditions of other rural areas if the traditional manufacture of batik,
or of cigarettes, or of other commodities is 'modernised' and centralised? The
high population density of rural Jawa which had been made possible until now
by sharing poverty and by manufacturing activities would turn into unbear-
able overpopulation. Rural areas would become more agricultural again at the
expense of the carrying capacity, and the surplus population would not have the
possibility of out-migration — which in western countries had contributed
considerably to solving the problem when similar difficulties emerged at the
beginning of the first industrial revolution.[25]

The task of transmigrating surplus population will then have to
include not only farmers with little or no land but also craftspeople,
manufacturing hands who have lost their source of income due to
modernisation. For better land use on the Outer Islands, more and more
people will have to be occupied in non-agricultural activities. It is up to
the politicians to take care that the introduction of highly sophisticated
and labour-saving technology will not block this possibility right from
the beginning.

NOTES

1. In 1983 North Sulawesi refused to accept any more transmigrants from outside the province but has had to resettle 4,000 of its own villagers in safer and more fertile areas. This is also the case for another 4,000 families returning from the southern Philippines, where they worked in plantations. (*Indonesia Times*, 25 Oct. 1983).

2. 'The people most disposed to the idea of migration tend to be landless agricultural labourers who, although they may spend part of the year as wage labourers in urban areas, are essentially farmers. . . . The "unskilled labourers from the rural areas and cities of Jawa" . . . are no different from the "padi farmers" who have been moving as transmigrants in the past, with the possible exception of a small percentage of city-born labourers.' (Hardjono, 1978, p. 108).

3. Arndt and Sundrum, 1977.

4. *Indonesia Times*, 25 May 1984.

5. Loc. cit., 28 May 1984; 10 April 1984; etc.

6. 'Environmental aspects of transmigration' (quoted by Egger *et al.*, 1981, p. 61).

7. 'East Kalimantan Transmigration Area Development Project', a project for technical cooperation between the Republic of Indonesia and the Federal Republic of Germany, general publications and working papers from 1976/7 onwards.

8. Egger *et al.*, 1981.

9. Such rules are mentioned in many studies with varying emphasis given to specific techniques. See, for instance, Ross (1980, pp. 75–7), Hanson (1981, pp. 231–5), Rieser (1975a, chapters 8 and 11), Suratman and Guiness (1977, p. 81).

10. Particularly recommended are *Koompassia spp.*, a legume that absorbs nitrogen and therefore produces litter rich in this nutrient (Egger *et al.*, 1981, p. 61).

11. Loc. cit., pp. 71–3.

12. See also Blom (1979, p. 19): 'It is felt that tree crops might become an important component of settlement farming both in new schemes and older ones where their cultivation was initially not foreseen.'

13. Leake, 1980, p. 74.

14. *Nachrichten für Aussenhandel*, 23 March 1984.

15. *Ceres*, vol.17, no.4 (1984), pp. 3–4; Stanton and Flach, 1979.

16. Hamzah, *Indonesia Times*, 24 and 25 Nov. 1983.

17. Donner, 1964.

18. Hamzah, *Indonesia Times*, 29 Sept. and 1 Oct. 1983.

19. These can well be transmigrants who introduce the techniques of their homelands alien to the local people. This may lead to rejection as well as adaption in both directions as the case may be. See for instance Zimmermann, 1975c, pp. 97–108; ditto 1975a, pp. 115–16.

20. Burger (1980), describes such a case in eastern Central Jawa.

21. Soewardi *et al.*, 1980, p. 61.

22. See Egger *et al.*, 1981, pp. 112–13.

23. Chernichovsky and Meesook, 1981, pp. i, ii, 47.

24. Bailey and Bailey, 1960, p. 277.

25. Horstmann, 1980, p. 101.

Part IV. ENVIRONMENTAL PROBLEMS OF NON-AGRICULTURAL LAND USE

12

INDUSTRY

Growing population pressure and the competition for land have led to environmental damage and a shortfall in agricultural employment (particularly in Jawa). This in turn has led to demands for an expansion in the industrial sector to help create more non-agricultural jobs and to relieve some of the population pressure from the countryside. In fact, though industry in the modern sense does not have a long tradition in Indonesia, this sector has shown remarkable growth rates in the last few years. Ranking second after the construction business, the annual growth rate of manufacturing industry, fluctuating strongly, reached 16% in 1981 but fell to 1.2% a year later.[1]

The spatial distribution of industrial enterprises is unbalanced, as is the case in most developing countries. Regionally, more than 80% of all industrial firms, except wood-processing, are located on Jawa, and of these about half are situated in or around Jakarta and West Jawa. Food-related industries, though, are concentrated around Surabaya in East Jawa.[2] Indonesian industry covers many sectors including construction materials and tools, vehicles, ranging from bicycles to heavy road-construction equipment, textiles, chemical fibres, electronics, paper and packing material, to name just a few which have a promising internal market. For various reasons, the industrialisation policy aims at encouraging the establishment of industrial enterprises on the Outer Islands, first, to exploit the locally available natural resources, and secondly, to offer work and income to the growing population.

In fact, industries based on natural raw materials such as wood and petroleum have developed well in the last few years, or are at least well advanced in planning; they include a chemical industry in South Sumatera based upon the petroleum of Palembang; a fertiliser plant in Samarinda based upon the petroleum of Balikpapan and several plywood factories making use of the still abundant rain forests of Kalimantan. However, such quite large-scale enterprises usually require heavy investment and result in comparatively few jobs. Therefore, the industrialisation policy favours the idea of establishing small and simple industries based on the agricultural sector, such as the processing of milk, cassava, leather, rubber and wood and the production and repair of farming tools. This would not only offer work and income, but would also contribute to the creation of new non-urban centres in generally rural areas.[3] But whether it would be realistic to envisage small-scale industries on the eastern islands of the archipelago supplying the bigger enterprises in West and Central Indonesia is doubtful, unless a cheap, reliable, integrated sea-transport system is established.[4]

Looking back upon the industrial development of Indonesia, it is clear that the rise in manufacturing production did not significantly increase the level of employment in industry, since the new investment after 1969 was very capital-intensive. If we take for instance the net effect of the transformation of the textile industry, which formerly employed almost one-third of all workers in manufacturing, the picture is depressing. During the period between 1966 and 1971, when production rose from 250 to 600 million metres, textiles as a whole lost more than half of its workforce. In the handloom and batik sectors, all but 100,000 of the 510,000 workers lost their jobs. The ILO reported that more than 70% of the workers in both sectors were unemployed.[5]

At the beginning of the industrialisation boom, there was virtually no investment made in the agricultural sector, and the majority of capital-intensive enterprises and foreign investment became concentrated around Jakarta and in West Jawa. It was only after 1973 that the government started to persuade potential investors to invest more in the poorer provinces, but this can in no way be considered a significant contribution to the reduction of unemployment.[6]

The subject of this study, however, is not the manpower question but the environmental impact of industrialisation. Industrial pollution is by no means confined to Indonesia nor to the countries of the Third World. In fact, the terrible precedents set by the industrialised countries of North America, Europe and Japan cannot be overlooked. Yet, when in 1972 the United Nations held the First World Conference on Environment, many of the developing countries were reluctant to follow the warnings. They argued that they were too poor to invest in protective techniques and that the industrialised countries were using the environmental question to deliberately retard their economic growth. In the meantime, there is hardly a developing country that would not acknowledge the dangers emerging from careless industrialisation. Nevertheless a combination of poverty, a growing population and the wish to absorb the maximum number of people in the industrial sector has led to environmental damage affecting those whom industrialisation was meant to help.

Indonesia has certainly done its utmost to direct public interest towards this matter. Quite a number of top-level specialists like Emil Salim and Otto Soemarwoto used every opportunity to call for action, and their efforts have been supported by hundreds of trained environmentalists engaged in research and fieldwork. Finally, in 1982, the Indonesian parliament promulgated an 'Act on the Basic Principles of Environmental Management', providing drastic penalties for those flouting environmental safeguards.[7] However, enforcing such controls in a country of Indonesia's size is extremely difficult. Even where polluting industries are concentrated in one area such as the environs of Jakarta, identifying the specific plants which are causing environmental damage is not easy.

In 1980 the Indonesian press reported that furious peasants had burnt down a government-owned chemical factory near Banjaran, not far from Jakarta. The reason for this was that for three years the waste waters of this factory had polluted the irrigation canals of the nearby rice farmers and finally destroyed their fields. Inspite of public impositions, the factory had not stopped directing its waste water to the fields, and thus had to pay an indemnity to the aggrieved party.[8] Events of this kind are rarely reported because, unlike in this case, the link between cause and effect is not normally so clear.

In general, air pollution caused by industry is not yet seriously affecting the health of the people of Indonesia for the simple reason that reverse winds blow day and night in the archipelago. However, as soon as industrialisation and urbanisation coincide, air pollution accumulates:

The degree of air pollution due to industrial development varies according to the economic structure of the region. Urban areas are more likely to suffer from air pollution because industries are concentrated in those areas. Urban residents are therefore subject to air contaminants arising from power production, gas, petrochemical industries and motorised vehicles, which may also affect local vegetation, the climate, as well as human health. . . . Jakarta for instance with its 4.6 million inhabitants and its rapidly growing industrialisation might have to face the resulting air pollution in the near future if no preventive measures are taken.[9]

The pollution of water, on the other hand, is visible everywhere:

Public waters are further polluted by industrial waste. Most of the industries discharge their waste into public waters. Some plants are already equipped with waste water treatment and waste recovery units. However, not all industrial waste waters especially of chemical industries, which may require expensive processes and operations are treated. The water bodies where such waste is discharged may after a short or long time become contaminated.[10]

In the town of Tangerang west of Jakarta (and part of the JABOTABEK planning area), there are 1.3 million inhabitants and approximately 500 industrial firms. It has become a new settlement area for people working in Jakarta and, whereas the population of West Jawa grows by 2.6% a year, the figure for Tangerang is 4.5%. Apart from traffic congestion, and a lack of clean water and proper housing, the Ci Sadane river passing through the town is heavily polluted by waste waters from these industries.[11] There is no doubt that the rivers and canals draining the area of Greater Jakarta and flowing into Jakarta Bay carry with them an unprecedented quantity of industrial waste. Though the specialists still haggle about the level of contamination in the bay, frequent cases of brain disorder among children seem to indicate clearly that mercury pollution from industrial waste waters is taking place.[12]

The situation is similar to that of the Surabaya river — the lower part of the Brantas river — which flows through the town of Surabaya, also an industrialised agglomeration on Jawa. Detailed studies revealed that during the wet season, when the water discharge is high, namely between 50 and 130 cu. m. per second, there is sufficient dilution of the inflowing wastes. Though the quality of the river water from Sonokarimun downstream is inappropriate for fishing, domestic, and even industrial purposes, it can still be used to irrigate the farmers' fields. In the dry season however, when the water discharge ranges between 5 cu. m. per second and less, dilution is drastically reduced. Then, the river water is so heavily polluted both with domestic and industrial waste that it is inappropriate for any use, quantitatively and qualitatively.[13]

On the Outer Islands, industrialisation based on petroleum and wood is on the increase. Though these enterprises are occasionally visited by a group from the Institute of Technology in Bandung (ITB) who study the environmental impact of the oil and the plywood industries, they can hardly be sufficiently controlled. The numerous plywood factories of the Mahakam river in East Kalimantan use the river not only as a means of transport for the wood, but generally also dispose their waste water into it.[14]

The influence of the official environmental policy, however, becomes visible only slowly. Public efforts directed towards greater environmental education seem to be bearing fruit, ably supported by the cooperation of many Indonesian universities.

When researching pertinent reports, we find a number of exemplary enterprises that are managed in a very considerate way as regards the environment. One case in particular may be cited. P.T. Kertas Bekasi Teguh, a privately owned pulp and paper mill in Bekasi, another industrial centre east of Jakarta, established a chemical recovery plant and now reuses about 90% of the chemicals otherwise lost with the waste water. In view of the necessary investment, this case is regarded as unique in Indonesia's current anti-pollution environmental programme. Moreover, to ensure a supply of raw materials, the mill distributes the seed of *Albizzia* trees to farmers. In this way, it supports the government's regreening campaign and offers an additional income to the peasants when they sell the wood after five years. The mill also buys paper waste, bagasse (dry sugar-cane refuse), sawmill waste etc., and thus contributes to the recycling of raw materials and the reduction of air and water contamination.[15]

Apart from such spectacular cases however, environmental consciousness seems to have spread among industrialists. Many entrepreneurs planning a new factory now request the advice of environmentalists in order to avoid undue pollution of air and water. Similarly, whole industrial complexes that have developed all over Indonesia or are at least in the planning stage, sign agreements with Environmental Study Centres in one of the nearby universities. In a significant breakthrough, twelve important industrial enterprises in Tangerang, the highly polluted town mentioned earlier, asked the University of Indonesia, Jakarta, to study their output of contamination and to advise them on the necessary remedies. Emil Salim, who was present at the ratification, hoped for more than just the action of a dozen firms: 'This cooperation needs the support of all elements in this area, because this precious cooperation will be useless without the participation of the people of Tangerang.'[16] This is not an isolated example (see Table 13).

During the United Nations' Second World Conference on Environment held at Nairobi in 1982, Emil Salim explained that Indonesia intends to build up an industry by integrating environmental concerns into all its development efforts, thus avoiding the same mistakes as the older industrialised countries. To this end, the country should abstain from simply copying Western countries and instead aim at pollution-free technologies as far as possible.[17] This is certainly a bold plan, and we might wonder

Table 13. COOPERATION BETWEEN INDUSTRY AND SCIENCE
IN THE FIELD OF ENVIRONMENTAL PROTECTION IN
INDONESIA, MID-1983

Industry/Industrial complex	*University (Province)*
Lhok Seumawe	Syah Kuala University (Aceh)
Indarung	Andalas University (W. Sumatera)
Palembang, Baturaja	Sriwijaya University (S. Sumatera)
Cilegon	Pardjadjaran University (W. Jawa)
Cibinong	Technological Institute Bandung (W. Jawa)
Bekasi, Karawang, Purwakarta	Bogor Agricultural Institute (W. Jawa)
Gresik	Airlangga University (E. Jawa)
Bangkalan	Airlangga University (E. Jawa)
Leces, Probolinggo	Brawijaya University (E. Jawa)
Tonasa, Gowa	Hasanuddin University (S. Sulawesi)
Cirebon	
Cilacap	} Under preparation
Kupang	

Source: 'Industries to seek ways to solve pollution', *Indonesia Times*, 4–6 June 1983.

how far it can be realised. The important point, however, is that
environmental protection is no longer regarded as a diversion designed to
slow down the industrialisation of the Third World. Another important
point is Indonesia's decision not to become the 'dustbin' for dangerous
waste from industrialised countries. Third World states with weak
environmental laws attract industries that had to be closed down at
home because of pollution. They are also a highly-prized dumping ground
for waste from the towns and dangerous industries of the North,[18] and it
is to be hoped that Indonesia remains strong in resisting such temptation.

NOTES

1. *Mitteilungen der Bundesstelle für Aussenhandelsinformation*, Jan. 1984.
2. Schmidt-Ahrendts, 1979, p.131.
3. *Indonesia Times*, 4 May 1981 and 9 Sept. 1981.
4. Loc. cit., 21 Dec. 1981.
5. Palmer, 1980, p. 227; see also Fremerey, in Nohlen and Nuscheler, 1978, pp. 178–97.
6. Palmer, 1980, p. 229.
7. *Indonesia Times*, 29 June 1982 and 10 July 1982.
8. *Frankfurter Rundschau*, 21 Feb. 1980.
9. Karimoeddin, 1973, p. 175.
10. Loc. cit., p. 174.

11.   *Indonesia Times*, 4 June 1983.
12.   Loc. cit., 5 July 1982.
13.   Wardoyo *et al.*, in Furtado, 1980, pp. 629–33.
14.   *Indonesia Times*, 19 March 1982.
15.   Loc. cit., 29 April 1983.
16.   Loc. cit., 4 June 1983.
17.   Loc. cit., 17 May 1982 and 21 June 1982.
18.   Loc. cit., 6 June 1983.

13

MINING AND ENERGY

Although mining is one of the economic activities which most severely violates the earth's surface, and usually produces an enormous quantity of waste in the form of removed overburden, waste dumps, slag heaps and tailings, there seems to be little awareness of the resultant environmental damage in Indonesia. In a UNESCO-sponsored seminar on ecology and environment held in Bogor in 1974, some very general guidelines for the exploitation of mineral resources were put forward:

1. The rate of exploitation of any mineral resource should be proportional to the amount of known deposits; this would spread benefits to future generations.
2. The effects of mining operations in terms of scenic values should be considered as one of the social costs of the project.
3. Plans for mining development should be subjected to a thorough ecological analysis.
4. Part of the revenue of mining should be devoted to the report or research and development of recycling technologies. This could be done through a depletion tax.[1]

If one follows the Indonesian press reports over the years, it is remarkable to observe that — despite a great involvement in the environmental problems of agriculture, urban areas, industry and the aquatic sector — the dangers emerging from mining are hardly dealt with. This is surprising since mines and processing units for basic raw materials are renowned as a source of highly poisonous water contamination caused by ore tailings and all kinds of chemicals. Cases are known from other parts of the world where such waste rendered irrigation and drinking water projects impossible, but 'notwithstanding their output of poisonous agents, industries gaining foreign exchange have always had priority to food production and public health requirements of the population.'[2]

Mining in Indonesia dates back to 1709 when the rich, predominantly alluvial tin deposits on the islands of Bangka, Belitung and Singkep were successively exploited.[3] Nowadays, these reserves show signs of exhaustion. Government efforts helped to increase the declining production up to a figure of 35,000 tonnes of tin concentrate (1981), but reserves are expected to be exhausted within forty years. At present, in pure metal production, Indonesia ranks third after Malaysia and Thailand with about 14% of the world production.[4] It is impossible to predict the longevity of Indonesia's tin production since this depends not only on the actual ore deposits and their metal content, but also on the available techniques and the price paid for tin in the world market. In the past, the ore on the islands of Bangka and Belitung had a tin content of 70%, and 1,000 cu.m. of washed sand produced 6 quintals of ore. Such deposits are now rare. Thus, tin mining proceeded on the tiny islands of Singket, Karimun and Kundor off the province of Riau, but the sands bring only 1.8 to 3 quintals of ore per 1,000 cu.m. with a comparable tin content.[5]

Since the environmental impact of mining, and particularly of tin mining, has obviously not yet received much publicity, we may instead refer to the cases of Thailand and Malaysia, where mining is carried out under similar conditions.

Tin mining needs vast amounts of water to wash out the ore, and this water, laden with mud, has to be disposed of. For many years, observers have repeatedly warned that these tailings from the tin mines consist of sterile material which, when washed over farmers' fields, can do great harm. Moreover formerly important ports have been silted-up and abandoned over a period of years. Even when care is taken to guide the tailings into the open sea (using specially dug channels) heavy rainfall can easily destroy these precautions, and upon arrival in the coastal waters, the sediment endangers the lives of marine creatures. Referring to observations in southern Thailand, where tin mining is a long-established and economically important industry, Pendleton states that:

One of the principal problems of the tin miners is to dispose of the 'slimes' or tailings, particularly the clays suspended in large quantities of water. These clays, being the result of profound weathering processes and having for the most part developed deep below the surface of the ground, contain at best almost none of the plant nutrient materials liberated from the weathering of the rocks nor any plant residues which would supply some plant nutrients. So not only is the plant-nutrient content of these clays very low, but their physical condition is not appropriate for upland crops. For paddy growing, the alternation of these slimes is relatively easy. The main danger here is that if the slimes get to going down in quantity with the water, they will rapidly bury the paddy soil too deeply; besides, the very fine mud will damage the growing paddy plants.[6]

In Malaysia, environmental damage as a result of rapid industrial development becomes more visible every day: 'The combination of a major shipping lane and the industrial zones makes Penang one of the most industrialised areas in Asia and one of the most polluted. . . . Rapid deforestation, indiscriminate mining practices, poor planning and accelerated expansion are taking a toll on the country and its resources. . . . Penang has a double environmental threat. In addition to industry and the tin mines, the Strait of Malacca is one of the world's busiest waterways. . .'[7] There is no reason to believe that the situation in Indonesia under comparable conditions is much different.

Ooi Jin-Bee has written a very detailed analysis of the environmental consequences of mining, particularly of tin mining, in the Kinta Valley upstream of Telok Anson, Malaysia, where mining has been going on for a long time already.[8] Mining usually starts with deforestation, and this may even exceed the area necessary to be worked, since additional wood may be needed to supply poles and timber for gangways, scaffolding, living quarters and many of the other features found in a mining landscape. There is an even greater demand for wood when the miners smelt the ore on the spot, and since hardwood provides an excellent raw material for charcoal, there is a great waste of valuable timber because large tracts of forest are cut down in order to extract the isolated species of hardwood. With the introduction of machinery, the forests are further denuded of hardwoods to be used as fuel for the engines employed in the mines. Thus, no matter whether mining is done in a traditional or a modern way, vast areas become denuded, and the consequences of this denudation will be all the more serious because mining is usually concentrated on hillsides and undulating land. Thus, soil erosion is initiated and accelerated on slopes formerly protected by a mantle of natural vegetation.[9]

Mining activities as such contribute to environmental destruction in a similar way by ground sluicing on hillsides and by the discharge of tailings from the mines into the drainage system of the valley. Washing out the ore from the waste into which it is embedded requires an enormous quantity of water to move the waste, inflicting severe scars on the landscape. Once the accumulated topsoil and most of the weathered material beneath is washed away — leaving only the core-boulders and partly weathered rock — the damaged landscape that results takes a long time to recover, but will probably never return to its original state. All these activities result in a constant discharge of large quantities of silt into the drainage system of the area, ranging in texture from coarse sand to fine slime, so that eventually the river beds are raised above the level of the surrounding country. Floods, formerly exceptional, now become the normal concomitant to heavy rainfall in the catchment basin, depositing extensive areas of alluvium in the lower reaches, and often creating almost permanent swamps.

Although efforts to restrict environmental damage in Malaysia date back to 1879, it is clear that these regulations were made solely to prevent the total devastation of the whole countryside by clearance of the protective forest-cover and by the haphazard discharge of silt. However the regulations have not led to a conflict between mining interests and the various government departments concerned with land, agriculture, forestry and drainage. We should therefore discuss two particularly sensitive phenomena.

First, one consequence of unrestricted mining is the reduction of once flourishing agricultural areas to desolate wastes not only by the conversion of agricultural land into mining concessions, but also because rivers distribute enormous quantities of sand and silt from the mines over the land. Ooi reports of a tributary of the Kinta river where a well-kept rubber estate had been reduced to long stretches of *belukar*, dead trees, and swampgrass, while the lower reaches of the river had turned into muddy lagoons. In a similar way, rice fields are rendered unproductive due to a layer of sterile slime being deposited via streams and canals from the mines. Secondly, the destructive impact of the waste waters and tailings from the mines by no means ends once the material reaches coastal waters. On the contrary, long-term observations on both the eastern and western coastlines of Thailand show that the mangroves as well as the corals are threatened and have already been partly destroyed by these wastes.[10]

Once mining has taken place, reclaiming the land for agriculture or at least for forestry is an extremely difficult and lengthy process. It also depends on the type of mining methods used. Where the land has been worked open-cast (and this is most often the case),

the resultant landscape has a hummocky appearance and consists of a series of abandoned mining pits, generally filled with water. It is usually difficult and economically impracticable to restore such land to its original level state by filling up these holes, although in places these pits have been filled with tailings discharged from neighbouring mines. This serves to restore the land as well as to absorb effluent that otherwise would find its way into the drainage system of the region. . . . On such land, denuded of its forest cover and often consisting of unconsolidated material, erosion is extremely rapid, especially when the mining site is situated on a steep slope. . . . Although coarse vegetation in time establishes itself on such land, the process of regeneration is so slow that the earth remains in a barren condition for many years.[11]

When the ore is produced by dredging (and this is the case in more than half of Malaysia's tin-mining), the environmental impact is no less dangerous. During a dredging operation, a large volume of alluvium from 15–45 metres in depth passes through the various stages of treatment from puddling to the final concentration of the ore. During this process 'all the organic matter in the original soil is lost, and the fine slimes jettisoned by the dredges contain very little vegetable matter, so that there is practically no humus in the soil of a newly dredged area. Such land is, therefore, of little value agriculturally. The natural succession of plant regeneration will eventually induce sufficient fertility to support a moderately good plant growth, but not for more than a century will this process be complete.'[12]

Obviously any form of mining, particularly open-cast and dredging, will seriously affect the environment near the mine, and though we have no access to recent studies undertaken in this field in Indonesia, we can assume that there are conflicts developing between the interests of mining on one side and forestry and agriculture on the other. The expansion of coal mining on the Outer Islands may easily threaten the expansion of agriculture under transmigration programmes (though the forest is the victim of both).

Mining coal on Sumatera dates back to 1846, and two of the earliest mines to be started (Ombilin in West Sumatera, 1892, and Bukit Asam in southern Sumatera, 1919) are still the most productive in Indonesia. In 1982, total coal production amounted to 481,000 tonnes for the whole of Indonesia, but the government's ambitious plans foresee an annual production of 17 million tonnes of coal during the 1990s.[13] This target is based upon more recent discoveries of coal deposits in Sumatera and Kalimantan (in particular) and also on expectations of Irian Jaya's potential. Estimates concerning coal reserves in Indonesia are continually being revised upwards and so are targets for coal production. In 1983, reserves of 22,500 million tonnes were envisaged, but at the same time the domestic demand for fuel coal grew in such a way that the authorities concerned feared the country would soon have to import coal unless the national production could be increased substantially. This situation has indeed encouraged foreign investors to engage in joint ventures for the mining of the newly discovered deposits in South and East Kalimantan. Since these mining activities will mostly be open-cast, enormous areas will be deforested and the ecology destroyed. The new contract for South and East Kalimantan coal mining refers to three separate areas covering together some 2.8 million ha.[14] Fortunately, before the agreement was signed, the Indonesian Parliament's Commission VI, responsible for industries, mining and energy, insisted on a number of conditions, one of them requiring that 'damage of environment and nature must be avoided' and 'nature conservation should be observed'.[15]

It is certainly encouraging that groups of scientists from time to time study what is going on in the new coal areas where the entrepreneurs want to make their profits, and the government urges them to fulfil production figures. Though it is not customary for exhausted mining areas to be regreened, hopefully the responsible authorities will not overlook the fact that devasted areas of such dimensions must inevitably have a severe impact on the surrounding area's vegetation, soil and hydrology.

Peat is another source of energy which although there is plenty available in Indonesia has to date hardly been used. The amount of peat in eastern Sumatera, southern Kalimantan and Irian Jaya has been estimated at no less than 17 million ha., and Indonesia ranks fourth in terms of world peat reserves. However this is only a rough estimate, for, in order to be commercially viable, peat layers must have an organic content of more than 65% and must be more than one metre thick.[16]

In view of energy requirements (in particular for the villages), the need for jobs, and the scarcity of land for transmigrants, Indonesia's peatland seems to offer opportunities that should at least be seriously considered. All the forms of energy being developed in Indonesia require high capital investment, high technology and a low labour input, and though peat can be developed in the same way, it is also possible to do it on a small scale, at moderate cost, with low technology and a high labour input. In this respect, peat offers an ideal basis to promote the energy development in the rural areas. Only about 15% of the 57,000 villages in Indonesia have electricity, and most of these are situated near the cities.[17] Experiments with peat-fired generators are already taking place with West German assistance, and Finland is to provide power plants running on gazified peat.[18]

The peat areas of Indonesia have been regarded as worthless in the past, and consequently have been avoided in transmigration plans. Nowadays however, the idea is gaining ground that if the peat could be extracted and used economically, not only could a valuable source of energy be exploited, but the land thus reclaimed could be used to settle transmigrants. Since this technique has been used for many years in the Dutch *veenkoloniën* (fen-colonies), it is practicable, but experience with tropical lowland peat is limited, and pilot schemes are necessary. This would at any rate be a form of mining that changes a landscape without destroying it completely.[19]

Oil production in Indonesia dates back to 1872 and had reached an important position in world terms before the Second World War. From the 1940s onwards the Indonesian oil industry experienced an impressive boom. This was achieved with the cooperation of more than forty foreign firms, mostly from the United States and Japan. Pertamina, the national trust for oil and gas, has developed as a 'state within the state' and runs its own fleet of tankers, aeroplanes, petrochemical plants and steelworks, and has many other business interests, to some extent carrying out its own independent development projects.[20] Today, petroleum concessions cover many islands, shelfs, and sea areas from the coastal waters of western Sumatera to the jungles of Irian Jaya, including parts of the South China Sea. Thus, Indonesia has developed as the largest petroleum producer east of the Persian Gulf, surpassed only by the People's Republic of China. In 1980, Indonesia produced 77.5 million tonnes of petroleum or less than 3% of total world production, but ranks thirteenth on the producer's list. Production and processing of natural gas has experienced a similar boom in the last few years.[21] In 1980, Indonesia produced 12,000 million net cu.m. of gas, less than 1% of total world production, but it has become one of the most important producers and exporters of liquified natural gas (LNG). Today, certain parts of Indonesia, particularly in Sumatera and eastern Kalimantan, are covered with drilling rigs, pipelines, refineries and oil and gas terminals. Production figures do occasionally stagnate though, and official policy, for good reasons, lays stress on the development of non-oil energy resources.

Oil extraction and oil processing industries in Indonesia have in the course of their development made use of more technology than other branches, especially when they started off-shore drilling. Legislation to protect the environment can be traced back to the year 1927 when the 'Storage of Oil Ordinance' came into force. This was followed in 1930 by the 'Mine-Police Regulation' and in 1967 by the 'Basic Mine Law'. Although this legislation is by no means perfect, it could serve as the basis for further improvements. Obviously the environmental impact of the petroleum and natural gas industries is monitored by various bodies. The Institute of Oil and Gas established a Study-Group on Pollution with the task of collecting data through surveys and research, while Pertamina set up an agency to coordinate all activities regarding oil pollution.[22] Some would doubt whether an agency established by the polluter itself can be very effective; on the other hand it is an interesting indicator of how environmental consciousness is spreading in this important field.

In concluding this section it should be reiterated that in general mining has an enormous impact on the environment, from the violation of the earth's surface and vegetation right down to the underground hydrology. Indonesia is well on its way to becoming an important mining state, and industrialised Asian countries like Japan and Taiwan as well as non-Asian states seem all too keen to plunder its non-renewable resources in the shortest possible time. These range from giant bauxite and aluminium smelter complexes like the Asahan Project in North Sumatera[23] to the working of simple quartz sand in East Kalimantan which the Taiwanese want to supply their glass industry with.[24] Certainly, Indonesia is happy to attract so much foreign capital, but it should remain careful not to sell out its natural inheritance for short-term financial gains.

In terms of energy supply, Indonesia has a multiplicity of resources. Petroleum, natural gas, coal and peat have already been mentioned, but there is another more uncommon source found there: geothermal energy. At the end of 1980 the Philippines had 446 megawatts (MW) of installed geothermal generating capacity, ranking second only to the United States in this type of power generation. In contrast Indonesia is lagging behind, though Jawa alone has some 112 volcanoes, 97 of which are regarded as inactive. The country has instead concentrated on the development of petroleum and natural gas, but since the government's policy now wisely aims at developing energy resources, the exploitation of geothermal energy has assumed much greater importance. It is conservatively estimated that a potential 10,000 MW of power could be derived from geothermal steam, 5,500 of which would come from Jawa, 1,400 MW from Sulawesi and 1,100 MW from Sumatera. Bali and Lombok are also thought to have potential geothermal reserves.

To avoid the dangers deriving from active volcanoes, drillings for underground steam are usually confined to so-called inactive volcanoes, particularly in areas of high rainfall — since rainwater is the source of groundwater and ultimately of underground steam. The only establishment now actually producing geothermal power is Kamojang, situated south-east of Bandung in West Jawa. It uses a crater of 14 sq.km., 1,500 m. above sea-level on the western slopes of Mt Guntur, a volcano that last erupted in 1885. After surveys had started in the early 1970s, the first production phase was inaugurated in 1983 with a capacity of 30 MW; expansion plans aim at 140 MW by 1987. The Mt Salak project is under way about 65 km. south of Jakarta, and will eventually produce 220 MW of power.

In terms of the environment, geothermal projects can be run without major problems apart from the dangers normally associated with volcanoes. One disadvantage, however, is that places with a geothermal steam potential are usually found in remote, less populated areas, so that the steam has first to be converted into electricity and then transported to the user. Fortunately, the electricity network on Jawa covers practically the whole island, so that each geothermal generator could feed its energy into it. However falling petroleum prices mean that cheap oil is available for domestic consumption, and under such conditions there is neither the means nor the political will to invest in geothermal energy. Rising petroleum prices, however, encourage the government to expand oil production and thus neglect geothermal energy once again. It seems that Indonesia has nevertheless decided to develop this free gift of nature as a source of energy that produces hardly any environmental damage.

Yet another possible source of energy which Indonesia could exploit is the sea. One such programme is being developed using 'Ocean Thermal Conversion (OTEC)', which attempts to generate electric power from the temperature difference in and on the water surface. Indonesia is making preparations for such an OTEC programme and two systems are being studied with Dutch and West German technical assistance to tap this source of energy that can be used without destroying the environment.[25]

As in many developing countries, the bulk of Indonesia's population lives in thousands of scattered villages, hamlets and individual farmsteads. Supplying them with energy is, therefore, a completely different task from electrifying a town. Since no developing country can afford to build a compound network connecting all villages to a central generating station, the only solution is to install decentralised units to supply a village or a group of villages; in many cases even farm-owned generating units are necessary. As part of Indonesia's drive over many years to bring energy to the farmers, most of the schemes developed have promoted 'alternative sources of energy'. Such techniques are usually based on the exploitation of renewable energy resources, particularly waste material that would otherwise be lost. With British and West German technical aid, often extended by volunteers living in the villages, various trials are under way to use solar energy for the production of drinking-water and ice and for cooking in remote places.[26] On a much wider scale, the use of the biomass for energy generation is also spreading.

It is generally agreed that most organic waste, agricultural as well as silvicultural, is simply squandered. In some countries, plant residues are fed to animals, but in most cases they are left *in situ*. This is particularly true for silvicultural waste. After logging, the branches and tops are left in the forests and all kind of residues from the saw mills are normally just left rotting on the spot. The volume of waste being produced in Indonesia is believed to amount to 25.4 million tonnes.[27] Various Indonesian research bodies like the Bandung Institute of Technology deal with the conversion of this organic waste into gaseous or liquid fuel for diesel or other engines, and it may be added that certain weeds growing in abundance in the tropics, for instance water hyacinths (*Eichhornia crassipes*) and even filamentous algae are good sources of energy production.

The development of wood-fired gasification schemes in pre-1939 Europe[28] is one example of the extent to which alternative energy resources can be found, given enough urgency.[29] Environmentally the use of vegetal waste for decentralised power generation does not seem to pose much of a problem — unless we consider the competing demands for residues to be composted and given back to the soil, fed to animals or used as fuel for generators. However this complex has not yet been given much thought.[30]

It is not only developing countries without fossil fuels like coal or petroleum, or with very small reserves of this kind such as Brazil, but also countries which do possess such natural wealth that think of developing alternative, renewable energy. Indonesia is one of these. At present, there are thirteen plants which convert sugar-cane into ethanol, mainly on the Outer Islands. In 1984 the first plant for the commercial production of alcohol was set up at Tulang Bawang, in the province of Lampung, Sumatera (with Japanese technical assistance), to convert cassava and sweet potatoes into ethanol, dry yeast solids, and high fructose syrup. This plant is designed to process 100 tonnes of tubers a day, and between 4,000 and 4,500 transmigration peasants living not far away are expected to supply the raw material. The planners have praised the advantages: a steady income for the farmers from a formerly 'worthless' crop, local production of fuel and the introduction of technology to the villages.[31] Given the fact that many transmigrants to Lampung hardly had a chance to grow rice and were forced to clear uplands for the production of less valuable upland crops, the arguments may be correct. Such a plant might, on the other hand, cause a certain amount of social disruption and unrest in the surrounding areas, and once the local people cannot produce enough food themselves, they will become dependent on their additional earned income to purchase food at higher prices from outside. An ESCAP study warned of drastic declines in food production and the raising of cash crops to bring in much needed foreign exchange. It explains that the large-scale cultivation of sugar-cane, cassava, maize and other cereals convertible to alcohol can lead to a reduction of farm inputs needed for food production; quite apart from this it means a diminution of the amount of land planted with food crops.[32]

According to ESCAP, the 1980s can be regarded as an 'energy transition period'. There is a growing realisation that in the future we will need renewable energy to replace shrinking reserves of fossil fuels, quite apart from the fact that burning these fuels places a heavy burden on our environment. In addition, developing countries with a scattered population need decentralised energy sources. This presents a challenge for the industrialised countries to support Indonesian technicians and scientists in their search for the right solutions.

NOTES

1. Sumardja, 1974, p. 25.
2. Schädle in Opitz, 1984, p. 224.
3. For the geological background see Kanayama, 1973.
4. *Metallstatistik*, 1983.
5. *Indonesia Times*, 16 Dec. 1983.
6. Pendleton, 1953, p. 225; see also Donner, 1978, p. 457.
7. Hollie, 1981.
8. Ooi, 1955.
9. Loc. cit., p. 46.
10. Donner, 1984, pp. 60–61; see also Soponkanaporn, 1979.
11. Ooi, 1955, p. 43.
12. Loc. cit., p. 44.
13. Katili, *Indonesia Times*, 18 Oct. 1983.
14. *Indonesia Times*, 3 Nov. 1981.
15. Loc. cit., 23 Aug. 1981 and 24 Sept. 1981.
16. Katili, loc. cit.
17. *Indonesia Times*, 24 Sept. 1981.
18. Loc. cit., 4 Feb. 1984.

19. Bronsgeest, 1979, pp. 60–67.
20. Röll, 1979, p. 165.
21. See for instance, Kositchotethana, *Indonesia Times*, 7 Nov. 1982.
22. Karimoeddin, 1973, p. 163.
23. Munthe, *Indonesia Times*, 9–11 Aug. 1983.
24. *Indonesia Times*, 24 May 1982.
25. Loc. cit., 27 Jan. 1983.
26. Loc. cit., 20 Nov. 1981 and 22 Jan. 1982.
27. Imanudin, *Indonesia Times*, 9 July 1981.
28. Wind, 1936, p. 445. It is remarkable that many Indonesian authors refer to the European experience of using wood gasification to drive lorries before World War II, when the difficulties of petroleum supply in case of war were being anticipated. In France, for instance, forestry publications pleaded in 1936 for *le gaz des forêts'* as motor fuel, and due to overproduction of fuel-wood the French government supported the conversion of trucks and tractors to wood-gas and even experimented with it to power locomotives on the state railways. In a *'Rallye Internationale Rome-Paris'* in 1935 over 3,500 km., wood-gas-driven cars overcame heavy rains in the Alps and extreme heat in the Appennines. 'The use of wood-gas', wrote a Dutch-Indonesian forestry magazine, 'does not pose any more problems. The experimental stage is behind us and we may assume that, in view of the growing danger of war, perfection will soon be reached.'
29. *Tectona*, 29 (1936), p. 782. It is astonishing to recall that when the International Forest Conference was held in London in 1936, the German delegation of three arrived having driven their wood-gas-driven car right from Berlin, and having paid no more than 9 marks per person for the whole trip.
30. Imanudin, loc. cit., Hadering, *Indonesia Times*, 29 March 1984.
31. Simanjuntak, *Indonesia Times*, 29 March 1984.
32. Vacharapongpreecha, *Indonesia Times*, 31 March 1984.

14

URBANISATION

The Malay, who comprise the main ethnic group of the Indonesian people, have hardly any urban tradition. Their traditional living space is the *kampung* (village) and not the *kota*, a word for town derived from Sanskrit. Even in the modern congested towns, a majority of Indonesians continue to live in their *kampung* on the outskirts, even when surrounded by concrete buildings. Historically speaking, we can distinguish three main roots of the modern Indonesian town. The conquest originating from the Indian sub-continent led to the creation of the early Hindu-Javanese empires with their residential towns centred on the *kraton* or princely court. The colonial Europeans developed western-style residential and administrative quarters and, finally, Chinese immigration, already documented in AD 414, resulted in extremely crowded housing and business quarters, often the only urban part of the settlement.[1, 2] For a long time, the towns grew on a moderate scale, and although centralising tendencies could be observed some 350 years ago, they never posed a problem. Even in 1930, there were only seven towns in the whole of Indonesia with more than 100,000 inhabitants; Jakarta (then known as Batavia) had somewhat more than 500,000.

European-style housing and urbanisation were introduced without considering the climatic conditions of Batavia.[3] As a result, this led to a series of epidemics over two centuries until Governor-General H.W. Daendels (1807–10) introduced a more healthy type of garden city, located south of the swampy coast on more elevated land. This led to a growing space consumption for European housing in all towns where foreigners constituted a certain proportion of the population. In Bandung, for instance, 12% of the inhabitants were European, 10% Chinese and 77% Indonesian. Their consumption of the available space for housing purposes in 1930 was 52%, 8% and 40% respectively.[4] This pattern remained largely the same until the end of the Second World War when, following the one-sided declaration of independence and the subsequent four years of fighting to expel the Dutch, an unprecedented rural exodus took place. Thousands of families moved to Jakarta, Surabaya, Bandung and other towns, and in the nation's capital, Jakarta, the population grew by 70% between 1949 and 1950.[5] Obviously the towns which became the targets of the migrants were in no way able to cope with the new problems. Till that time the administration had had to deal only with upper and middle-class residents who earned a decent income, paid taxes and did not pose serious quantative problems. After the first wave of immigration was over, however, the flow of rural people to the urban centres, and in particular Jakarta, did not diminish and has continued to the present day. Consequently, there has been a slow but steady growth of the urban population in Indonesia (see Table 14). Before going into the environmental details of this process, it is interesting to compare the Indonesian situation with that of South-east Asia in general.

Table 14. POPULATION TRENDS IN MAJOR INDONESIAN
TOWNS 1930–83
(× 1,000 inhabitants)

	1930	1961	1971	1980	1983[a]
Jakarta	533	2,973	4,567	6,503	7,636
Surabaya	342	1,008	1,269	2,028	2,289
Bandung	167	973	1,154	1,463	1,602
Medan	77	479	620	1,379	1,966
Semarang	218	503	633	1,027	1,269
Palembang	108	475	614	787	903
Ujung Pandang	85	384	497	709	888
Malang	87	342	429	512	560
Padang	52	144	187	481	726
Yogyakarta	137	313	394	399	428
Banjarmasin	66	214	277	381	437
Pontianak	45	150	194	305	355
Bogor	55	154	186	247	274
Cirebon	54	158	187	224	273

[a] Official estimates for 31 Dec.

Sources: Kantoor voor de Statistik, Indisch Verslag, 1938; Population Censuses 1961, 1971; The Far East and Australia 1984–5, London, 1984.

South-east Asia is characterised by a low overall level of urbanisation, but at the same time by a high degree of primacy. This means that only a comparatively small percentage of the total population of a country lives in towns, whereas the metropolis may be many times the size of the second largest city. Greater Bangkok's population is 34.90 times the size of the second largest town (Chiang Mai), and 62.7% of Thailand's urban population live in the capital. This picture is not valid for Indonesia. In 1980, Greater Jakarta was only 3.2 times the size of the second largest town, Surabaya, and only some 20% of Indonesia's urban population live in the capital city.[6] This seems to be a tremendous advantage, since it means that the problem of urbanisation is spread more evenly over several towns, resulting in more or less manageable problems in each of them.

In fact, until the early twentieth century the urban population was well distributed. According to the 1920 census, 6.6% of Jawa's population lived in cities. In 1930, 8.7% lived in 102 communities of a 'more or less urban character' and over half of the urban population lived in cities with more than 100,000 inhabitants. This pattern changed however when the urban population of Jawa started to grow more quickly than the total population of the island, and it has to be stated that the larger the city, the quicker the growth.[7] Moreover, during the 1960s and 1970s interprovincial migration also began to increase. The main targets of the movements were, after transmigration projects, the Javanese towns, and in particular Jakarta, to which 40% moved immediately, while others reached the capital city after first settling in smaller towns. 'While the population of Jakarta is only one-third of the total urban population of Jawa, Jakarta received 75% of the lifetime migrants, and 69% of the total migrants going to urban areas in Jawa, Jakarta has therefore received a preponderant share of migration to urban areas.'[8] Earlier it was mentioned that Indonesia has a fairly evenly distributed urban population. However this advantage is only relative.

The problems which Thailand is facing in Bangkok are now beginning to confront the Indonesian government in a similar way in Jakarta and other big towns. The number of people living in Jakarta grew six-fold within no less than thirty-three years, reaching a population density of 11,000 persons per sq.km. This has led on the one hand to social disequilibrium and criminality, poverty and juvenile delinquency, and on the other to a concentration of capital and economic activities, thus increasing the attraction of the city.[9]

This study is not the place to deal in detail with town planning and housing in Indonesia, but rather with the environmental problems resulting from urban land use. Foremost among these is the need to provide the newly arrived masses with accommodation and services like a clean water supply, and waste water and rubbish disposal. These problems are exacerbated by air, water and noise pollution and, as the case may be, natural calamities such as floods, volcanic eruptions or earthquakes.

It is evident that none of the Indonesian cities was prepared to properly absorb the newly arrived rural migrants as far as employment (income) and housing (including public utilities) were concerned. Once the natural absorption capacity was exhausted, the newcomers established themselves in makeshift huts and slums located along railway tracks and waterways. 'Growing urban congestion has led to death rates much higher than in rural areas where the food is more varied and healthful and sanitary conditions are better. The sheer impossibility of providing employment, shelter, potable water, and other necessities to the new-comers has created social and economic stresses which have manifested themselves in high crime rates, prostitution, and juvenile delinquency.'[10] In response to this overcrowding, a determined family planning campaign was launched among townspeople and migrants which yielded spec-tacular results. The birth rate fell from 44 to 33.5 per 1,000 and the death rate from 17 to 10.5 per 1,000 (as the result of better general health care programmes run in conjunction with family planning schemes), while the number of family planning acceptors surpassed 500,000. Yet the head of the Jakarta Family Planning Coordination Board 'describes the population problem in Jakarta as intractable. The urbanisation process had speeded up and according to a recent census, the population in the cities had increased by 24% in the last year. Jakarta was still a favoured destination and its population increased by 3.99% per year.'[11]

Various surveys have shown that the motive behind the rural-urban migration is primarily the lack of employment in the village (push factor) and not so much the attraction of city life (pull factor). The growing rural population faces the exhaustion of virgin land suitable for culti-vation and the lack of alternative employment in industry or services. Though such 'migrants on the average are not contributing more than their share of poverty in the city'[12], they feel that they can make a living in the town rather than in their village, though no doubt for some time the quality of life would be worse, the air would be less clean, they would probably have to sleep on the pavement and there would be no relatives or friends close at hand.[13] 'The most positive [aspect] of the slum community', writes Krausse, 'is to provide the city with an abun-dant supply of manpower'; and it is certainly true that migrants often spend more time at work than the native citizens,[14] since they are despe-rate to gain a firm foothold in the city. The picture of the migrants' socio-economic absorption is an extremely heterogeneous one. Papanek even writes of a 'culture of poverty' that developed in Jakarta. He distin-guishes between the poorest of the poor (scavengers who live by collect-ing discarded paper or cigarette butts), and the richest poor (traders,

Even in the modern, congested towns, the majority of Indonesians continue to live in their *kampung*.

peddlers, construction workers), with many variations in between. When, in the early 1970s, the municipality tried to make the town 'beautiful and orderly', it was estimated that there were 45,000 street vendors, 30,000 loafers or *gelandangan* (see Litahalim, 1983) and 10,000 prostitutes forming part of the 'informal sector'.[15] It is therefore not usually the 'formal sector' — meaning properly established employers or any other registered business — that absorbs the newcomers, for the simple reason that the number of formal working places is not growing at the same speed as the number of job seekers. In addition, most of them cannot offer any qualifications whatsoever. Rather it is the 'informal sector' that absorbs the bulk of unemployed migrants without capital or professional qualification. But the city does offer a chance to those with a better education who have no prospects in the rural area. And since the migrants leave their villages not for higher wages but because they have no prospects at all at home, most of them will never return even if their situation is initially worse.[16]

As soon as a migrant family succeeds in establishing itself with a hut and a tiny piece of land in a *kampung* another form of employment emerges. 'Subsistence production' is the third way of earning a living, whether it be by vegetable-growing, raising chickens, fishing, housing construction, or waste collection. This income, though of the utmost importance to the urban population, is not recorded and therefore never appears in any official statistics.[17] These hidden earnings created in the informal and the subsistence sector explain why, despite low calculated per capita incomes, nobody actually perishes. We have described this situation in some detail to hint at the enormous potential and the will to survive among the urban masses, since urban environmental problems will never be solved unless this potential is made part of the efforts. There is little hope that this pattern of urbanisation will change in the foreseeable future. On the contrary, in 1983 analyses by ESCAP concluded that by the year 2000, six out of every ten Asian city-dwellers will live in slums or squatter areas, unless governments curb the massive rural exodus by taking the necessary measures in the countryside.[18] Emil Salim, Minister of State for Population and Environment, summarised the Indonesian situation as follows:

First of all, Indonesia's population continues to grow at a rate that does not show any drastic abatement in the forseeable future. Our population growth-rate at present stands at 2.3% per annum and even if we could manage to lower this to 1.7% per annum between now and the year 2000, by that time our population will be an estimated 217 million. This means that within the next two decades we must provide for the livelihood, food, and housing for an additional 50 million people. On top of that, at an annual rate of 4% (higher in some places) our cities are growing faster than the rest of the country. Up to now the large majority of our population lives in the rural areas but this pattern is bound to change and we anticipate that by the year 2000 one-third (about 71 million) will be city dwellers and the remaining two-thirds will remain in the villages. But even now the increased flow from the rural areas into the cities is noticeable.[19]

This uncontrolled influx into the towns has led to an expansion in the amount of densely populated sub-standard housing. In Jakarta, 43% of the housing area is illegally occupied and thus without public utilities. A survey in 1968 showed that 47.9% of the newly erected homes were temporary buildings, 28.5% semi-permanent and only 23.6% were built to last. More than two-thirds of the city's inhabitants live in the first two categories of housing, and an estimated 100,000 have no fixed abode whatsoever. They sleep under bridges, on the pavement and the like.[20] The congestion of these quarters is often underestimated. A survey in Yogyakarta showed that densities — referring to administrative boundaries — fluctuated between 3,600 and 25,900 inhabitants per sq. km. In the built-up areas, this figure ranges from 7,500 to 43,300.[21] In attempting to solve the problem of low-cost housing, the authorities have been unable to meet the growing demand either quantitatively or qualitatively. It cannot be overlooked that construction programmes meant for low-income families have actually been occupied by the middle-classes while the living conditions of the poor masses have hardly changed.[22] The reason for this is the rather costly design — even if the authorities decided to switch to cheaper multistorey buildings with flats which the slum-dwellers usually do not accept.[23]

A far more promising scheme that is being developed is the Kampung Improvement Programmes (KIP). Up to 70% of Indonesia's city-dwellers live in *kampungs*, which despite the heterogeneous origins of their inhabitants usually have a strong community spirit. The programmes aim at mobilising people concerned for self-help activities to become more aware of the *kampung* as a social unit of human beings. '*Kampungs* are more than just residential areas', states Ismartono, 'they are also places of work for the street vendors, store-keepers and serviceman and employees of small industries. This rich mixture of diverse socio-economic groups is made possible by a sense of interdependence, reinforced by the spirit of *gotong royong* — mutual aid and cooperation — and is therefore not only accepted as a reality, but also as a desirable form of urban settlement in Indonesia. The government's policy on *kampungs* in the overall urban development plan has been to maintain and improve them, rather than demolish and replace them with new settlements.' And 'although the results of the self-help approach had sometimes fallen short of city planers' expectations, it was not only cheaper, but the sense of pride and achievement that resulted from the community's participation often motivated residents into embarking on other projects. They improved their homes, built community centres, or organized social activities to improve the *kampung's* communal well-being.'[24] According to the Director of Housing, the KIP had affected over 3.5 million people living in 200 towns throughout the country by early 1984. The provision of housing, water-supply and sanitation for low-income groups is as far as possible synchronised with the government's policies on agriculture, population, budgeting, credit, employment opportunities, technological development and social welfare.[25] Certainly, self-help cannot cope with major investments such as the supply of clean water, but it can provide useful support for public efforts. It is well known that clean water is the prerequisite for healthy living since water-borne diseases constitute the main hazard for children as well as adults. Therefore the urban administrations do try their best to expand their water-supply systems, but despite all efforts the population growth continues to outpace the construction of water-treatment plants and pipes. Between 1969 and 1982, the supply of clean water financed by the national budget grew from 9,000 to 33,888 litres per second and now reaches approximately 36% of the urban population. If we add non-governmental supply as provided by industries or ports under provincial budgets, about 16.9 million people or roughly half of the urban population is reached.[26]

Clean water depends not only on processing and distribution, there must also be enough 'raw' water available. This has proved to be a serious environmental problem in Jakarta. The possibility now exists that its water supply will dry up sometime in the future. During heavy rainfall the canals cannot carry the water away quickly enough, main thoroughfares are flooded to a depth of 40 cm., traffic breaks down, life and property are threatened — nobody would ever think that the waterworks could fall short of its raw material having secured a reliable water supply. But in Jakarta the canals and waterways are often so full of rubbish that during heavy rainfall the water overflows and causes flooding, thus putting an intolerable strain on the already overburdened water supply.[27]

However, since piped water does not reach every part of the town, industry and housing projects rely more and more on their own water supply. This means that open wells are dug or artesian wells sunk — sometimes to a depth of more than 150 m. — to be independent of the municipal supply. According to the *Indonesia Times*, Real Estates, 'which are constructing new houses and residential areas in Jakarta, have resorted to sinking artesian wells for their water-supply. . . . There are more than 2,500 such wells in Jakarta.'[28] Unfortunately, with the increasing use of deep artesian wells, the volume of underground water is gradually decreasing. Every now and then the water-supply dries up at certain levels, forcing deeper and deeper drilling for water. The problem is that as soon as the water-table dries up, sea water seeps in.

The Directorate of Geology, Bandung, in a study of Jakarta's fresh water resources, concluded that whereas during the Dutch colonial period fresh water could still be pumped from the Tanjung Priok area near the coast, the fresh water limit has now receded some 10 km. inland. In 1979 the salt water had penetrated to a line from Pedongkelan in the west to Ujung Krawang in the east of Jakarta (see Fig. 27).

Usually, groundwater pumped to the surface is replaced from the surrounding layers in due course. It seems, however, that environmental destruction is already hampering this process. Quite apart from the fact that the immediate surface of the capital has been covered with asphalt and concrete and thus lost its absorption capacity for rainwater, the area that supplies most of Jakarta's water has been damaged in another way. The forests in the district of Bogor are subject to human negligence and

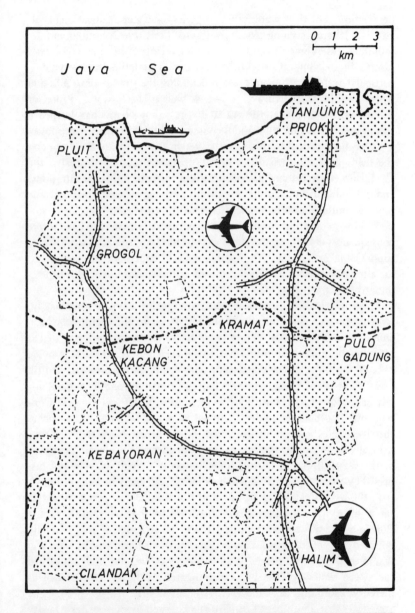

Fig. 27. APPROXIMATE LINE OF SALT WATER INTRUSION INTO
JAKARTA'S WATER-TABLE

Source: Indonesia Times, 25 Jan. 1982.

destruction. The river water which supplies the water purification instal-
lation in Jakarta, coming from the Bogor forest, the Ci Lutung and the
Jatiluhur reservoir, is constantly contaminated by industrial and
human waste and littered with rubbish.[29] The Jatiluhur reservoir in
particular suffers both from erosion denuding the surrounding hills and
from the influx of fertiliser residues as well as waste water. Professor
Soemarwoto stated that the water in the storage basin has reached a
condition of eutrophication and is consequently unfit for many purposes.
This has been caused by the waste water from the city of Bandung and
the fields used for agriculture in the upstream catchment area of the
Ci Tarum. Nitrates and phosphates have reached high concentrations,
and in combination with a large amount of organic erosion material, the
oxygen content has been reduced to zero.[30] This is an interesting example of how geographically remote environmental damage can have severe
consequences elsewhere. It seems that in the long run, the clean water
supply of Jakarta can only be secured if a safe catchment area is reserved
for a treatment plant and the processed water then transported via pipelines to the capital.

The similarly difficult task of urban sewage disposal is, unfortunately,
largely left to local residents to deal with. Plans for this sector have been
considered only recently, and it is indicative of the low priority that is
given to this field that, until 1990, the government is aiming to provide
only 20% of the urban population with waste water disposal. Up till
now, all types of waste water including human excreta have been discharged into natural or artificial watercourses, rivers and canals.[31] Consequently, both household and industrial waste water usually reaches
the rivers and the open sea untreated, where it may be diluted and thus
neutralised, as long as the inflow does not surpass the cleaning capacity
of the water. In some coastal areas, however, where water is not dispersed by the action of wind and currents, it can easily become polluted.

In the second half of 1983, the deaths of a number of people who had
eaten fish caught in Jakarta Bay caused great public concern. Earlier, a
Burmese newspaper had reported that: 'Jakarta Bay is more heavily
polluted by metal wastes, such as mercury, than Japan's Minamata Bay
was in the 1950s.' It referred to an Indonesian environmentalist who had
found the waters of the bay contaminated by heavy metals such as
mercury, cadmium, zinc and lead. A Japanese specialist observed at the
same time that a number of fishermen's children living near Jakarta Bay
exhibited symptoms similar to 'Minamata disease'.[32]

However, scientific opinion in Indonesia was divided over the precise source of the pollution — though it was agreed that the situation was potentially catastrophic. As well as heavy industrial residues, there are dangers posed by the dumping of non-industrial waste and household rubbish into rivers. The effects of this are skin irritation, damage to the waste water system and the fishing grounds, and last but not least the drinking water supply, since public wells are often sunk only 10 or 20 metres away from the river.[33] Cleaning of waste water is only done in exceptional cases, though in the tropics a system of biological waste treatment, using the prevailing high temperatures, could be practicable.[34] In Jakarta, some 17,000 cu.m. of organic and non-organic rubbish are produced every day, exceeding the disposal capacity of the municipal cleaning service (only about 12,000 cu.m. can be collected by the city's 3,000 workers, and the 400 or so waste disposal trucks often get stuck in traffic jams). An analysis of the city's rubbish revealed that 76.4% was domestic, 14.4% came from markets, and a mere 3.6% from industry. Though this composition would be ideal for conversion into fertiliser, such techniques are used to only a minor extent; the bulk of the rubbish is eventually dumped into disposal pools.[35] It would be worthwhile considering methods of composting urban rubbish, possibly combined with sewage sludge from waste water treatment plants, in order to support the recycling of valuable components of urban waste back to nature.[36] Since Indonesia hopes to provide 60% of the rural population with rubbish disposal by 1990,[37] and may invite private companies to take over at least the processing of urban waste,[38] there is hope that in the long run the more dangerous parts of it may be neutralised and the rest converted into fertiliser, building materials or electric energy.

Finally, road traffic is another important source of pollution as far as noise, air pollution and immediate danger to life is concerned. Traffic jams nowadays occur on all roads and at all times of the day in Jakarta. In January 1982, the capital had 876,784 registered motor vehicles, a figure which grows by 15% per annum, whereas the road network (3,600 km.) expands by only 0.4% a year. In Jakarta there are 4 million commuters out of a population of 6.5 million, and traffic jams are said to cause a loss of 7 million man-hours a day. Consequently, traffic moves extremely slowly in the built-up areas of the city, emitting far more carbon monoxide and noise than would be the case at normal speed. To give an example, the Intercity bus between Jakarta-Cililitan and Bogor covers the distance of some 50 km. in 15–20 minutes. A city bus, on the other hand, covering a distance of some 7 km between Cililitan and the centre (Merdeka Selatan) takes almost one hour.[39]

Traffic moves extremely slowly in the built-up town, producing more carbon monoxide and noise than in the case of normal traffic flow.

The congestion of motor vehicles results in a severe concentration of air pollution, particularly during peak hours, and especially in the main thoroughfares (Jl. M.H. Thamrin, Jl. Jendral Sudirman and Jl. Kramat). Public buses are another source of air pollution. When overburdened, they emit greater amounts of carbon monoxide, and though most buses keep within the legal limits, their simple weight of numbers makes the air in Jakarta oppressive. The Minister of Health admitted that some public transport passengers fainted or became sleepy due to carbon monoxide,[40] and the minister, Emil Salim, warned in 1982 already that air and water pollution in Jakarta will reach a dangerous level, unless appropriate measures are taken immediately.[41] Traffic accidents are also numerous in Indonesia though there are only 263 road deaths per one million registered vehicles compared with 385 in the USA and 500 in West Germany.[42]

Indonesian as well as foreign observers agree that the environmental problems in the towns and particularly in Jakarta require enormous public investment in water pipes, sewage disposal, rubbish collection and the necessary processing plants. But there is also agreement that individuals as well as neighbourhood groups could do much to ease the situation. S. Hamzah has written in the *Indonesia Times*:

Everyone is a rubbish producer, how careful he is with his littering will depend on his social status, individual discipline and his neat living habits. Now everyone is aware that careless littering can pose dangers to human life. When rubbish is disposed of at improper places, many problems will occur. Flood and diseases are just a few examples of how dangerous the problem is. But, most of the Jakartans and other cities' residents are likely to be tolerable in living amidst scattered rubbish, murky water and polluted air. Such sights are common in most sectors of these cities. Usually, trash is regarded as useless, and so tiny attention is paid to handle it properly. . . . Rubbish is scattered everywhere and just mounted on river banks and in canals. Despite of his bad habit, everyone should be fully aware of how poor his discipline is in taking care of rubbish that he produces.[43]

Another Indonesian writer has concluded that:

An important instrument is the creation of public awareness of the harmful effects of pollution. One has to bear in mind that pollution has not only direct consequences upon the life of man, plants, animals, corrosion of buildings, but more important are the indirect consequences upon social nature. Technical problems though sometimes costly, are not so difficult to solve, but it is the social aspect of the problem that is really difficult to tackle. One way of doing so is to create public consciousness at all levels. It is in that field also that we want the cooperation and support of the developing countries either in the method or means of communicating.[44]

However the authorities have had some success in raising people's awareness of environmental issues by organising competitions with an environmental theme.[45]

The technical starting-point for any improvement is the construction of a basic infrastructure, particularly in the *kampungs* and the slum areas. This was done over a ten-year period in Jakarta and the World Bank acknowledges that 'nearly two-thirds of the slum areas have been improved by means of infrastructural investments leading to a reduction of overpopulation and a better provision with public services'.[46] Critical observers, however, ask what really has been done for the poor. 'It is the question whether the rapid increase in the number of motorcycles after the improvement of the *kampung* can really be regarded as an indicator for better living conditions since it means noise, air pollution and a danger for children as well.'[47] They also criticise the fact that the affected social group had not been activated to identify itself with the improvement. People were not ready to take care of the water-taps and drainage systems, but rather suffered from the loss of their cheap accommodation during the process of improvement.

This helps illuminate a very crucial point. Lasting improvements can only be attained if the people concerned participate right from the start, and if the planners abandon western patterns as a model. Therefore, as mentioned earlier, *kampungs* or similar units may be the right framework to work from, but it is important to educate people and to strengthen inner structures such as *kampung* councils and *gotong royong* which are threatened by modern influences.[48] Another important fact, however, is that an improvement in the living conditions depends more on employment (income) than on the environment. When looking at the *kampungs* and slums we tend to forget that the quality of housing in the western sense does not have such a high priority in hot countries.[49] Some observers criticise any form of slum improvement and instead demand the removal of the social factors which gave rise to the slums.[50] This is easier said than done. Officials who are responsible for the survival of the poor in the towns have practically no influence over the reasons that make the migrants come. The officials could 'close' the towns, and in 1970 the Governor of Jakarta actually did so, banning newcomers who did not have work and accommodation. This proved completely unsuccessful: the daily average of 648 new migrants did not diminish and this law is no longer enforced.[51]

The administration is finally coming to terms with the fact that the process of urbanisation cannot be stopped. The development of 'urbanised regions' has been admitted,[52] and knowledgeable experts now construct theoretical models of a 'Javanese megalopolis' which, by the year 2000, might comprise an urban population of 90 million living in a network of cities along the northern coast of the island.[53] To some extent, a step in this direction can already be observed. In view of the meagre relief that transmigration projects may bring, the urban development strategy for Indonesia also aims at incorporating smaller towns into the inevitable urbanisation process: 'The middlesized and smaller communities also have a vital role to play. If handled correctly, they can become a vehicle for the development of their hinterlands, centres for the expansion of employment opportunities in activities such as agrobusiness which do not require the environment of the metropolis. More than anything else this may help to ease the pressure on the larger cities.' In fact, between 1971 and 1980 towns like Medan and Palembang in Sumatera, Balikpapan and Samarinda in Kalimantan, and Bandung, Cirebon, Semarang and Surabaya in Jawa showed a considerable annual increase.[54] Little by little, the network of towns becomes denser (see Fig. 28).

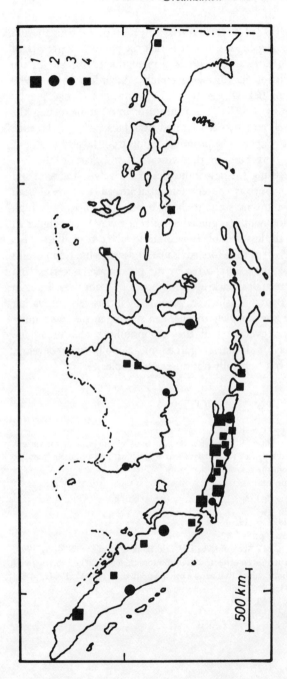

Fig. 28. DISTRIBUTION OF THE MAIN URBAN GROWTH CENTRES IN INDONESIA, 1980

1 = Cities with more than 1 million inhabitants
2 = Cities with 500,000 to 1 million inhabitants
3 = Towns with 250,000 to 500,000 inhabitants
4 = Other towns with central and growth functions

Base map: Rutz, 1976 (simplified).

298   *Land Use and Environment in Indonesia*

The JABOTABEK plan for Jakarta is yet another contribution towards a Javanese megalopolis. It is true that the population growth of Jakarta City was greater in the 1960s than in the 1970s, but in the environs the population has increased rapidly. According to the 1980 census, the town of Bekasi grew by 3.60% annually, Tangerang by 4.07% and Bogor by 4.59%.[55] In a measure aimed at increasing the amount of industrial employment located near Jakarta City, the then Governor sought to expand the planning region from Jakarta to Bogor in the south, to Tangerang in the west and to Bekasi in the east (JABOTABEK), settling the new industries exclusively in the west and the east (leaving the south as a purely residential area and a source of clean water). Since these towns seem to be growing together, it is high time that this development was guided in a planned way.[56] The bigger an urban unit grows, the less possible it becomes to take care of all its needs in a centralised way. It is therefore advisable to decentralise responsibilities into manageable units, supported by the population concerned and with financial and technical back-up being provided from the centre. In the same way that a rural smallholder survives by intensively cultivating his plot of land, so the *kampung* dweller can survive in the town quite decently if he takes over part of the responsibilities that urban life requires. Ultimately, a Javanese 'garden megalopolis' may develop, inhabited by poor, but hopefully not miserable people.

NOTES

1. Lehmann, 1936, pp. 109–10.
2. Wijoyo, *Indonesia Times*, 9 July 1982. Because the Chinese element for various reasons wants to disappear in the Malayan crowd, Chinatown still has a character quite different from the rest of the city.
3. See Cobban, 1976.
4. Lehmann, loc. cit., p. 119.
5. Messmer, 1979, p. 44.
6. See also Yeung, 1978, p. 19.
7. Van der Kroef, 1956, p. 747.
8. Sundrum, 1976, p. 81. According to a World Bank estimate in 1984, by 1990, Jakarta, Surabaya and seven other towns with more than 500,000 inhabitants will together house about half of Indonesia's urban population. (IBRD, 1984a, p. 8).
9. Atmavidjaja, 1983, pp. 74–5.
10. Pelzer in MacVey, 1963, p. 20.
11. *Indonesia Times*, 20 Jan. 1984.
12. Krausse, 1979, p. 67.
13. Temple, 1975, p. 80.
14. Krausse, loc. cit.
15. Papanek, 1975, pp. 1–27.
16. Sethuraman, 1975.

17. Evers, 1981*a* and 1981*b*.
18. *Indonesia Times*, 3 Sept. 1983 and 9 Sept. 1983.
19. Salim, *Indonesia Times*, 1 June 1983.
20. Grimm, 1976, p. 89.
21. Yunus, 1978, p. 35.
22. Fremerey, 1978, p. 192.
23. Litahalim, *Indonesia Times*, 30 May 1983.
24. Ismartono, in *Indonesia Times*, 6 July 1982, describes such activities in Surabaya and Yogyakarta.
25. *Indonesia Times*, 17 Jan. 1984.
26. *Business News*, 6 April 1982. According to the World Bank, in 1980 the urban water supply was provided in the following ways (percentage figures): pipe 26.4, pump 11.7, well 52.7, spring 2.7, others 4.8 (IBRD, 1984*b*, p. 42).
27. *Indonesia Times*, 16 May 1984.
28. Loc. cit., 25 Jan. 1982.
29. Loc. cit.
30. Thijsse, 1974, p. 9.
31. In 1980, 29% of Indonesia's urban population had private toilets with septic tanks, and another 17% had toilets without septic tanks. The rest, 54%, 'share public or other facilities' (IBRD, 1984*b*).
32. *The Guardian*, Rangoon, 5 Aug. 1983.
33. *Indonesia Times*, 10 Sept. 1983; 29 Oct. 1983; 2 Feb. 1984.
34. Suwarnarat, 1979.
35. Hamzah, *Indonesia Times*, 8 Aug. 1983.
36. The advantages of such techniques for tropical countries has been described by Reichling in Egger *et al.*, 1972, pp. 242–60.
37. *Business News*, 6 April 1983.
38. *Indonesia Times*, 2 Aug. 1983.
39. Ryanto, *Indonesia Times*, 1 July 1982.
40. *Indonesia Times*, 2 Oct. 1981.
41. Loc. cit., 31 May 1982.
42. Ryanto, loc. cit.
43. Hamzah, *Indonesia Times*, 8 Aug. 1983.
44. Karimoeddin, 1973, pp. 162–3.
45. See for instance *Indonesia Times*, 29 April 1983 and many current reports.
46. IBRD, 1979, p. 95.
47. Messmer, 1979, p. 49.
48. Messmer, loc. cit.; Lindauer, 1972.
49. Messmer, loc. cit.; p. 54.
50. Seidensticker in Egger *et al.*, 1972, pp. 238–41.
51. Yeung, 1978, pp. 22–3.
52. Horstmann, 1982.
53. Thijsse, 1975*b*. This compares with the megalopolis in the eastern United States where, between Boston and Washington DC, 40 million people crowd together; or Japan, where there are 40 million between Tokyo and Osaka; or the megalopolis spanning London, Hamburg, the Ruhr and Paris, with its 60 million people. Even the water supply for these people presents a serious problem.
54. *Indonesia Times*, 7 Jan. 1984.
55. *Neue Zürcher Zeitung*, 6 March 1981.
56. *Indonesia Times*, 23 June 1983.

15

THE AQUATIC SECTOR

Though there are old Indonesian songs which celebrate the exploits of brave seafaring ancestors, the Indonesians of today are not very attached to the ocean. This seems strange if we consider that the country has a coastline of some 81,000 km., and that it has more sea than land in its national territory. Indonesians frankly admit that the country suffers not only from the lack of an efficient maritime network between the many islands, but also from the lack of the necessary marine science and technology; the country has hardly any personnel trained in marine biology. To remedy these deficiencies — of which the government is well aware — a National Committee for Ocean Technology has been established which prepared a master plan for REPELITA V (1984/5–1989–90) dealing with subjects like food, energy, minerals and the control of marine pollution. Substantial assistance has been extended by the United States[1] and the Netherlands.[2] One exception is off-shore oil drilling. In this field, the Indonesians have accepted modern technology, and whereas in 1970 oil was only produced from land-based units, in 1981 no less than 34% of Indonesian oil came from off-shore drillings 30–150 m. deep.

The traditional way of using aquatic potential is fishing. Indonesia has access to considerable stocks of fish of various species. The country's territorial waters cover vast shelf areas as well as deep seas, coastal swamps, brackish water stretches and substantial inland waters. After declaring an Exclusive Economic Zone (EEZ) on 21 March 1980, the government claimed territorial rights over some 5 million sq.km., maintaining that this zone could yield around 5.8 million tonnes of fish per annum, providing work and income for roughly 1 million fishermen.[3] So far, however, no more than 1.7% of this potential has been used.

The prime fishery zone, the so-called upwelling areas, cover roughly 750,000 sq.km. Its annual potential is estimated at 4.2 million tonnes of fish — a substantial proportion of the total resources. These coastal waters have always been the main fishing grounds of the artisanal fishermen. Not surprisingly, the warm tropical waters support a multiplicity of at least 200 varieties of fish. The important species for fisheries are sardine, mackerel, Chinese herring, skipjack and jack mackerel. Others include bluefin tuna (*Thunus thynus*), milkfish (*Chanos chanos*), coral fish

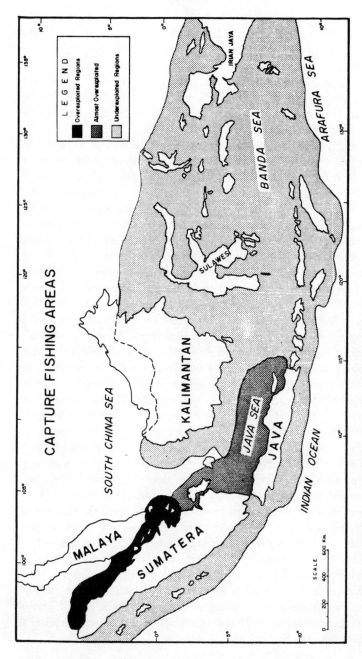

Fig. 29. UTILISATION OF THE INDONESIAN FISHING AREAS

Source: Collier et al., 1977.

(*Pomacanthus*) and sea bream (*Abramis brama*), but there are also sharks (*Selachii*) and rays (*Raja spp*.). Many tuna species are migrating to the Maluku and Banda seas, especially in the wide upwelling areas, whereas the Indian Ocean south of Jawa seems to be the spawning ground for the bluefin tuna. Among the crustacea, shrimps are the most important species. Shrimp grounds stretch along the coasts of Sumatera, Kalimantan, Jawa and Sulawesi; however substantial resources can also be found along the south coast of Irian Jaya.

Fishing is an important economic activity in Indonesia for the simple reason that it provides a source of protein that is cheaper and more evenly available than meat, and because it provides work and income for a fishing-dependent population of approximately 5.6 million people. There are about 1 million fishermen, and 90% of them pursue so-called 'artisanal fishing' in simple boats with simple equipment, whereas the remaining 10% are 'commercial fishermen' who work in larger boats with more modern technology.[4] Because the demand for fish and crustacea is growing, and because the fishermen are technically confined to a rather limited sea area, parts of the Indonesian waters have already been overfished;[5] the bulk, however, is still underexploited (see Fig. 29). To make full use of these areas, it is necessary to improve professional training and to equip the fishing boats accordingly. In order to establish a major fishing industry, much more investment is needed, particularly in cooling facilities for the storage and transportation of the fish. The representatives of Indonesian fishing interests are well aware of these constraints and have, from time to time, invited foreign counterparts to participate in the exploitation of their maritime wealth. Moreover, in the 1970s the government tried to encourage Indonesian fishermen to use larger, more efficient trawlers.[6] The danger signs, however, became apparent very quickly. Foreign (mainly Japanese) fishing firms were only too keen to exploit Indonesian waters to supply the insatiable Japanese market with fish and shrimps. Japan, which was then operating 82 fishing boats in Indonesian waters, requested a license for 300.[7] However a presidential decree, no. 39/1980, prohibited all foreign trawlers from operating in Indonesian waters after 1982, and stiff penalties were introduced for those ships caught violating the new laws.[8] An agreement was reached with Japan for joint efforts to develop new fishing nets capable of reducing hauls of fish other than shrimp.[9] It seems that the government wants to prevent overfishing by commercial interests and, at the same time, protect the livelihoods of the artisanal small-scale fishermen and those who run fish and shrimp farms.

It is obvious that the growing demand for fish and seafood in the industrialised countries — and Japan is one of the biggest consumers — encourages all sorts of entrepreneurs to procure fish and shrimps in whatever way they can, poaching not excluded. It is to be expected that such entrepreneurs, notwithstanding the fact that they form joint ventures with Indonesian firms to satisfy the law, are not very interested in sustained yields from foreign fishing grounds, but merely in immediate catch and profit. It is not surprising that the Indonesian Minister of Justice has thus urged caution in jointly exploiting the nation's maritime resources with foreign parties. He admitted that his country's ability to handle the fishing technology is still low and that Indonesia needs technical assistance from abroad. He also added a word of caution: 'We must learn from experience in dealing with foreign parties', he said and, to cite a sad example, he mentioned the deplorable exploitation of Indonesia's forests in the past: 'At the time, Indonesia was known for its abundance of dense forests; as a seemingly inexhaustible source of timber which on the international level has been dubbed *green gold*. Cooperation with foreign parties has resulted in the mass destruction of our forest, while reafforestation went very laxly.' Therefore, Indonesia does not want to incur incalculable maritime losses by opening its archipelago to all interested parties.[10]

But quite apart from foreign interests, the growing Indonesian demand for fish requires an increase in production figures. Attempts have been made to introduce Western techniques into Asian fisheries, but in many cases they ended in failure. The reason is that basic Western conceptions as to the goals of fisheries development are often inappropriate for Asian conditions, the techniques introduced are not suited to the local situation and not enough sensitivity is employed in dealing with the local community. D.K. Emmerson, a World Bank fisheries specialist, maintains that too many *feasibility* studies were carried out when *desirability* studies were needed. Instead of following the western tendency towards maximum economic yields — which does not always suit developing countries with poor, malnourished, unemployed people — it would be better to attain maximum sustainable yields with traditional labour-intensive fishing instead of introducing the latest labour-displacing technologies. The optimum sustainable yields would be achieved, according to Emmerson, by 'a deliberate melding of biological, economical, social and political values, designed to produce the maximum benefit to society from stocks that are sought for human use, taking into account the effects of harvesting on dependent or associated species.' It is, in short, a 'bioanthropological approach' taking into account the interests of humans as well as those of the biotopes.[11]

In view of the worsening situation of the Indonesian fishing population, these ideas are worth considering. The development of sea as well as inland fisheries should be included in regional policies, since artisanal fishing is often the only source of income for a large section of the population in vast areas of Indonesia.[12] Unfortunately, efforts in the transmigration of fishing families have not fulfilled expectations, and though there are 420 registered fishing cooperatives, observers have generally deplored their moderate impact in the fishing sector.[13]

Inland fisheries account for about one-third of the total registered catch in Indonesia. They include open inland waters such as rivers and lakes, freshwater and brackish water ponds, swamps, and rice fields. The government lays great stress on the development of fishponds and shrimp farming since the total area of brackish water and freshwater suitable for inland fisheries is estimated to be no less than 9.4 million ha. The fish production potential, based on intermediate management techniques, is estimated at around 1.4 million tonnes a year. At present, about 70% of the inland catch comes from rivers and lakes, and 8% from rice fields.[14] In 1982, the country's existing fishponds covered an area of 150,000 ha., to which another 30,000 ha. will be added during REPELITA IV. Government policy aims at increasing shrimp production through shrimp farming, after the banning of trawlers resulted in a fall in production figures.[15] Emphasis is laid on the establishment of shrimp and fish farming in the coastal waters of the Aru archipelago, south of Irian Jaya, where extremely favourable conditions prevail.[16]

It is still true to say that the maritime potential of Indonesia has only been used to a minor extent. In the province of East Kalimantan, for example, which includes a maritime area of 120,000 sq.km., only 20% of its potential has been used so far. The potential of the coastal waters is believed to be 68,000 tonnes a year, but only 13,000 tonnes are landed; the deep sea waters can possibly yield 250,000 tonnes a year, but only 80,500 tonnes are caught. Thus, another 169,500 tonnes a year seem to be at the province's disposal.[17]

However, environmental problems in the maritime sector are caused mainly by other human activities. Some years ago it was officially reported that severe conflicts between fishing, oil extraction and other industries concerning waste disposal had not yet been settled. However, the same report revealed that 'marine pollution has drawn the attention of the government, particularly the Directorate General of Fisheries'. A number of tanker accidents have caused concern not only among the fishing families affected, but also among the authorities. Dwindling fish stocks caused by contamination of the coastal waters threaten the livelihoods of fishermen living in a number of coastal areas along northern Jawa.[18] Bagansiapiapi on the coast of Riau (Sumatera) was once 'the world's biggest supplier of *teburuk* fish (*Clupea spp.*)' which has now completely disappeared due to the pollution of the coastline.[19] A similar development is reported from Bengkasi island off the Riau coast, once renowned for its Chinese herring; the stock is now depleted. Effluents from fertiliser plants in East Jawa have polluted brackish water ponds near Gresik and killed milkfish fry and young shrimp,[20] whereas pollution of the coastal waters by metal elements and coli bacteria has not only caused a decrease in shrimp production, but also made the buyers reluctant to purchase further Indonesian shrimps.[21] These few examples help to show that fish and shrimp production, both on and off-shore, is increasingly threatened by the environmental impact deriving from modern economic factors.

In view of the fact that the total volume of the earth's oceans amount to 1,360 million cu.km. (or 326 million cu.miles) of water, it seems unlikely that the effluent of a minor quantity of waste material — the 940 industries of Singapore discharge some 30,000 cu.m. daily into the sea — could actually do any harm. It is known that every natural body of water has its own self-cleaning ability, and over the millenia this has worked perfectly well in safeguarding water against contamination. From this we may conclude that the oceans are large enough to digest or dilute any amount of pollution poured into them. Unfortunately, this is not the case, since the oceans are not a single entity where water constantly circulates, but are formed by separate bodies of water kept apart by the varying landscape of the sea-floor. In addition, minor entities like bays are often only marginally a part of the larger body of water, and permanent winds or streams may hold masses of water back so that they cannot mix. Under such conditions — and they only prevail along coasts — the discharge of waste can easily surpass the self-cleaning capacity of the water. Hence such local pollution disasters like the one in Minamata Bay may occur, when, in 1953, mercury poisoning killed or maimed thousands of Japanese.[20] It is, therefore, of great importance to identify the type of pollution at its source and undertake every step to stop it, or at least try to reduce it as much as possible.

An archipelagic country like Indonesia, situated between West and South Asia on one side and East Asia and the American west coast on

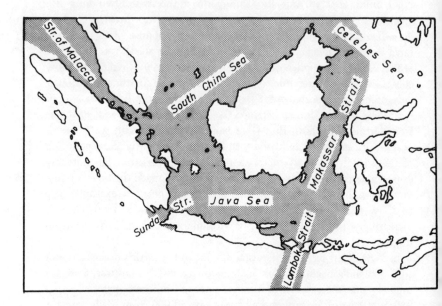

Fig. 30. INDONESIAN WATERS MOST AFFECTED BY POLLUTION FROM SHIPS AND OFF-SHORE OIL DRILLING

Source: various press reports.

the other, is necessarily crossed by many shipping lanes. Heavily-used shipping lanes and straits, particularly those used by Japanese super-tankers plying between the Middle East and Japan, are the South China Sea, the Celebes Sea and the Java Sea. However the busiest shipping lanes are the Malacca Strait, the Makassar Strait and the Lombok Strait (see Fig. 30). A particular concentration of shipping and hence pollution can be found in the shallow and narrow Malacca Strait between Malaysia and Sumatera, through which more than 37,000 ships and supertankers pass every year. This means a high risk of supertanker accidents and a potential threat of pollution to the adjacent coasts of Indonesia, Malaysia and Singapore. The pollution of the Malacca Strait in 1975 by oil from the stricken Japanese supertanker *Showa Maru* is still fresh in people's mind.[23]

It is however not only the danger of supertanker accidents that jeopardises the marine environment. When water ballast from tankers is discharged into the sea it leaves a floating petroleum residue on the sea surface. This residue is a threat both to the environment and to local fishermen's nets. An ASEAN/UNEP discussion in Jakarta in 1983 was to make arrangements for continuous monitoring of the oil content of the seas, oil-tar floating on the sea surface, sea-tar deposited on seashores and the size of oil slicks.[24]

The environmental impact of international maritime traffic has been recorded and analysed for many years. When, in 1969, the Applied Oceanographic Division of Dillingham Corporation undertook a survey of all the events that led to pollution of the sea between 1956 and 1969, it was found that thirty-eight major oil spills had occurred. About two-thirds were caused by tanker accidents, 8.33% by refineries, 5.55% by off-shore drilling operations, 7.5% by general cargo vessels and 11% by other sources. The increasing volume of oil and shipping activities in Indonesian waters means that Indonesia is one of the countries in the world most vulnerable to marine pollution by oil spillage.[25] In the meantime, the off-shore activities of the oil companies have greatly increased, and with them the threat of marine pollution.

We have already dealt with effluents from land-based industries that discharge their waste waters through the rivers and thence into the sea. This should be remembered in the present context, and it should be added that this marine pollution is by no means confined to industrial agglomerations such as Jakarta or Surabaya, it can also be found in remote areas. According to the Industrial Development and Research Centre (Ambon), the Ambon Bay in the province of Maluku is heavily polluted,[26] and waste from numerous palm oil mills discharged into the sea has a detrimental effect on the marine environment, as has been reported from Malaysia.[27]

The general public are not aware of the problems caused by the interaction of the agricultural and the aquatic sector. More recently, however, it has become obvious that 'development' or 'modernisation' in agriculture can lead to unwanted side-effects for inland waters as well as for the sea. 'Malaysia, one of the more developed of the developing countries . . . is now experiencing the hazards of pollutants which are the by-products of intensive development activities, including the excessive use of pesticides and insecticides in agriculture, the silting and erosion resulting from land development and mining activities.'[28] This remark can also be applied·to Indonesia. For example, fish were poisoned by the side-effects of pesticides used in the rice fields of East Jawa (1969) and West Sumatera (1970).[29] 'The uncontrolled use of pesticides', states Karimoeddin, 'especially if sprayed from the air, might unintentionally contaminate rivers, lakes, pools and other open waters. The contamination of open waters is particularly harmful for Indonesia where land fishery constitutes about 40 per cent of the total fish production in the whole country (1968/69). Thirty per cent is produced in Jawa, of which a substantial part is produced from wet-farming fields.'[30] In a similar way the excessive use of chemical fertiliser charges river water and eventually sea water with plant nutrients which leads to eutrophication, increasing the demand for oxygen and reducing the amount of 'clean' water available.[31]

When dealing with soil erosion elsewhere in this study, we have explained how Indonesian and in particular Javanese rivers carry large amounts of silt into the sea. This has proved catastrophic for the fishermen living and working along the north coast of Jawa. They claim that the heavy siltation of several rivers has resulted in the growth of mangrove thickets at the river-mouths and the development of mud-banks along the coast. The fishermen are now forced to traverse heavily silted areas 3 to 4 km. in width in order to reach the open sea where they can throw their nets or lines, and they complain that, without dredging, large boats can no longer enter the river-mouths to reach their landings.[32] Similarly, land reclamation work is often accompanied by silting for at least as long as it takes the new land to stabilise. Unfortunately, during this period, substantial quantities of silt may find its way into coastal waters where it has a detrimental effect on coral colonies. This has been observed in such zones on Singapore island, although the corals do re-colonise an area once the conditions improve.[33] For the time being, however, the outlook for Asia's coral reefs is rather bleak:

Acres of sea floor littered with dead coral, scummed by green slime, or bleached like bones in the sun. Elegant staghorn and elkhorn colonies, once teeming with fish, decapitated. Entire reefs, which took millenia to rise, crushed into powder in minutes. Such scenes of human devastation below the waves are increasingly common in tropical Asia, where some of the world's richest and most stunning coral reefs are rapidly becoming vast graveyards of the sea.[34]

According to A.H.V. Sarma, an ESCAP oceanographer, coastal degradation in Asia and the Pacific basin is reaching proportions not only dangerous to human life, but also detrimental to development:

The degradation is caused mainly by inland and coastal pollution due to discharge of human and industrial wastes, agricultural run-off, sand mining, unscientific dumping of rocks on the coastline, maritime commerce, tanker traffic, harbour development, tourism, the breaking-up of corals and the destruction of mangrove swamps. . . . But there are no straight answers to

Mangrove belts not only protect the coast from erosion by the sea; they also constitute a habitat for fish and shrimps and a spawning-ground for many marine species.

these problems. As in the case of coral mining in Sri Lanka, the livelihood of villagers is involved, and they would only cooperate if alternative employment is provided to them. . . . People must be made to understand the situation. They must realise, for instance, that if they hack out the corals which act as a buffer against the fury of the ocean, their houses will be washed away in a storm. Nor do the fishermen know that when they destroy the corals and mangrove swamps, they also destroy the breeding grounds of fish. Our approach is to manage the coastal environment in such a way that it can be sustained for future use.[35]

E.D. Gomez, a Filipino marine biologist, found that in his country only 5.5% of 632 reef sites surveyed were in excellent condition. In two reports he stressed that saving coral is not merely a matter of aesthetics but also one of safeguarding vital food supplies, especially since there remains a great deal of ignorance about what coral reefs can do for mankind. K.F. Jalal, chief environmentalist of ESCAP, believes that the Asian Pacific region will not be able to meet its food requirements from the sea if the degeneration of reefs and other areas continues.[36]

Such an area under threat is the mangrove belt, which not only protects the coast from being eroded away by the sea, but also constitutes a habitat for fish and shrimps and a spawning and nursing ground for many marine species. In Indonesia, the authorities are well aware of this impending danger, and as early as 1970 measures were taken to replant mangrove stretches, together with firewood plantations, along the coast north of Jakarta. This stabilised the area and led to a gradual increase of one hectare of land a year.[37] However, as in other countries in South-east Asia, mangrove forests in Indonesia are cut down for firewood and the production of charcoal, and are even used as the raw material for paper mills and chipboard factories; they are also converted into fishponds and areas of human settlement. The Indonesian authorities would be well-advised to establish their own centres for marine research to give their policy a sound scientific basis. The Marine Biology Centre at Semarang — the result of cooperation between the Diponegoro University in Indonesia and the University of Newcastle-upon-Tyne in England — deals, in particular, with coastal fisheries and the possible breeding of fish and shrimps in fishponds and paddy-fields, and with problems linked to silt-accumulation in rivers and pollution connected with paddy-fields, ponds and coastal lagoons. The Centre believes that it can perform an important role in monitoring and treating the effects of pollution on marine life in South-east Asia.[38]

Within the framework of the United Nations there are various regional agreements designed to protect the oceans. The five South-east Asian countries recently announced a regional sea programme, dealing scientifically with oceanography, pollution and coral and mangrove eco-systems. In the field of environmental management its main targets are oil pollution control, waste disposal and information exchange.

There is no doubt that the Indonesian authorities are keen to protect the country's environment, particularly at sea. The local governments of Bali and West and East Nusa Tenggara for example, have decided to protect sea mammals such as the Irrawaddy dolphin (*Orcaella brevirostris*) and the sea cow or dugong (*Dugong dugong*), commonly found in the Gulf of Kupang (Timor).[39] The problem, however, is not so much creating the legal framework as enforcing it. Indonesian tropical ornamental fish — just to name one example — are continuously in demand from Western countries for aquaria and fish-breeders, and the main buyer and distributor for the international market is Singapore. Freshwater ornamental fish are caught in the inland waters of Sumatera, Kalimantan and Sulawesi, whereas marine ornamental fish are drawn from the waters surrounding Jawa, Bali and Flores. There are reports of poaching in Indonesian territorial waters around We island at the northern tip of Sumatera, one of the country's national marine parks, from where the fish are then smuggled to Singapore.[40] This practice raises the danger of over-fishing and the extinction of certain species, and, at any rate, disturbs the ecological equilibrium.

The sad case of two species of aquatic creatures (frogs and sea-turtles) has already caused concern among the international conservation community.

A country with extended swamps and wet-rice fields usually has quite a large frog population, and Indonesia, in common with a number of developing countries, has seized the opportunity to earn foreign exchange by exporting frogs' legs, — regardless of the cruel way in which the animals are deprived of their legs. *Rana monodon*, a rather small frog, is the variety normally exported, although increasing overseas import duties are making this less profitable. The farmers concerned have called for the development of natural frog breeding in irrigated rice fields and for strict controls on the use of DDT to protect frog breeding.[41] In contrast, the freshwater fish-farms located in several regions of West Sumatera foresee a bright future in developing *Rana macrodon*, a giant frog weighing one kilogram or more — compared to 100 grams for *Rana monodon*. Experts maintain that this giant species is a speciality of the region and cannot be found elsewhere, but could be multiplied rapidly through controlled breeding practices.[42] From the environmentalist's standpoint, harvesting frogs, if limited to maintain the population, may be tolerable, but the experience of countries such as Bangladesh gives cause for concern. Some 70 million American bullfrogs (*Rana catesbliana*) and comparable species are exported every year, and this has led to a tangible ecological impact. The frogs used to eat many harmful insects, but the catching of frogs on such a large scale has led to an increase in the number of insects in the rice fields, so that insecticides have had to be applied.[43]

Sea-turtles belong to the indigenous animal population of the warm tropical seas and their eggs as well as their meat have always been part of the coastal population's diet. This is true for Indonesia as well, where people used to collect turtle eggs for their own consumption and for sale in the market. For example, there are a few small islands, Sumama, Sangalik and Belambangan in the Strait of Makassar off East Kalimantan, of which the latter two are known for their turtle egg production. Over the years, 2.2 million eggs have been collected annually, representing not only a considerable marketable commodity but also a source of revenue for the local administration. However, the production of this area decreased considerably from 63.4 tonnes in 1978 to 51.4 tonnes in 1982. The authorities in the district concerned revealed:

An estimated 2,000 sea-turtles in the Berau waters in East Kalimantan were illegally caught and killed for their meat and shells last year. . . . Two fishing boats were apprehended in the last few months for poaching turtle in the Berau waters. One of the fishing vessels, each of 20 tonnes capacity, originated from East Jawa while the other was from Bali, where turtle meat is popularly consumed. . . . The district head . . . has expressed concern over the possible extinction of species unless the poaching is halted. Worse still, the disappearance of turtle in the Berau waters might eventually upset the ecosystem, aside from depriving the livelihood of local inhabitants who earn their living by picking and selling turtle eggs.[44]

In the meantime, the Berau district administration has declared Sunama island a protected habitat for sea-turtles, and continues breeding them under its own supervision. No less than 12,000 animals are hatched every year and released into the sea once they are strong enough not to be devoured by the large number of sharks in the waters near the islands.[45]

This brief outline gives some idea of the problems involved in protecting the marine environment, particularly where valuable species such as turtles are involved. Two species, *Chelonia mydas* (green turtle) and *Eretmochelys imbricata* (hawk's-bill) are not only transported and eventually killed in a most cruel way, they are also collected illegally wherever they can be found. Indonesia is a signatory of the Convention on International Trade in Endangered Species of Wild Fauna and Flora (CITES), yet in 1981 for instance, a so-called 'turtle breeding station' on Bali illegally processed some 10,000 sea-turtles which had been caught in remote areas such as Irian Jaya. These were then prepared for the meat and tortoise-shell to be exported to other signatory states such as West Germany. In 1981 West Germany imported no less than 216 tonnes of tortoise-shell, either directly with falsified papers or indirectly via Italy or Singapore, and about 230 various turtle products, brought back as souvenirs by tourists, were confiscated.[46] Indonesian authorities, such as the Bali Nature Protection Centre, frankly admit that 'mass slaughtering of sea-turtles is continuing in Bali during the past few years', and that 'these turtles were slaughtered for human consumption and tourism promotion purposes. . . . Most of these turtles were brought to Bali from Nusa Tenggara, Timor, South and South-east Sulawesi, Maluku and Irian Jaya, because the sea creatures in Bali waters are nearing extinction.' According to their figures, between 20,000 and 23,000 turtles a year were being killed in the early 1980s.[47]

It is certainly very difficult for a country as geographically diverse as Indonesia to enforce its own protective laws, unless the local authorities cooperate. To help maintain the shrimp stocks of Irian Jaya for instance, the government has regulated the number of vessels that are permitted to operate under strict licensing,[48] and the Minister of Justice announced that fish poachers and those who pollute the sea will be watched by patrol boats and punished according to the law. On the other hand, the Marine Legal Aid Institute reports again and again that foreign ships pollute Indonesian waters and then escape scot-free, because Indonesia is 'very generous' compared to Singapore (where any ship that is caught throwing rubbish into the sea or causing an oil-spill will be fined S$5,000 on the spot).[49] However it is immeasurably easier to enforce these regulations in the territorial waters of Singapore than in the vast maritime expanses of Indonesia.

Unless environmental awareness can be raised among all the potential polluters, legislation will never yield the desired results. This means that eventually the local authorities will have to be supplied with the necessary patrol boats and aeroplanes, radio communications and so forth, and that they will have to place more stress on the protection of the environment than on the financial rewards to be gained by its destruction. At the time of writing, these hopes have not yet been realised.

NOTES

1. Simanjuntak, *Indonesia Times*, 4/5 Nov. 1983.
2. *Indonesia Times*, 7 Jan. 1984. — Here, the 'Second Snellius Expedition' (from mid–1984 to mid–1985) may be mentioned, in which more than 400 Indonesian and Dutch scientists participated, studying the eastern Indonesian waters under criteria such as geology, biology, fisheries and pollution and other subjects. (*Indonesia Times*, 16 March 1984.)
3. *Indonesia Times*, 29 March 1984.
4. Zurek, 1980.
5. Not all the techniques used by local people respect environmental requirements. Some fishermen on Irian Jaya apply a vegetable poison to narcotise the fish, but since they spare spawning grounds, overfishing is avoided. More recently, however, dynamite fishing has become popular, a technique that kills all marine creatures indiscriminately within a certain radius. West German Overseas Service volunteers have attempted to introduce the *brush park method* (which provides an artificial environment in which the fish can breed) to increase fish production without furthering the ruinous exploitation of the stocks (von Wahlert, 1985.).
6. Röll, 1979, p. 162.
7. *Indonesia Times*, 24 March 1983.
8. *Business News*, 12 Jan. 1983.
9. *Indonesia Times*, loc. cit.
10. Loc. cit., 17 May 1983.
11. Emmerson, 1980.

12. Zurek, 1980, p. 281.
13. Loc. cit., pp. 284–8.
14. 'The present status of fisheries in Indonesia', 1974, pp. 412.
15. *Indonesia Times*, 3 June 1982.
16. Loc. cit., 8 Nov. 1983.
17. Loc. cit., 17 Jan. 1984.
18. Loc. cit., 4 Nov. 1983.
19. Loc. cit., 4 March 1982.
20. 'The present status . . .', loc. cit., p. 424.
21. *Indonesia Times*, 21 Sept. 1983.
22. Hadering, *Indonesia Times*, 22 March 1983; for the Minamata case see Gunnarsson, 1971.
23. *Malaysia Business Times*, 18 June 1981; *Indonesia Times*, 11 Aug. 1983.
24. *Indonesia Times*, 11 Aug. 1983; 'The present status . . .', loc. cit.
25. Karimoeddin, 1975, p. 176.
26. *Indonesia Times*, 17 Feb. 1983.
27. *Malaysia Business Times*, loc. cit.
28. Loc. cit.
29. 'The present status . . .', loc. cit.
30. Karimoeddin, 1975, p. 170.
31. A concentration of pig-rearing and the keeping of poultry can have a similar effect, for the whole complex see SIDA/FAO, 1972.
32. Djajanto, *Indonesia Times*, 2 Nov. 1983.
33. *Malaysia Business Times*, loc. cit.
34. Gray, *Indonesia Times*, 11 Nov. 1983.
35. Waningasundara, *Indonesia Times*, 3 Oct. 1983.
36. Quoted by Gray, loc. cit.
37. *Indonesia Times*, 19 May 1981; 3 June 1982.
38. Loc. cit., 24 Dec. 1981.
39. Loc. cit., 24 June 1982.
40. *Business News*, 2 Feb. 1983.
41. *Indonesia Times*, 25 Nov. 1983.
42. Loc. cit., 24 Feb. 1984.
43. *Globus*, German television, 7 Oct. 1984.
44. *Indonesia Times*, 14 Aug. 1981; 12 Jan. 1983.
45. Loc. cit., 12 Jan. 1983.
46. *Die Zeit*, 14 Jan. 1983; 9 Sept. 1983.
47. *Indonesian Observer*, 14 Nov. 1983.
48. 'The present status . . .', 1974, p. 424.
49. *Indonesia Times*, 28 July 1981.

16

TOURISM

Though the Indonesian press reports frankly on environmental damage caused by agriculture and industry, and does not conceal the detrimental impact of growing population pressure and certain economic activities on nature, tourism still enjoys a special status as a growth industry. This is also true for scientific studies — if they touch the subject at all. However scientific institutions covering the South-east Asian region and those dealing with tourism on a global scale envisage many problems. In its 1982 Annual Report, ESCAP notes that 'the impact of tourism upon the traditional social fabric and the physical environment of the historical communities has caused concern in many countries, but several have attempted to control such effects rather than restrict the flow of tourists.' In other words, 'they risk everything just to earn dollars.'[1] And the 'Manila Declaration on World Tourism' of 1980 stated:

Tourism resources available in the various countries consist at the same time of space, facilities and values. These are resources whose use cannot be left uncontrolled without running the risk of their deterioration, or even their destruction. The satisfaction of tourism requirements must not be prejudicial to the social and economic interests of the population in tourist areas, to the environment or, above all, to natural resources, which are the fundamental attraction of tourism, and historical and cultural sites. All tourism resources are part of the heritage of mankind. National communities and the entire international community must take the necessary steps to ensure their preservation.[2]

In 1983, at its fifth General Assembly in New Delhi, the World Tourism Organisation (WTO) in its *Delhi Statement* pleaded for the promotion of larger movements of people both within their own countries and internationally; for the study and gradual elimination of pollution problems caused by tourists; and for the working-out of a precise programme for the conservation of heritage and the environment.[3]

It is interesting to note that these organisations, next to dealing with the social and economic impact of tourism, lay considerable emphasis on the environmental aspect. This is hardly the case with scientific studies, most of which are carried out by economists or social scientists whose spheres of interest are financial results (the input/output approach), or changes in society and behaviour (the social approach). This is all the more disappointing since in many countries of Asia — not to mention the Mediterranean basin — tourism has led to the deterioration of the environment to such an extent that it can no longer be overlooked. A.H.V. Sarma, the ESCAP oceanographer quoted earlier, refers mainly to Sri Lanka, but also to comparable countries when he deplores the haphazard building of hotels and guesthouses, the plague of pollutants resulting from human habitations and business, the turning of beach resorts into asphalt jungles, thus destroying their natural beauty, and the presence of coliform bacteria indicating that there are disease-carrying bacteria lurking. According to him, coastal zones in 'near-critical conditions' where oxygen deficiency is already present are found, to name a few, in the upper Gulf of Thailand, Manila Bay in the Philippines, and the Hooghly estuary in India, south of Calcutta.[4]

The Mediterranean basin is an excellent though shocking testing ground for what tourism can do to the environment when it is regarded exclusively as a growth promoting enterprise:

Because tourism entails a concentrated impact of people in a well-defined corridor along the shoreline, the ecological consequences of this growing population movement are intensified. Large numbers of tourists use large quantities of coastal space. In the process, the touristic uses of space compete with other uses that contribute to the quality of the coastal environment. Open beaches, clear swimming waters, seaside parks and sand-dunes give way to hotels, recreational facilities and a multitude of boats that crowd the narrow strip of the Mediterranean coastline. Pollution of the sea as well as the shore itself is often the case. The fact that tourism is a fast growing industry earning foreign currency is related to its potential as an ecological problem. To this reason one should add also the quick spread of tourist activities from one locality to another. The fast growth of the industry encourages entrepreneurs to develop tourist facilities in a manner that is detrimental to the coastal environment. Local and national governments, in their eagerness to attract more tourists in order to develop the local economy, as well as to improve the balance of payments, are often ready to disregard zoning regulations and other qualitative standards. The results of such indiscriminate development are often ruined beaches, polluted coastal waters and overloaded public services.[5]

The way in which the environmental beauty of the Mediterranean basin has been sacrificed in order to attract mass tourism as a source of income is certainly a discouraging example. The process has continued because the more affluent Northern Europeans are constantly attracted south by the prospect of relatively cheap sunny holidays. A similar process can be observed between North America and its southern neighbours and, more recently, between Japan and South-east Asia, though not yet on such a scale. The more the southerly and not so rich countries compete for the affluent tourists from the north, the more some of them are ready to go to sometimes scandalous lengths to ensure the preparation of holiday resorts at the expense of local people and the environment.

In Venezuela, for example, the authorities used highly poisonous defoliants to clear a beach of noxious algae. They thus killed not only the algae but also the fish and other marine creatures and brought disease to the local people. The enormous water consumption of a tourist — some 600 litres per day — often leads to a conflict of interests between a hotel or holiday resort and the local farmers who have cultivated the soil for generations. This is a particular problem in areas where there is a dry season and water is scarce. In southern Tunisia, for example, near the oases of Nefta and Tozeur, where people make their living from highly intensive irrigated horticulture, some luxury hotels were built, together with deep-well drilling. The result was that many of the local wells, vital for the oasis farmers, dried up, while a tourist consumed as much water in a day as could have irrigated 200 sq. m. No wonder that there are cases of violence in such countries directed against the authorities as well as against tourists. In Nepal, a Himalayan country suffering from extreme deforestation and soil erosion, trekking tourists contribute to this process by using additional firewood on their way to enjoy the mountain scenery. The local porters who accompany them neglect their fields because the trek offers them more money in the short-term. Thus, the cultivation and stabilisation of a delicate agrosystem suffers because of such an intrusion. There is no doubt that tourists in large numbers, or when concentrated in a few points of interest, have a negative impact on the surrounding area. Landscape, plants and animals become endangered or are destroyed, the quietude is gone, motor traffic, noise and air pollution increases and the coastal and mountainous environment suffers.[6]

Such general remarks are perhaps best illustrated by cases reported by journalists or scientists. The Filipino marine biologist E.D. Gomez, when dealing with the coral colonies of South-east Asia, states that tourists have disfigured many once spectacular underwater gardens. This process can be verified in many diving resorts. In the Caribbean state of Haiti, the use of sophisticated spear-guns by sports divers has severely reduced the population of fish and lobsters, but unfortunately local fishermen have adopted the technique and are also depleting their fishing grounds.[7]

The environmental impact of tourism certainly changes from place to place. Conflicts over land use do not occur if tourist buildings are constructed in areas that are of no interest to other parties. Sandy coastal strips are not normally used for agriculture, though they may well be communal land. Occasionally, fishing villages are considered to be 'in the way' of tourism planners. When development takes place, access to the sea is restricted and the fresh breeze that once made life in the villages more agreeable is blocked by hotel buildings. The local population gradually moves away till only a minority remain, often living in slums near the hotels. A once well-balanced picture of man and nature has been destroyed by the spread of tourism. There are also cases where peasants are pushed out onto marginal land to make way for a holiday resort. And where the fishermen are tolerated, the impact of tourism on the coastal environment will make their former habitations worthless. Protected areas, so-called National Parks, were originally established to give a refuge to threatened species. However, increasing areas of such land are being regarded from a tourism viewpoint, especially for hunting, in order to maximise foreign exchange earnings.[8]

Attention should be paid to the fact that tourism is not necessarily a great foreign exchange earner for the developing countries. If we consider that the establishment of a touristic centre or zone requires much foreign investment and offers only a very limited number of jobs for the local population, that the government often has to subsidise the costs of investment, and that the dependency on foreign interests increases, the idea of tourism as an important developing factor seems doubtful. In many cases it furthers inflation, aggravates social disparities and destroys the local economic equilibrium, the social order and the standards of values — in short, it certainly brings change, but not necessarily to the advantage of local people, let alone the destruction of the environment.[9]

As far as Indonesia is concerned, the country has hardly been discovered by mass tourism except for the island of Bali. In 1983 Indonesia received less tourists than any other ASEAN nation:[10]

	millions
Singapore	2.8
Malaysia	2.3
Thailand	2.2
Philippines	0.9
Indonesia	0.6

Though the country has seventeen major points of entry by air or sea, most visitors arrive in Jakarta and remain there for a few days. The country's ten main tourist destinations in order of importance are: Jakarta, Bali, Yogyakarta, West Jawa, North Sumatera, East Jawa, South Sulawesi, Central Jawa, West Sumatera and North Sulawesi. In 1983, a tourist stayed on average for ten days and spent about US$58 per day. The annual overall total comes to roughly US$381.2 million. In early 1984, foreign visitors came from many countries, the biggest groups from Japan (15.2%), Singapore (12.4%), Great Britain (9.3%), the United States (7.9%), and West Germany (5.2%).[11] For the whole of Indonesia, the number of tourists entering the country developed in the last few years as follows:[12]

		Growth %
1978	468,614	
1979	501,430	7.0
1980	561,578	12.0
1981	600,151	6.9
1982	592,046	- 1.4

It is obvious that the general world recession is reflected by decreasing numbers of tourists, yet the minister in charge of this sector still believes in tourism as the main future foreign exchange earner when oil and gas resources are depleted.[13] However for many years the authorities regretted the fact that they did not succeed in attracting mass tourism to Indonesia, and critics blamed them for having relied for too long on the natural beauty of the country instead of attracting comfort-loving tourists by building luxury hotels all over the country.[14]

By the year 2000, the tourism authority hopes to push the number of visitors up to 1 million or so a year, offering new tour possibilities in West and East Nusa Tenggara, East and West Kalimantan and Riau (Sumatera). The tourism authorities in Bali alone hope for some 400,000 tourists a year, particularly since there are now direct flights to the airport of Denpasar, thus avoiding Jakarta.

Indonesia's various provinces have rather ambitious ideas for the development of their touristic potential. Riau, situated on the east coast of Sumatera opposite Malaysia and Singapore, thinks that no less than ninety locations are capable of being developed as tourist sites. The first priority, however, will be given to marine tourism with boating, yachting and regattas, centred around Trikora Beach in Tanjungpinang on the island of Bintang, where an old tradition of boat-racing exists.[15] East Kalimantan, known for its petroleum and timber industries, intends to attract tourists through its special wildlife reserve for the orang-utan, located 1,000 km. from Samarinda. This area can be reached on one of the great rivers; in addition, there is a natural reserve of 50 ha. of orchids containing 50,000 species.[16] In East Kalimantan, the local authorities announced the development of Tanjung Beach near Balikpapan 'to attract more tourists to the region as well as to provide healthy recreation for the local people. The beach has white sand, and off-shore can be found sandy islands surrounded by sea-gardens with beautiful corals where lovely fish live.' It seems that they indulge in the hope that tourism will preserve the coastal ecosystem. A local environmentalist said this, and added that he hoped that the decision would lessen the commercial exploitation of sand and live corals all along the coast.[17]

So far, studies about the impact of tourism on society in Indonesia have been confined to Bali, where roughly half of the visitors stay. In the 1920s, when the first few tourists arrived on the island, there was hardly any impact to speak of; in 1975 however, 250,000 people visited Bali. In the meantime, two small towns have been developed to accommodate tourists: Sanur, east of Denpasar, with comfortable hotels for group travellers with a fixed programme, and Kuta, to the south-west, with simpler lodgings for individual travellers. In the southern peninsula, at Nusa Dua, a tourist resort is planned which will accommodate a great number of tourists in what is presently a thinly populated region.

In contrast to any other tourist destination in Indonesia, Bali offers the chance to study the consequences of mass tourism on the indigenous population over a comparatively long period with respect to behaviour, the standard of values and so forth. However, these findings cannot be generalised, for the Balinese constitute a very specific group among the Indonesian family of peoples. They are Hindus and live in very tightly-knit village communities. All the studies undertaken by Indonesian and foreign social scientists agree that 'many Balinese are able to think in the same rational way as Western businessmen in the field of economy; this, however, does not prevent them from integrating themselves into the mystic ground conception of family and village after their work . . . the sound and strong village community has succeeded in absorbing the new influences, and gaining new impulses from them.'[18] Another academic concluded that: 'The social life is so dominating in the Balinese society that every Balinese, in the first place, is a faithful member of his society; only in the second place is he holding any other position. In a case of cultural conflict, the Balinese will, without hesitating, stay with his community and, without visible regret, leave any other organisation or institution to which he belonged so far.'[19] Nobody can predict though what will happen following decades of mass tourism. Moreover, studies about non-Balinese groups affected by mass tourism have not yet been prepared. The Director-General for Culture, Mrs. Haryati Soebaio, maintains that Indonesia intends to preserve its cultural values (meaning obviously the historical monuments as well as the standard of values in society): 'To earn as much foreign currency as possible does not mean we should sacrifice our cultural values.'[20] And yet, just the effort of turning cultural values into hard currency can easily create social and environmental conflicts.

Borobudur, situated not far from Yogyakarta in the centre of Jawa, one of the greatest Buddhist sanctuaries, erected in about A.D. 800 and saved from decay through a substantial financial contribution by UNESCO for its restoration between 1975 and 1982,[21] is an outstanding tourist attraction. However it is surrounded by palm plantations, rice-fields, and villages inhabited by people who have lived there for generations. Some of them have cultivated pieces of land that formerly were infertile but are now productive thanks to their efforts. Borobudur is their's, and they feel that they belong to Borobudur. Unfortunately,[22] the authorities developed ideas of how to 'market' the sanctuary in the most direct way, namely by clearing the palm plantations and evicting the peasants from their fields and villages in order to build a giant 90 ha. tourist complex around the stupa of Borobudur. Four hundred families in four villages would be affected. The head of the Buddhist Society of Indonesia protested against the plans, because he feared the spread of tourist attractions such as night clubs. This is certainly not the proper environment for one of the world's most celebrated sanctuaries. How seriously the project is meant is another question. It would, however, be just another example of how tourism irreverently seizes hold of holy and historical places round the globe just to attract foreign exchange. Similarly, there are risks inherent in the idea of 'marketing' natural sanctuaries by opening them up to hunters in exchange for hard currency. When carefully managed, the protection of nature can be compatible with gains from tourism, but, in the same area, there is always the danger of conflict between tourists' hunting and the indigenous people's protein supply.[23]

Economically, and indirectly ecologically, the impact of mass tourism is very diverse and may have positive as well as negative consequences. If it leads to an increased demand for labour, it may well reduce unemployment in the region, but if better salaries lure farmers' sons and daughters away from their villages, agricultural production may suffer. A growing demand for food, especially fruit and vegetables, may encourage the farmers in the vicinity to grow more and to increase their income, but in times of scarcity the hotels are able to pay higher prices for produce than local consumers can afford. Tourists also stimulate demand for fish and seafood which can benefit the fishing industry, but there is always the danger that certain areas will be overfished, and shrimp and lobster stocks may decline rapidly. In Bali, it is reported that the growing demand for fish has encouraged the development of artisanal fishing methods which have provided an income for about a thousand people.[24]

These few remarks will have shown that the consequences of tourism and particularly mass tourism are by no means confined to charter company arrangements and the construction of hotels. Tourism is a very complex phenomenon whose effects can vary from land use conflicts to the destruction of the environment. If not properly planned, it can easily lead to disaster. In the context of this study, the problem is not so much who actually gains from tourism — foreign enterprises or local people — nor whether a society of proud but poor peasants might lose their identity by becoming waiters, beggars, or even prostitutes and criminals. What is important here is whether tourism — as a type of land use — destroys the potential of an area which, if properly maintained, could yield sustained income to a great number of people.

To achieve these targets, a cautious policy, based upon proper planning, is necessary. It must not deal with tourism alone, but should involve a range of measures such as regional planning, environmental policy, the development of agriculture and other economic spheres and, last but not least, cultural policy. 'One of the most pressing tasks, therefore, has to be to find out the carrying capacity of the various touristic regions and to take care that it is not surpassed.'[25] Indeed, 'tourist carrying capacity' has to become the key word in tourism planning:

Carrying capacity and saturation of tourism areas is a reality, not just a concept. There is a definite need to establish acceptable levels both for use in initial planning and in cases where saturation exists. The problem lies in defining and quantifying these levels into a usable form. Statistical measurement is possible with much plant and infrastructure, but many criteria can never properly be evaluated and compared in such precise terms. These obvious difficulties should not be allowed to divert attention from the importance of carrying capacity. It is important that a body of knowledge on the subject should be built up through experience to provide guidelines for those concerned with tourism planning and management. Assumptions based on these guidelines (standards), can then be tested and modified to suit the particular circumstances of individual cases. Impact studies complement the understanding of carrying capacity and saturation levels; they are not an alternative. Carrying capacity standards can be used as a guideline to the development of project proposals, which can then be tested by impact studies. There is a need to establish a body of knowledge on carrying capacity and saturation which can act as a general guide to governments and others having different situations in time and space. At present such information is limited, but it can be built up from both theoretical concepts and from practical experience.[26]

This should be given the fullest consideration by those who centrally, regionally, and locally are responsible for what will happen to the land, vegetation, animals and the aquatic realm sacrificed to tourism.[27]

It has now generally been acknowledged that a given natural environment can only be used by man to a certain degree without being violated or even destroyed. We have seen that agricultural land use and transmigration have to respect such limits, and that surpassing them would lead to environmental destruction, or at least a severe reduction of the area's natural potential. Similarly, we have shown that the aquatic environment cannot be overburdened with pollutants without losing its self-cleaning capacity and its function as a habitat for creatures. Tourism is no different. In this case, the carrying capacity or saturation point of the environment is determined by the number of visitors to a certain area, and the degree of physical development that is needed to accommodate them without reducing the attraction of the place for tourists or the quality of life of the resident community.

Experience in many parts of the world shows that market forces alone will never ensure that the carrying capacity is not surpassed as long as tourists accept the deteriorating environment, and the local population puts up with declining living conditions. Therefore, it is the responsibility of the government at national, regional and local levels to ensure that the quality of the environment is maintained in the interests of tourists, the local community and the nation.[28] Indonesia has not yet experienced any major environmental catastrophe caused by tourism. Fortunately there is still time to ensure that this does not happen.

NOTES

1. *Indonesia Times*, 7 June 1983.
2. UNEP, *Industry and Environment*, vol. 7, no. 1, p. 1.
3. *Indonesia Times*, 15 Oct. 1983.
4. Wanigasundara, *Indonesia Times*, 3 Oct. 1983.

324   *Land Use and Environment in Indonesia*

5. Gonen, 1981, pp. 380–1.
6. Loc. cit.; Mäder, 1983, Vorlaufer, 1984.
7. Donner, 1980, p. 355, see Gray, 1983.
8. Vorlaufer, 1984, pp. 218–32.
9. Beller *et al.*, 1980, pp. 251–85.
10. *Indonesia Times*, 19 March 1984.
11. *Sunday Times*, Singapore, 14 Oct. 1984.
12. Various press reports, for instance *Business News*, 28 Sept. 1983. — The Ministry of Tourism's figures are more optimistic.
13. *Business News*, loc. cit.
14. *Neue Zürcher Zeitung*, 24 Feb. 1977.
15. *Indonesia Times*, 2 May 1984.
16. Loc. cit., 10 Jan. 1984.
17. Loc. cit., 2 May 1984.
18. Leemann, 1978, p. 23.
19. Soemardjan, in Kötter *et al.*, 1977, p. 201.
20. *Indonesia Times*, 20 Oct. 1983.
21. *UNESCO Kurier*, vol. 24, no.2 (1983), pp. 8–23.
22. The following is taken from a leaflet entitled *Tourism Project Borobudur in Indonesia and the Consequences* (in German), issued by the Initiative for Human Rights of all Citizens of the ASEAN States (IMBAS) in West Germany. It refers to two press articles, namely *Mutiara* (4–17 March 1981) and *Buang Minggu* (27 June 1982). Early in 1985, part of the restored monument was blown up by persons unknown. This was initially attributed to extremists hoping to block the revival of Buddhism, but the possibility that it was carried out by those who feel threatened by tourism plans has not been ruled out. (*Die Welt*, 23 Jan. 1985.)
23 *Indonesia Times*, 5 Nov. 1981; Vorlaufer, 1984, pp. 223–32.
24. Vorlaufer, 1984, p. 125.
25. Gutzler, 1978, p. 84.
26. World Tourist Organisation, 1984, p. 36.
27. Gonen, 1981, p. 381. Tourists can hardly be regarded as protectors of the environment, and although the International Union of Official Travel Organisations (IUOTO) has been concerned about the accusation that tourism 'causes pollution and ruins cities', it had to acknowledge the continuing deterioration of the coastal environment in the Mediterranean despite all efforts of the tourist economy.
28. World Tourist Organisation, loc. cit.

Part V. CONCLUSIONS

The outcome of this study, based upon local research, discussions with experienced specialists, and the examination of technical literature published during the last 100 years, is rather discouraging. Knowing from my own experience the hopeless struggles of technical advisers in many developing countries to convince the authorities of the urgent need to halt environmental destruction, one feels frustrated to discover that in Indonesia our colleagues fought in vain for the same measures over 100 years ago. Leafing through old technical papers, we can find accurate predictions of what is happening today in the Sahel, in Ethiopia, in the Amazon basin and partly also in Indonesia. However, in many cases the experts underestimated the scale of such disasters, being unaware of the technical means to destroy the earth's surface. If we compare the situation 50 or 100 years ago with today, we can see that not much has changed — except that the earth now has more people and less arable soils. And yet we cannot simply capitulate; we must continue to fight for the survival of our earth — and that means for our own survival.

It is not only warnings from the nineteenth century that are relevant today. Often we find that the same technical proposals for soil conservation which we are trying to introduce now were being made then — with the same discouraging results. At present, however, there are more mouths to be fed from each hectare, and the costs of urgent technical measures are many times higher than before.

Economic as well as ecological criteria are decisive for land use. But if we give the economic viewpoint exclusive priority, that means if we use a certain piece of land only because we expect to maximise financial returns, we are in danger of eventually destroying our source of income. Therefore, priority or at least equal importance should be given to ecological or environmental criteria. This means that we should not develop any form of land use if we can anticipate future environmental destruction. This may well sound utopian at a time and in a society where input/output thinking guides nearly all decisions, and material gain is the only thing that counts. For centuries, Western people around the globe have followed the maxim 'What's wrong with profit?', and we now can see the result: the destruction of natural resources, often irretrievably lost. This is the result of the myth of the free gifts of nature.

There are extremely rich countries that have exploited the natural resources of the poor countries for centuries, and we have extremely rich élites in those countries doing the same with nature and the poor masses. The poor masses, whose numbers continue to increase at an alarming rate, have little choice in the matter. They live on the soil, the forest, and the water that is left over to them, and as their population expands more and more, they subsist on their 'capital' — the environment — by consuming it.

If we believe UNEP's figures — and there is no reason not to do so — we cannot help but be worried. Every year, some 6 million ha. of land turns into desert, and another 21 million ha. of arable land are lost for agricultural production.[1] The annual loss of 16 million ha. of tropical forest[2] is not simply a depletion of natural beauty and wood reserves, it also leads to deterioration of the micro-climate and the loss of species. Experts believe that at present, our planet loses one species every day, but that figure could easily become one per hour by the year 2000 if nothing substantial is done.[3] If mankind is to survive, we must invest in the protection of our environment and our resources. If this means making consumer goods more expensive and profits smaller this is only to be expected. We have to bear this in mind when analysing the situation of a specific country. Fortunately, in Indonesia the authorities are conscious of the dangers posed by a steadily growing population and the environmental damage that results. The national population policy, which includes family planning, is supported by practically all social groups, and is beginning to produce measurable benefits for the country. It is vital that this policy is continued, especially in the Outer Islands where the myth of unlimited space still prevails, particularly in the transmigration areas.

It is, however, obvious that the impact of the most successful family planning policies will not lead to a significant decline in population pressure on the soils for some decades yet. Thus in view of the situation on Jawa, other policies are required. These have in fact existed for a long time in the form of transmigration from Jawa and other small islands with a high population pressure to the Outer Islands. Unfortunately, experience shows that transmigrating thousands of families is technically and financially limited and corresponds to only a minor fraction of the annual population increase on the overpopulated islands. Moreover, the absorption capacity of the Outer Islands is limited and the unsettled land there is very susceptible to destruction by inconsiderate land use.

Thus Indonesia is facing the following situation. It has to accept a population increase for another few decades; it can transmigrate only a small fraction of the annual increase from Jawa to the Outer Islands; it has to take care of the remaining and growing Javanese population; and it has to remember that most of the Outer Islands' surface is unsuited for growing field crops — on pain of environmental and ecological destruction. In view of these factors, Indonesian policy in the field of population, land use and environment should comply with the following strategies. The present population policy should be continued and strengthened; the carrying capacity of Jawa needs to be increased both in terms of agricultural and non-agricultural land use; and the Outer Islands need to be developed in an integrated way while respecting their economic circumstances.

These proposals are rather general and need to be discussed in more detail. The ideal policy, namely to shift and settle people according to the natural potential of an area, is impracticable for both political and practical reasons and because there is no more room left on Jawa; even the victims of volcanic eruptions have nowhere to go. Thus, people on Jawa will have to remain where they are, but they can save their environment or at least reconstruct it so as to prevent any further deterioration. Quite a number of technical projects are planned to further this end.[4]

The appraisal of the carrying capacity of a certain area, preferably a watershed (the catchment area of a river system) requires an enormous amount of work in the fields of geology, hydrology, pedology, botany and so forth, let alone economics and marketing. It is quite possible though to reduce this task to those studies which are environmentally crucial in avoiding further or future deterioration. The results of these studies should then be brought to the knowledge of the local people. This is a very important point, for it has been found in many developing countries, Indonesia included, that without the understanding, cooperation and participation of the population concerned, no change, however moderate, will be accepted, to say nothing of measures that would tangibly alter the living conditions.

One may argue that it is hoping for too much to expect simple peasants to understand complicated environmental interrelations they have never heard of before, and make decisions that may fundamentally change their future life. This is true to a certain extent, but all farmers in Indonesia, except those living under extremely favourable conditions, have experienced the deterioration of their environment, though they cannot see a way of avoiding it. Next to general environmental education, which should begin in the primary school, the 'model area' is an excellent means of instructing and persuading farmers to care more for their environment.

Since efforts in the field of watershed management (already envisaged by the Ministry of Forestry) can be successful only where people cooperate, public support should concentrate preferably on areas where such cooperation is guaranteed. Then, hopefully in the course of time, a number of model watersheds will come into existence where protection and production are in equilibrium and environmental deterioration has come to a halt. Under such conditions, field, cattle and forest products would be produced, partly under irrigation, then processed and marketed, and local craftsmen would provide the necessary services to maximise local income. Such 'model' areas could then serve to demonstrate what the policy is aiming at and help to persuade community councils, peasants' representatives and the like. Once other barriers — such as an unfavourable tenurial system — are overcome, it may show that more people can be settled on the soils of Jawa, or the same number can subsist on them without further destroying the environment. But expectations should not be exaggerated: a growing number of people will have to be accommodated outside agricultural production, be it in rural or urban occupations. And, we are well aware that even now, many more hands than are actually needed are used to produce certain agricultural products. But as long as we have an excessive supply of labour, we have to face — if at all possible — further 'involution' and 'shared poverty' on Jawa. Industrialisation, the service sector — not to forget the 'informal sector' — and migration slightly reduce population pressure on the soil, but they do not offer a permanent solution.

The enormous surface of the Outer Islands, still mostly covered with lush green jungle, is misleading in that it seems to be an area of unlimited potential. Decades of transmigration efforts on various islands show however that indiscriminate cultivation of soils previously covered with forests, swamps, or tropical grass easily leads to their exhaustion and ecological destruction. This, together with the ruthless commercial exploitation of forests, has caused great concern among the authorities. Science, already sceptical of the myth of fertile tropical soils hidden under forests a century ago, has in the meantime discovered the mechanisms and the secrets of a luxuriant vegetation growing on virtually sterile soil. Undoubtedly the Outer Islands do possess areas suitable for the production of rice and annual crops, but they are limited and often scattered in small areas. Moreover clearing the jungle is hardly the way to gain land for the continuous economic production of rice or other annual crops.

Paradoxically, natural tropical forest, when properly managed, is able to provide sustained yields of forest products, but if the area is cleared and the ground turned into fields, it may soon be exhausted and not even able to carry forest any more. We have seen that tropical grassland is a soil reserve that has hardly any value of its own, except as a ground cover. Its preparation for cultivation is difficult and needs technical support from outside, but it is a potential that can be realised without destruction. On the other hand, if soils turn out to be less suitable for crop production, they may be used for perennial crops or reforestation. It is particularly useful to make trials with the planting of valuable tree species on such soils instead of considering clear-cutting tropical forest for tree monoculture.

Swampland ought to be given secondary priority for transformation into cultivated land, since it has its own economic and ecological merits. Swampland produces various interesting tree species, and provides a natural habitat for a large number of animals, and thus is well worth protecting. Careful management is thus required when opening-up swampland for cultivation in order to keep as many ecological values as possible untouched.

One typical swamp plant, the sago palm, is drastically underused as a source of basic food. In terms of agricultural policy it provides an interesting subject for research since it would be a mistake to force wet-rice cultivation on people who have previously lived on sago as their indigenous staple food (growing naturally in the biotope they inhabit). However substantial parts of the swampland may be suitable for wet-rice production.

Least recommended for being turned into cultivated land are the tropical forest reserves. After Brazil, Indonesia is the world's most forested country and nearly 10% of the earth's tropical forests are found there; however, large areas of forest are threatened by destruction. In view of their outstanding importance in supporting the ecological, hydrological, climatological and pedological equilibrium and — if properly managed — their vast sustained wood yields, the forest soils ought to be the last to be taken under cultivation, except in cases where it can be done without causing much harm.

It cannot be denied that peasants in the tropics do have an understanding of how to produce food without destroying the basis of their existence, but this understanding has often been lost because of the

growing demands for various products grown on the soil and the intervention of world market interests. Their impoverishment has forced
them to think in terms of their own survival, even at the expense of the
environment and of future generations.

It is, therefore, most important to improve extension services all over
the country, and pass on the results of the. many experiments with
protective land use techniques such as agro-forestry, the planting of
shelter-belts, regreening, contour-farming, terracing and the use of
mulch and covering crops to the peasants. They have to understand that
the protection of the natural potential in the tropics requires cultivation
techniques which are as close as possible to the natural biotope, namely
the rain forest. Bearing these factors in mind, the ecological way of converting tropical land into permanent crop land would be to establish
permanent crops such as rubber, coffee, cocoa, cloves, etc. — provided
the soil is properly protected by litter or cover crops.

But since the aim is to produce at least enough food for local requirements, other forms of ecologically suitable land use have to be developed.
Next to agro-forestry, mixed farming including home gardens, fodder-
crops and the intensive use of green manure through legumes and composting offers one way forward. Moreover mixed farming, including
field crops and livestock, presupposes careful ecological land use.

Obviously, the various landscapes in a country of Indonesia's dimensions offer different potentials and thus require different approaches
towards planned land use. Where irrigation is practicable, wet-rice may
be the most suitable basic crop; where swamps prevail, the sago palm
deserves more attention, and in the drier islands of the east, various palm
species may constitute the ecologically most suitable crop.[5] But where
the growing of upland crops on rain-fed fields is the only way to make a
living — and this is the case in most areas — extreme care is required if
this potential is not to be completely lost. The most recommended way
to use such soils is to introduce the home garden.[6]

Over a period of more than 100 years, science has collected a great deal
of information from observations and field experiments. Now is the time
to bring these old reports up to date with new observations and the
results of more recent experiments and turn them into a consistent
policy. Everybody is now aware that the Indonesian authorities and a
great part of the public have become environment-conscious thanks to
the untiring efforts of some Indonesian specialists and politicians, and to
the readiness of the Indonesian press[7] to disseminate information in this
field. When discussing the Bill of Environment Management Provision

in early 1982, the various political parties in parliament all supported the proposals made by the minister, Emil Salim. The Partai Democrasi Indonesia requested measures to protect the forests by cancelling the existing forest concessions and to fight pollution no matter what its source; the Armed Forces faction blamed industry and shipping interests for adversely affecting the environment, indicated that environmental deterioration, particularly in the field of forestry and hydrology, would enter a critical stage, and said that greater care was required in using chemical fertiliser and pesticides; the Karya Pembangunan pointed to the fact that the proposals would fit perfectly with the philosophy of Panca Sila and that man should protect the environment by training the village population in the necessary skills; lastly, the Persatuan Pembagunan faction recalled God's command to manage the earth properly. Referring to the noise and the pollution caused by factories and vehicles, the representative 'hoped that minister Salim would continue to sing his song of environmental information to the public'.[8]

Although making speeches in parliament does not change anything in the field, it is good to know that the public representatives have leant their support to the pertinent measures. Both government and university scientists — particularly in the faculties of geography and agriculture — are heavily involved in environmental efforts. Here again the question that arises is, to what extent do these results actually influence existing projects and policy? Well-informed observers deplore the fact that the East-Asian fear of losing face often prevents research findings from being applied in practical projects.[9] Another handicap is the ethnic multiplicity of Indonesia. There was, for many years, suspicion of Javanese guardianship of the Outer Islands. More recently, however, it seems that the army has become an integrating and stabilising element, and the fact that representatives of all of Indonesia's ethnic groups are now trained in Javanese or regional universities has helped to reduce this feeling.[10] The transmigration of millions of Javanese to the Outer Islands may yet require further measures to help soothe possible animosities. Giving decisive rights to the lower levels may help to overcome difficulties of this kind and strengthen the readiness of landowners to care for their own future.

Many of the obstacles which lie in the way of a successful transmigration policy have been dealt with in detail earlier. Unsatisfactory selection and preparation of sites, insufficient support and the virtual lack of extension services for peasants who try to adapt themselves to an alien environment result not only in poverty and often in resignation, but also in ecological destruction which could have been avoided. On the other hand, some of the successful transmigration areas are beginning to resemble Jawa in terms of population pressure and environmental damage.[11]

The problems of land use and environment are, as we have shown, by no means confined to the rural sector. Mochtar Lubis, a knowledgeable Indonesian observer of economic development and of the environmental scene, is greatly worried by the tendency prevailing today in Indonesia to indiscriminately accept Western standards to guide the future of a people and a society that is actually Asian. By accepting these standards, he argues, we are in danger of repeating the same mistakes made by Europe, the United States and, in particular, Japan. He insists on strict controls to guard against the exploitation of national resources by foreign interests and against environmental destruction:

I would not like to see Indonesia go the way Japan has gone: advance industrially but destroy its own nature. If that is the fate of everyone, if all Latin American, African, all Asian countries are trying to reach the same standards of industrial development as the West, I think we are going to pollute the whole world very soon. We must be happier with a simpler style of life. We must find our own directions, develop our own goals . . . I think what has happened in Japan, in America, is simply not good. It does not satisfy the real needs of man . . . We don't need those big, enormous, frightening cities like New York, Tokyo, Osaka, where people are like ants . . .

He does not see why Indonesia should accept the system of mass production (economies of scale) only to maximise profits, to import more and more cars and to have foreign firms producing goods in Indonesia with a high level of capital investment, thus destroying small-scale national industries:

I believe we can strike into our own direction. We can really develop something new, find relations within our society which are more harmonious and less oppressive. But I don't know whether we have the time to do this, or whether the world will give us the time . . . We are being colonised by these things . . .[12]

This is indeed the crucial question: whether the world market will permit a country, large or small, to go its own way?

We can also apply Lubis' legitimate demands to the rural sector. An ecologically well-balanced agricultural development, under tropical conditions, in the first place requires emphasis on subsistence production. Market production should only be undertaken when conditions permit.

Subsistence production in this sense does not mean a miserable existence; on the contrary, a mixed form of agriculture with food crops, fruit, fowl, cattle and possibly fish guarantees a much more balanced diet than most Indonesians have today. Moreover there will always be a surplus for the market to finance the purchase of necessary items from outside and to supply food to the urban population and the rural craftsmen who do not produce their own.

This certainly contradicts those programmes which think in terms of the money economy and monoculture, which serves the interests of a world market whose rules work much to the detriment of Third World farmers. Producing primarily for the world market has, to date, certainly brought benefits for well-to-do farmers, traders and speculators, but at the same time has led to impoverishment, indebtedness and finally loss of land and property among the rural masses.

This does not mean that export production should be completely abolished. It does mean, however, that this sector is not suited to offering a source of income for the mass of rural people. And as long as the non-agricultural sector is unable to offer alternative employment, the majority of the population will depend on its own subsistence production. Agricultural policy, therefore, should devote a substantial part of its work and funds to promote subsistence production which incorporates environmental safeguards.

Dissociation from the world market is, however, hardly feasible given the many vested interests of influential figures in foreign trade, the manufacturing of Western consumer goods, and world markets, let alone the impact this would have on the country's social and environmental situation. Yet the ecological deterioration of Indonesia will only be halted if the rural development policy follows certain priorities. The first is to support family farming,[13] primarily on a subsistence basis with *wherever possible* a strong protective component to satisfy the basic needs of society; the second is to support labour-intensive small and medium-scale industries and services which take care to observe environmental safeguards.

The tragedy of the environmental future of Indonesia lies in the fact that all Indonesian authorities, colonial as well as post-colonial, have been aware of the dangers for many years. One of the many observant Dutch scientists, J.H. Kievits, who was very concerned with the country's physical future, wrote the following in 1891 in his long essay entitled 'Waar gaan we heen?':[14]

Where are we going? Will the Dutch in future reap the sad reward of having changed Insulinde's emerald girdle into nothing but barren heaps of stone? Is there no way to stop the extravagant destruction of the arable land from where a treasure of organic matter is continuously washed unused into the sea? How should agriculture be practised in order to preserve the land as a living space for a well-to-do, happy population? How can the government contribute to meet this end?

Since this was written, Indonesia has gained independence and is now responsible for the same problems. It is a vast country with great potential, and with an industrious population and authorities aware of what needs to be done. Should it not be possible to save Indonesia from internal and external destruction?

NOTES

1. *Generalanzeiger*, Bonn, 16 May 1984.
2. Guppy, 1984, p. 929.
3. Myers, 1979.
4. Ramsey and Wiersum, 1974; Faber and Karmono, 1977.
5. For details see Fox, 1977.
6. The home-garden technique, together with *sawah*, has probably been used in some areas of Jawa since the tenth century without causing excessive environmental deterioration, largely because it is a kind of natural forest transformed into a 'harvestable forest'. It is dominated by a mixture of perennial trees with underplantings of annual and other herbaceous plants. There is often a closed, multiple storey canopy which is similar to natural forest and protects the soil against sun and rain. The ratio of nutrients in the vegetation to those stored in the soil is high, and the deep-rooted tree crops can take up nutrients from deeper soil layers. These ecological facts make the home-garden resistant to erosion and the leaching of nutrients. Furthermore, home-gardens are mostly situated on low slopes, which when combined with their intensive management, reduces erosion to a minimum. (Terra, 1958; Missen 1972; Ramsey and Wiersum, 1974.)
7. Tinker, 1974, p. 302.
8. *Indonesia Times*, 4 Feb. 1982.
9. Since no 'inferior' researcher dares to advise a 'superior' project manager or even a minister about his findings, unless they reflect the opinions of the superior, it is difficult to change a policy or to abandon a wrong one, unless the responsible superior is replaced. Snodgrass (1979, p. 19) quotes the Indonesian saying *'sedang bapak senang'* which roughly translated means, 'As long as the boss is happy, why bother him with bad news?'
10. Frenerey, 1978, p. 176.
11. Kronholz, *Asian Wall Street Journal*, 7 Sept. 1983, reports the latest developments in the transmigration programme.
12. Jon Tinker (1974, pp. 304–5) interviewed Mochtar Lubis, environmentalist and editor of the influential Jakarta daily *Indonesia Jaya*, which has been fighting since

the early 1970s against the exploitation and destruction of the Indonesian environment and society.

13. This does not exclude the collective efforts of peasants in cooperatives. Cooperatives, which are European in origin, have so far not been very successful in changing agricultural practises in developing countries.

14. Kievits, 1891, p. 712.

SELECT BIBLIOGRAPHY

Adisoemarto, S., and E.F. Brünig, *Transactions of the Second International MAB-IUFRO Workshop on Tropical Rainforest Ecosystems Research*, Hamburg-Reinbek, 1979.

Aditjondro, G.J., 'Mud-bank farming: a fast vanishing tradition in Irian', *Indonesia Times*, 29 Dec. 1983.

Aggarwala, N., 'Millions in Indonesia move to the Outer Islands', *Bangkok Post*, 21 Oct. 1983.

——, 'Falling output threatens the big dream', *Times of Papua New Guinea*, 18 Nov. 1983.

Alamgir, M., 'Programmes of environmental improvement at the community level: Bangladesh, Indonesia, South Korea', *Development Discussion Papers*, 103, Cambridge, Mass., 1980.

——, 'Programmes of environmental improvement at the community level,' *Economic Bulletin for Asia and the Pacific*, 32, 1 (1981), pp. 56–99.

Alphen de Veer, E.J. van, and A. Vink, 'Bestrijding van *alang-alang* met mechanische en chemische middelen,' *Tectona*, 42 (1952), pp. 97–116.

Appleman, F.J., 'De practische mogelijkheden der natuurbescherming in Nederlandsch-Indië', *Tectona*, 30 (1937), pp. 799–807.

Arakawa, H. (ed.), *Climates of Northern and Eastern Asia*, Amsterdam, 1969.

Arndt, H.W., 'Transmigration: achievements, problems, prospects,' *Bulletin of Indonesian Economic Studies*, Canberra (hereafter *BIES*), 19,3 (1983), pp. 50–73.

——, and R.M. Sundrum, 'Transmigration: land settlement or regional development?', *BIES*, 13,3 (1977), pp. 72–90.

Atmawidjaja, R., 'Jakarta und das Problem der Urbanisation', *Materialien zum Internationalen Kulturaustausch*, 18 (1983), pp. 73–6.

Baharsjah, S., 'Indonesia', in Ensminger (ed.), 1978, pp. 73–92 (1978a).

——, 'Population and food production in Indonesia', in Dams *et al.* (eds), 1978, pp. 130–9 (1978b).

Bailey, K.V., and M.J., 'Causes and effects of soil erosion in Indonesia', *Symposium on the Impact*, 1960, pp. 266–78.

Bakker, A.J., 'The Hydrology of the Anten river (West Java)', *OSR News*, 3 (1951), pp. 14–28.

——, and Ch. L. van Wijk, 'Infiltration and run-off under various conditions on Java', *OSR News*, 3 (1951), pp. 56–9.

Baren, F.A. van, 'A century of soil science in Indonesia', *OSR Publications*, 24, Surabaya (1950), pp. 3–13.

Battuta, Ibn, *Reisen ans Ende der Welt, 1325–53*, Tübingen, 1974 (mainly pp. 226–34).

'*Bebossing van geerodeerde gronden op Java: Literatuurstudie*', Wageningen, Landbouwhogeschool, Werkgroep Tropisch Houtteelt, 1973, vol.I.

Becking, J.H., *De Djaticultuur op Java*, Wageningen, 1928.

Beers, H.W. (ed.), *Indonesia, Resources and Their Technical Development*, Lexington, 1970.

Beller, H., *et al.*, 'Problemfeld Ferntourismus am Beispiel eines Projektes auf Bali in Indonesien. Nusa Dua — dem Fremden dienen', in Dennhardt and Pater, 1980, pp. 251–85.

Bemmelen, J.M. van, 'Über die Ursachen der Fruchtbarkeit des Urwaldbodens in Deli (Sumatra) und auf Java für die Tabakkultur und die Abnahme dieser Fruchtbarkeit', *Die Landwirtschaftlichen Versuchsanstalten*, 37 (1890), pp. 374–408.

Bemmelen, R.W. van, *The Geology of Indonesia*, The Hague, 1949.

Berger, L.G. den, 'Landbouwscheikundige onderzoekingen omtrent de irrigatie op Java', *Proefschrift*, Delft, 1915.

——, and F.W. Weber, 'Verslag van de water en slibonderzoekingen van verschillende rivieren op Java', *Mededeelingen van het algemeen Proefstation voor den Landbouw*, no.1, Batavia, 1919.

Bergsma, E., 'Field boundary gullies in the Serayu river basin, Central Java', in *Serayu Valley Project, Final Report*, vol.2, pp. 75–92, Enschede, 1978.

Bernet Kempers, A.J., 'Indonesien', *Historia Mundi*, vol.8, pp. 87–118, Bern, 1959.

Bintaro, R., 'Pressure of population in Klaten and its obvious results', *The Indonesian Journal of Geography*, 1,1 (1960), pp. 45–50.

Birowo, A.T., 'BIMAS: A package programme for intensification of food production in Indonesia', *SSIP Bulletin* (Basel), no.42 (1975), pp. 83–104.

Birowo, A.T., *et al.*, 'Landwirtschaft', in Kötter *et al.*, 1979, pp. 382–447.

Biswas, M.R., *United Nations Conference on Desertification in Retrospect*, Laxenburg (Austria), 1978.

Blaut, J.M., 'The nature and effects of shifting agriculture', *Symposium on the Impact . . .*, 1960, pp. 185–98.

Blom, P.S., 'Agricultural production and socio-economic problems in Indonesia, in particular on Java: An overview', *Abstracts on Tropical Agriculture*, 5,12 (1979), pp. 9–20.

Bodenstedt, A., 'Ländliche Sozialstruktur und angepasste Agrartechnologie in Südostasien. Das Beispiel Indonesien (Java)', in Röll *et al.*, 1980, pp. 1–8.

Boeke, J.H., *Ontwikkelingsgang en toekomst van bevolkings-en ondernemings-landbouw in Nederlandsch-Indië*, Leiden, 1948.

——, 'Dualism in colonial societies', in Evers (1980), pp. 26–37 (repr. of 1966 edn).

Boerema, J., 'Typen van den regenval in Nederlandsch-Indië', *Verh. Kon. Magn. Obs.*, 18 (1926).

Bohlander, K.R., 'Forestry and wood processing', East Kalimantan Transmigration Area Development Project (hereafter TAD), Report no. 5, Hamburg, 1978.

Bois, E. du, 'Het kalkgehalte onzer koffiegronden', *Koffiegids*, 1 (1899/1900), pp. 410–18.

Booth, A., 'Irrigation in Indonesia', *BIES*, 13,1 (1977), pp. 33–74; 13,2 (1977), pp. 45–77.

Boserup, E., *The Conditions of Agricultural Growth: The economics of agrarian change under population pressure*, London, 1965.

Brammah, M., 'Planning new rural settlements in Indonesia', *Ekistics*, 43,257 (1977), pp. 199–203.

Brendl, O., *Bericht über eine Studienreise nach Südostasien*, Vienna, 1982.

Bronsgeest, J., 'Nieuw leven in compagniesland', in *Spectrum Atlas van de nederlandse landschappen*, Utrecht and Antwerp, (1979) pp. 60–7.

Brookfield, H., 'On man and ecosystems', *International Social Science Journal* (UNESCO), 34 (1982), pp. 375–93.

Brown, L., 'Eine neue Bodenethik', *Forum der Vereinten Nationen*, 5,8 (1978), pp. 1–2.

Brüning, K., 'Asien', *Harms Erdkunde*, vol.3, Frankfurt/Main, 1954, pp. 471–99.

Bruijnzeel, L.A., 'Hydrological and biochemical aspects of man-made forests in south-central Java, Indonesia,' *Serayu Valley Project, Final Report*, vol.9, Amsterdam, 1983.

Bryant, N.A., *Population Pressure and Agricultural Resources in Central Java: The Dynamics of Change*, Michigan diss. 1973, Ann Arbor, Michigan, 1977.

Buku Saku Statistik Indonesia (Statistical Pocketbook Indonesia), Jakarta, var. years.

Burbridge, D., *et al.*, 'Land allocation for transmigration', *BIES*, 17,1 (1981), pp. 108–13.

Burger, G., 'Agrare Intensivierungsprogramme in Mittel-Java und Probleme ihrer Realisierung', *Geographische Rundschau*, 27,4 (1975), pp. 151–61.

——, 'Traditioneller Landbau und Regionalentwicklung im Konflikt: Das Beispiel des Rawa-Beckens in Zentral-Java', in Röll *et al.*, 1980, pp. 149–60.

Bus de Giesignies, L.P.J. du, *Rapport van de Commisaris-general Dubus over het stelsel van kolonisatie*, Batavia, 1827.

Büsgen, M., 'Die Forstwirtschaft in Niederländisch-Indien', *Zeitschr. für Forst- und Jagdwesen*, 36 (1904), pp. 1ff.

Callick, R., 'Tale of the transmigrasi', *Times of Papua New Guinea*, 18 Nov. 1983.

Carol, H., 'The calculation of theoretical feeding capacity for tropical Africa', *Geographische Zeitschrift*, 61,2 (1973), pp. 81–94.

Carson, R., *Silent Spring*, Boston, 1962.

Carter, V.G., and T. Dale, *Topsoil and Civilization*, University of Oklahoma Press, 1976.

Casey, J.H., 'Selling agroforestry', *Ceres*, 16,6 no.96 (1983), pp. 41–4.

Chandrasekaran, C., and S. Suharto, 'Indonesia's population in the year 2000', *BIES*, 14,3 (1978), pp. 86–93.

Chapman, V.J., 'Mangrove vegetation', 'Pre-Congress Conference . . .', 1971, pp. 172–85.

Charras, M., *De la forêt maléfique à l'herbe divine: La transmigration en Indonésie: Les Balinais à Sulawesi*, Paris, 1982.

Chernichovsky, D., and O.A. Meesook, 'Regional aspects of family planning and fertility behaviour in Indonesia', *World Bank Staff Working Paper no.462*, Washington DC, 1981.

Chia, L.S., and C. MacAndrews, 'Environmental Problems and Development in South-east Asia', in Mac Andrews and Chia, 1979, pp. 3–14.

Chorley, R.J. (ed.), *Water, Earth and Man: A Synthesis of Hydrology, Geomorphology and Socio-Economic Geography*, London, 1969.

CIDA, 'East Indonesian Regional Development Study', Ottawa: Canadian International Development Agency, quarterly reports, 1974 ff.

Clauss, W., 'Subsistence and commodity production in the Simalungung Highlands of North Sumatra', working paper, no.5, Bielefeld, 1981.

——, and J. Hartmann, 'Agrarentwicklung in Indonesien. Zwei Beispiele aus Nord-Sumatra und Java', working paper, no.7, Bielefeld, 1981.

Cobban, J.L., 'Geographic notes on the first two centuries of Djakarta', in *Changing South-East Asian Cities*, Singapore and London, 1976, pp. 45–57.

Collier, W.L., *Declining Labour Absorption (1878–1980) in Javanese Rice Production*, Bogor, 1980.

——, *et al.*, *Income, employment and food systems in Javanese coastal villages*, Southeast Asia Series, no.44, Athens, Ohio, 1977.

——, *et al.*, 'Acceleration of rural development of Java', *BIES*, 13,3 (1982), pp. 84–101.

'*Comptes Rendus du Congrès International de Géographie, Amsterdam, 1938*', vol.2, *Travaux de la Section IIIc — Géographie Coloniale*, Leiden, 1938.

Cornelius, P., 'De ecologische combinatie-methoden bij aanleg en verbetering van plantages en bossen', *Bèta*, 13,5 (1978).

——, 'Een ecologische kombinatiemethode voor landen bosbouw in tropische gebieden met droogteperioden', *Bèta*, 13,4 (1978).

Coster, Ch., 'Typen van stervend land in Nederlandsch-Indië: no. 1. Bevolkingstheetuinen in de Preangar', *Tectona*, 29 (1936), p. 781.

——, 'Typen van stervend land in Nederlandsch-Indië: no. 2. Het areaal van de zuikerfabriek Boedoean, Besoeki', *Tectona*, 29 (1936), pp. 961–2.

——, 'Typen van stervend land in Nederlandsch-Indië: no.3. Het Tjikeroe-gebied', *Tectona*, 30 (1937), pp. 155–7.

——, 'Typen van stervend land in Nederlandsch-Indië: no.5. Veeweide in het Sindang-palai-gebied,' *Tectona*, 30 (1937), pp. 316–18.

——, 'Bovengrondse afstroming en erosie op Java', *Tectona*, 31 (1938), pp. 613–728.

Dahm, B., 'Indonesien — ein historischer Rückblick', in Kötter *et al.*, 1977, pp. 65–115.

——, *Indonesien. Geschichte eines Entwicklungslandes (1945–71)*, Leiden and Cologne, 1978.

Dam, H. ten, 'Cooperation and social structure in the village of Chibodas' in *Indonesian Economics: The Concept of Dualism in Theory and Policy*, The Hague, 1961, pp. 347–82.

Dames, T.W.G., *Korte nota in zake enkele landbouwkundige problemen op Sumba, met een advies tot hulp op korte termijn*, Dienstreisbericht, 1952.

——, 'The soils of east central Java', *Contributions to the Central Research Station*, no.14, Bogor, 1955.

Dams, T., *et al.* (eds), *Food and Population Priorities in Decision-Making*, Westmead, 1978.

Daneš, J.V., 'Das Karstgebirge des Goenoeng Sewoe in Java', in *Sitzungsbericht der Königlich Böhmischen Gesellschaft der Wissenschaften*, Prague, 1915.

Danhof, G.N., 'Bijdrage tot oplossing van het *alang-alang* vraagstuk in de Lampongsche districten', *Tectona*, 33 (1940), pp. 197–225.

Daroesman, R., 'An economic survey of East-Kalimantan', *BIES*, 15,3 (1979), pp. 43–82.

——, 'Vegetative elimination of *alang-alang*', *BIES*, 17,1 (1981), pp. 83–107.

Daryadi, L., 'Indonesian forest resources', *The Malaysian Forester*, 41,2 (1978), pp. 118–20.

Deinse, H.A. van, 'Aufstieg, Blütezeit und Untergang der Europäischen Agrarwirtschaft in Indonesien', in *Für Erich Selbach*, Krefeld, 1975, pp. 70–82.

Dennhardt, J., and S. Pater (eds), *Entwicklung muss von unten kommen*, Reinbek, 1980.

Dequin, H., 'Agricultural planning for regional development: Case study West Sumatra, Indonesia', *Zeitschrift für ausländische Landwirtschaft*, 10,4 (1971), pp. 305–32.

——, 'Land resources in Indonesia', *Geoforum*, 6,3–4 (1975), pp. 257–60.

——, *Indonesien — 10 Jahre danach — Agrarwirtschaft und Industrie in der Regionalentwicklung einer tropischen Inselwelt*, Riyadh, 1978.

Devas, N., 'Indonesia's Kampung Improvement Program: an evaluative case study', *Ekistics*, 48,286 (1981), pp. 19–36.

Dick, H., 'The oil price subsidy, deforestation and equity', *BIES*, 16,3 (1980), pp. 32–60.

Dietrich, G. (ed.), *Erforschung des Meeres*, Frankfurt/Main, 1970.

Dijk, J.W. van, and V.K.R. Ehrencron, 'The different rate of erosion within two adjacent basins in Java', *Mededeelingen van het algemeen Proefstation voor de Landbouw*, no.84 (1940), pp. 3–10.

——, and W.I.M. Vogelzang, 'The influence of improper soil management on erosion velocity in the Tjiloetoeng basin, West Java', *Mededeelingen van het algemeen Proefstation voor de Landbouw*, no.71 (1948), pp. 3–9.

Djajonto, W., 'No escape from poverty for Indonesian fishermen', *Indonesia Times*, 2 Nov. 1983.

——, 'Village self-help becomes reality through unique method', *Indonesia Times*, 16 Jan. 1984.

Dobben, W.H. van, and R.H. Lowe-MacConnell (eds), *Unifying Concepts in Ecology*, The Hague, 1975.

Dobert, M., 'Operation transmigrasi', *Development Forum*, 9,5 (1981), p. 4.

Dollee, A.W., 'Sedimentproductie ten gevolge van regenval in het Iratunseluna gebied om Centraal Java', unpubl., n.d.

Domrös, M., 'Über das Vorkommen von Frost auf Java/Indonesien, insbes in den Pengalengan Highlands', *Erdkunde*, 30 (1976) pp. 97–108.

——, 'Klima' in Kötter *et al.*, 1979, pp. 21–37.

Donner, W., 'Die "*laghi collinari*". Eine neue Technik für die Entwicklung semiarider Landschaften', *Perrot-Bibliothek*, no.12, Calw, 1964.

——, *The Five Faces of Thailand: An Economic Geography*, London, 1978.

——, *Haiti. Naturraumpotential und Entwicklung*, Tübingen, 1980.

——, *Thailand ohne Tempel. Lebensfragen eines Tropenlandes*, Frankfurt/Main, 1984.

Dubois, E., *Pithecanthropus erectus, eine menschenähnliche Übergangsform aus Java*, Batavia, 1894.

Dudal, R., *Soil Survey and its Application in Indonesia*, FAO/ETAP Report no.1509, Rome, 1962.

Dürr, H., 'Regionalentwicklung in Indonesien 1974–79', *Geographische Rundschau*, 27,4 (1975), pp. 169–78.

——, 'Regionalberichte und -pläne für Indonesien. Problemhintergrund, Dokumentation, künftige Aufgaben', *Erdkunde*, 31 (1977), pp. 146–56.

——, 'Raumentwicklung im Dilemma zwischen Wachstums- und Gleichheitszielen. Indonesien als Beispiel', *Geographische Rundschau*, 34,2 (1982), pp. 49–57.

Edelman, C.H., 'Studiën over de bodemkunde van Nederlandsch-Indië', *LEB Fonds Publicatie*, no.24, Wageningen, 1947.

Egger, K., 'Zehn Jahre nach Stockholm. Das Umweltprogramm der Vereinten Nationen (UNEP) in seinem politischen Umfeld', *Vereinte Nationen*, 30,4 (1982), pp. 113–16.

——, *et al.*, *Ökologische Probleme ausgewählter Entwicklungsländer*, Hamburg, 1972.

——, *et al.*, Ecological recommendations for transmigration, TAD Working Paper no.3, Hamburg, 1981.

Eichelberger, R., 'Regenverteilung, Pflanzendecke und Kulturentwicklung in der ostindischen Inselwelt', *Geographische Zeitschrift*, 30 (1924), pp. 103–16.

Emmerson, D.K., 'Rethinking artisanal fisheries development: Western concepts, Asian experience', *World Bank Staff Working Paper*, no.423, Washington DC, 1980.

Engelen, G.B., 'The Serayu Valley Project, Central Java, Indonesia (1973–80)', *Geo-Journal*, 4,1 (1980), pp. 71–5.

Ensminger, D. (ed.), *Food Enough or Starvation for Millions?*, New Delhi, 1978.

Eppink, L.A.A.J., and J.P.A. Palte, 'Some socio-economic aspects of soil erosion in the Jatiluhur area', Project Report no.12, Bandung/ Wageningen, 1980.

Ernst, K., and G.R. Zimmermann, 'Ein Beitrag zur Steigerung der Nahrungsmittelerzeugung, der Beschäftigung und der Einkommen. Das landwirtschaftliche Produktionsmittelprojekt Klaten/Indonesien,' in *Zeitschrift für auslandische Landwirtschaft*, 13,2 (1974), pp. 169–81.

Escher, B.G., 'De Goenoeng Sewoe en het probleem van de karst in de tropen', in *Handelingen van het XXIII. Nederl. Natuuren Geneeskundig Congres 1931*, Haarlem, 1933.

EUROCONSULT, *Report Joint Review Mission Rawa Sragi Project*, Arnhem, 1980.

Evans, J., 'The growth of urban centres in Java since 1961', *BIES*, 20,1 (1984), pp. 44–57.

Everingham, J., 'Poverty in Paradise', *Asia Magazine*, 21, X26 (25 Sept. 1983), pp. 3–9.

Evers, H.D., *Sociology in South-east Asia*, Kuala Lumpur, 1980.

——, 'Subsistence production and wage labour in Jakarta', Working Paper no.8, Bielefeld, 1981*a*.

——, 'The contribution of urban subsistence production to incomes in Jakarta', *BIES*, 17,2 (1981*b*), pp. 89–96.

——, and J. Hartmann, 'Erklärungsversuche zur Krise der Agrarentwicklung Javas', Working Paper no.3, Bielefeld, 1981.

——,*et al.*, 'A survey of low income households in Jakarta: Selected summary tables,' Working Paper no.17, Bielefeld, 1982.

Faber, Th., and M. Karmono, *Serayu Valley Project, Final Report*, vol.1 (*General Report*), NUFFIC, 1977.

FAO, *Soil erosion by water: Some measures for its control on cultivated lands*, FAO Agr. Dev. Paper no.81, Rome, 1965.

——, *Forestry in Indonesia*, Report TA 2984, Rome, 1971.

——, *Termination report, Upper Solo Watershed Management and Upland Development Indonesia*, Rome, 1976.

FAO/IBRD, *Indonesia, smallholder cattle and buffalo development project* (preparation mission), Rome, 1979.

FAO/Soil Research Institute, *Project document: Land resources evaluation with emphasis on the Outer Islands'*, Bogor, 1978.

Fachurozzi, S., and C. MacAndrews, 'Buying time: Forty years of transmigration in Belitang', *BIES*, 14,3 (1978), pp. 94–103.

Fisher, C.A., *South-east Asia: A Social, Economic and Political Geography*, London (repr. 1971).

Flathe, H., and D. Pfeiffer, 'Grundzüge der Morphologie, Geologie und Hydrologie im Karstgebiet Gunung Sewu (Java)', *Geologisches Jahrbuch*, 83 (1965), pp. 532–62.

Fokkinga, J., 'Boschreserveering en inlandsche landbouw van overjarige gewassen op Java en Madoera', *Tectona*, 27 (1934), pp. 142–89 and 338–85.

Forbes, D., 'Mobility and uneven development in Indonesia: A critique of explanations of migration and circular migration', in *Population Mobility and Development: South-east Asia and the Pacific*, Canberra, 1981, pp. 51–70.

Fox, J.J., *Harvest of the Palm: Ecological Change in Eastern Indonesia*, Cambridge, Mass., and London, 1977.

Franke, R.W., 'The green revolution in a Javanese village', unpubl. Ph.D. thesis, Harvard University, 1972.

Frauendorfer, R., 'Das Rawa Sragi Project. Sumpfentwässerung im Süden Sumatras,' in Brendl (ed.) 1982, pp. 673–88.

Fremerey, M., 'Indonesien', in Nohlen and Nuscheler (eds), vol. 4.1 (1978), pp. 169–200.

Fromberg, P.F.H., 'Over de guano. Hare oorsprong, chemische samenstelling en werksamheid als meststof', *Natuurk. Tijdschr. Ned.-Indië*, 6 (1854), pp. 63–84.

——, 'Eerste verslag van den uitkomsten der bemesting van koffijboomen met guano en andere stoffen in de proeftuin te Genteng', *Tijdschr. Nijverheid Ned.-Indië*, 4 (1858), pp. 298–307.

Fruin, Th.A., 'Overpopulation and the emancipation of the village', in *Indonesian Economics*, The Hague, 1961, pp. 333–43.

Fuchs, F.W., 'Moderne Kolonisation in Niederländisch-Indien', *Koloniale Rundschau*, 29 (1938), pp. 316–42.

Furtado, J.I., 'The status and future of the tropical moist forest in South-east Asia', in *Developing Economies and the Environment*, Singapore, 1979, pp. 73–120.

——, 'Freshwater swamps and lake resources: A synthesis', in Furtado (ed.), 1980, pp. 797–8.

—— (ed.), 'Tropical ecology and development', in *Proceedings of the Fifth International Symposium of Tropical Ecology, 1979*, Kuala Lumpur, 1980 (2 vols.).

Fyfe, W.S., 'The environmental crisis: Quantifying geosphere interaction', *Science*, 213 (3 Sept. 1981), pp. 105–10.

Ganssen, R., and F. Hädrich (eds), *Atlas zur Bodenkunde*, Mannheim, 1965.

Gauchon, M.J., 'Some aspects of watershed management economics: Upper Solo Watershed Management and Upland Development Project', *Field Termination Report* no.7, Solo, 1976.

Geertz, C., *Agricultural Involution: The Process of Ecological Change in Indonesia*, Berkeley, 1963.

Gerlach, J.C., 'Bevolkingsmethoden van ontginning van *alang-alang* terreinen in den Zuider en Oosterafdeeling van Borneo', *Landbouw*, 14 (1938), pp. 446–50.

The Global 2000 Report to the President, Washington DC, 1980.

Gómez-Pompa, A., *et al.*, 'The tropical rainforest: a nonrenewable resource', *Science*, 177 (1972), pp. 762–5.; see also readers' correspondence, loc. cit., 181 (1973), pp. 893–5.

Gonen, A., 'Tourism and coastal settlement processes in the Mediterranean region', *Ekistics*, 48, 290 (1981), pp. 378–81.

Gonggrijp, G., 'Bescherming van Indië's bosschen. Vernietiging van toekomstige welvaart', *Tectona*, 30 (1937), pp. 682–3.

Gonggrijp, J.W., 'De snelheid van devastatie van den bosschen in de buitengewesten', *Tectona*, 30 (1937), pp. 793–9.

Gonggrijp, L., 'Het erosie-onderzoek', *Tectona*, 34/35 (1941), pp. 200–20.

Goodland, R.J.A., and H.S. Irwin, *Amazon Jungle: Green hell to red desert?*, Amsterdam, 1975.

Gourou, P., *L'Asie*, Paris, 1953.

Grandstaff, T., 'The development of swidden agriculture (shifting cultivation)', *Development and Change*, 9 (1978), pp. 547–79.

Gray, D.D., 'Asia's coral reefs facing rapid destruction', *Indonesia Times*, 11 Nov. 1983.

Grimm, K., 'Problems of urban development planning', *Intereconomics*, 11,3 (1976), pp. 88–92.

Groeneveld, S., 'Agrarberatung und sozialökonomische Feldforschung am Beispiel der Agrarberatungszentren in West-Java. Ein Plädoyer für Aktionsforschung', in Röll *et al.*, 1980, pp. 87–100.

Groeneveldt, W., 'Bosvernieling in Nederlandsch-Indië, speciaal in de buitengewesten', *Landbouwkundige Tijdschrift*, 49 (1937), pp. 318–36.

Guiness, P., *et al.*, *Transmigrants in South Kalimantan and South Sulawesi: Inter-island government sponsored migration in Indonesia*, Yogyakarta, 1977.

Gunnarsson, B., *Japan's Harakiri*, Stockholm, 1971.

Guppy, N., 'Tropical deforestation: A global view', *Foreign Affairs*, 62,4 (1984), pp. 928–65.

Gutzler, H., 'Tourismuspolitik zwischen Wirtschaft, Umwelt und Kultur', *Materialien zum Internationalen Kulturaustausch*, 7 (1978), pp. 80–5.

Haaf, G., 'Noch sind die Tiere nicht verloren', *Die Zeit*, 14 Jan. 1983.

Haan, F. de, *Priangan: De Preanger Regentschappen onder het nederlandsch bestuur tot 1911*, Batavia, 1910–12.

Haan, de, 'Beschouwingen over hydrologie en erosie', *Bergcultures*, 9,11 (1935), pp. 1113–25.

Haan, J.H. de, 'Erosie, theorie en praktijk', *Tectona*, 34/5 (1942), pp. 55–63.

——, 'The economic and social aspects of erosion control', *OSR News*, 3 (1951), pp. 69–72.

——, 'Silt transport of rivers in Java', *Flood Control Series*, 3 (1952), pp. 179–84.

——, 'Boswezen en erosie-bestrijding', *Tectona*, 43 (1955).

Haar, Ch., 'A program for land registration and land transfer in Indonesia', *Ekistics*, 41,244 (1976), pp. 155–7.

Hadad, I., 'Large and small-scale industry development in Indonesia. Problems and prospects of an industrialisation process minimizing 'loss' of indigenous culture', in *Culture and Industrialisation*, 1978, pp. 148–66.

Hadering, K., 'The regional seas approach, a focal point for environmental cooperation', *Indonesia Times*, 22 March 1983.

——, 'Fisheries development as maritime community assistance', *Indonesia Times*, 23 June 1983.

——, 'Raw materials for biogas plants', *Indonesia Times*, 29 March 1984.

Hadisumarno, S., 'Coastline accretion in Segara Anakan, Central Java, Indonesia', *Indonesian Journal of Geography*, 9,37 (1979), pp. 45–52.

Hagreis, B.J., 'Ladangbouw', *Tectona*, 24 (1931), pp. 598–631.

Hall, J.J.H. van, 'Uitputting van den grond', *Tijdschr. Ind. Landbouw Gen.*, 3(1873), pp. 192–201.

Hamzah, S., 'Mangrove forest saves the coastal ecology', *Indonesia Times*, 7 March 1983.

——, 'Individual discipline key to solving rubbish problem', *Indonesia Times*, 8 Aug. 1983.

——, 'Tubewell, a water crisis breakthrough', *Indonesia Times*, 29 Sept. to 1 Oct. 1983.

——, 'Toraut Dam, an integrated irrigation project', *Indonesia Times*, 24 and 25 Nov. 1983.

——, '*Galungung* victims return to "dangerous areas" ', *Indonesia Times*, 16 Jan. 1984.

——, 'Lombok Selatan demands high price', *Indonesia Times*, 11 April 1984.

Hansen, G.E., 'Indonesia's green revolution. The abandonment of a non-market strategy towards change', *SEADAG Papers*, 1972/6.

Hanson, A.J., 'Transmigration and marginal land development', in *Agriculture and Rural Development in Indonesia*, Boulder, Colorado, 1981, pp. 219–35.

——, *et al.*, 'Settling coastal swamplands in Sumatra: A case-study for integrated resource management' in *Developing Economies and the Environment*, Singapore, 1979, pp. 121–75.

Hardjono, J.M. *Indonesia: Land and People*, Jakarta, 1971; particularly pp. 91–105.

——, *Transmigration in Indonesia*, Kuala Lumpur, 1977.

——, 'Transmigration: a new concept?', *BIES*, 14,1 (1978), pp. 107–12.

Hardy, F., 'Some ecological aspects of tropical pedology', *The Malayan Journal of Tropical Geography*, 2 (1954), pp. 1–8.

Harloff, Ch. E.A., 'Over het kruipen van de bodem in het noordelijk deel van het Regentschap Bandjarnegara', *De Mijningenieur*, 11 (1930), pp. 96–101.

Hartge, H.H. and H.J. Wiebe, 'Die Böden sind an allem schuld', *Umschau*, 78,20 (1978), p. 636.

346   *Land Use and Environment in Indonesia*

Hartmann, A., 'Sozialökonomische und agrartechnische Aspekte der "Grünen Revolution" in Indonesien', *Vierteljahresberichte der Friedrich-Ebert-Stiftung*, no.51 (1973), pp. 61–76.

Hartmann, J., *Subsistenzproduktion und Agrarentwicklung in Java/Indonesien*, Saarbrücken, 1981.

Harts-Broekhuis, A., and H. Palte-Gooszen, *Demografische aspekten van armoede in een javaans dorp*, Utrecht, 1977.

Hatzfeldt, H., *Economic Development Planning in Indonesia*, Bangkok, 1969.

Heberer, G., and W. Lehmann, *Die Inland-Malaien von Lombok und Sumbawa*, Göttingen, 1950.

Heine-Geldern, R.von, 'Urheimat und früheste Wanderungen der Austronesier', *Anthropos*, 27 (1932), pp. 543–619.

Hendry, P., 'Where the desert stops', *Ceres* 17,2 (1984), pp. 20–4.

Hesmer, H., *Der kombinierte land- und forstwirtschaftliche Anbau*, vol.II *Tropisches und Subtropisches Asien*, Stuttgart, 1970.

Hill, R.D., and J.M. Bray, *Geography and the Environment of South-east Asia*, Hong Kong, 1978.

Holle, K.F., 'Iets over bemesting van koffijtuinen', *Tijdschrift Nijverheid en' Landbouw in Ned.-Indië*, 12, New Series 7 (1866a), pp. 88–91.

——, 'Een groot gevaar, dat sluipend nadert', *Tijdschrift Njiverheid en Landbouw in Ned.-Indië*, 12 (1866b), pp. 122–34.

Hollerwöger, F., 'The progress of the river deltas in Java', *Scientific Problems of the Humid Tropical Zone, Deltas and their Implication*, Proceedings of the Dacca Symposium (UNESCO), 1966.

Hollie, P.G., 'Malaysia starts to suffer ills of industrialisation', *International Herald Tribune*, 4 Feb. 1981.

Holttum, R.E., 'Adinandra Belukar: A succession of vegetation from bare ground in Singapore island', *The Malayan Journal of Tropical Geography*, 3 (1954), pp. 27–32.

Homan van der Heide, J., 'Eenige gegevens nopens het slibgehalte van irrigatiewater', *Archief Java Suiker Industrie*, 122/II (1904), pp. 669–705.

Hope, G.S., et al. (eds), *The Equatorial Glaciers on New Guinea*, Rotterdam, 1976.

Horstmann, K., 'Indonesien. Bevölkerungsproblem und Wirtschaftsentwicklung', *Geographisches Taschenbuch*, 1958/9, pp. 410–23.

——, 'Die Erhöhung der Tragfähigkeit des ländlichen Javas durch nichtbäuerliche Erwerbstätigkeit', in Röll et al., 1980, pp. 101–10.

——, 'Stadtregionen auf Java? — Erste Annäherung', *Forschungsbeiträge zur Landeskunde in Süd-und Südostasien* (Festschrift für Uhlig), 1982, pp. 146–56.

Huges, C., and D., 'Teeming life in a rain forest', *National Geographic*, 163,1 (1983), pp. 49–65.

Huizer, G., 'Peasants mobilization and land reform in Indonesia', *ISS Occasional Papers*, no.18, 1972.

Hull, T.H., 'Fertility decline in Indonesia: A review of recent evidence', *BIES*, 16,2 (1980), pp. 104–12.

——, *et al.*, 'Indonesia's family planning story: Success and challenge', *Population Bulletin*, 32,6 (1977).

Hunter, L., 'Tropical forest plantations and natural stand management: A national lesson from East Kalimantan?' *BIES*, 20,1 (1984), pp. 98–116.

IBRD, *World Development Report 1979* (in German) Washington DC, 1979.

——, *World Development Report 1983* (in German) Washington DC, 1983.

——, *5th Urban Development Project — Indonesia*, Washington DC, 1984a.

——, *Indonesia — Urban Services Sector Report*, Washington DC, 1984b.

——, *Indonesia — Second Swamp Reclamation Project*, Report no.4645–IND, Washington DC, 1984c.

ILO, *Poverty and Landlessness in Rural Asia*, Geneva, 1980.

Imanudin, 'Energy from agricultural waste', *Indonesia Times*, 9 July, 1981.

Imber, W., and H. Uhlig, *Indonesien*, Bern, 1979.

Inansothy, S., 'The change of vegetation and soil properties after logging of natural forest', BIOTROP Internal Report (unpubl.), 1975.

Institut für Asienkunde, *Wirtschaftspartner Indonesien*, Hamburg, 1979.

Institut für Baustofflehre, *Marginalsiedlungen in Ländern der Dritten Welt*, Stuttgart, 1979.

Iskander, N. *et al.*, 'The building of demographic competence in Indonesia', *Studies in Family Planning*, 5,9 (1974), pp. 289–93.

Ismartono, Y., 'On the way to improvement', *Indonesia Times*, 6 July, 1982.

Jacobs, J., 'Diversity, stability and maturity in ecosystems influenced by human activities', in van Dobben and Lowe MacConnell, 1975, pp. 187–207.

Jones, P.H., 'Lamtoro and the Amarasi model from Timor', *BIES*, 19,3 (1983), pp. 106–12.

Joosten, J.H.L., 'Ontwikkeling en problemen van de bemoeienis van den Landbouwvoorlichtingsdienst met de bodenbescherming in West Java', *Landbouw*, 17,12 (1941), pp. 1063–80.

Junghuhn, F., *Topographische und naturwissenschaftliche Reise durch Java*, Magdeburg, 1845.

——, *Java, seine Gestalt, Pflanzendecke und innere Gestalt*, Leipzig, 1852–4 (3 vols.).

——, *Licht und Schattenbilder aus dem Binnenlande von Java*, Leipzig, 1855.

Kali Konto Project, 'Forestry for rural communities', Perum Perhutani/Staatsbosbeheer, working papers, 1979.

Kälin, J., 'Die ältesten Menschenreste und ihre stammesgeschichtliche Deutung', *Historia Mundi*, vol.1, pp. 33–98, Bern, 1952.

Kahirman, M., and A. Mardjuki, 'The role of aerial photos in the management of Indonesian forests', *The Malaysian Forester*, 41,2 (1978), pp. 125–7.

Kanayama, S., 'Tin bearing granites and tin placers in Bangka and Billiton islands in Indonesia', *Tonan Ajia Kenkyu* 11,3 (1973), pp. 321–7 (English summary).

Karimoeddin, T., 'National report of Indonesia: Creating consciousness of the public', in *One World Only: Industrialisation and Environment*, Tokyo, 1973, pp. 161–82.

Kartawinata, K., 'The tropical forest research programme of BIOTROP and recommendations', in Kartawinata *et al.*, 1978, pp. 169–79.

——, 'The Kalimantan Timur Project', in Adisoemarto and Brünig (1978), pp. 223–7.

——, 'East Kalimantan: A comment', *BIES*, 16,3 (1980), pp. 120–1.

——, and R. Atmawidjaja (eds), *Coordinated study of lowland forests in Indonesia*, Bogor, 1974b.

——, *et al.*, *East Kalimantan and the Man-and-Biosphere Program*, Samarinda, 1978.

Katili, J.A., 'Coal and peat in Indonesia, potentials and prospects', *Indonesia Times*, 18 Oct. 1983.

Keleny, G.P., 'Soil organization and land use pattern with special reference to Indonesia', *Symposium on the impact* . . ., 1960, pp. 127–32.

Keyfitz, N., 'The ecology of Indonesian cities', *American Journal of Sociology*, 66 (1960/1), pp. 348–54.

Khan, H., 'Indonesia', in *World Atlas of Agriculture*, vol.2, pp. 220–52, Novara (Italy), 1973.

Kievits, J.H., 'Waar gaan we heen?', *Teysmannia* 2 (1891), pp. 701–12; 3 (1892), pp. 1–10, 79–89, 121–34.

——, 'Het nut van terrassen', *Tijdschrift Nijverheid en Landbouw in Ned.-Indië*, 45 (1893), pp. 280–303; 46 (1894), pp. 136–49; 263–70.

Kinzelbach, W., 'Umweltprobleme in China', *Umschau*, 81,24 (1981), pp. 754–8.

Klar, schön war's, aber . . ., *Tourismus in die Dritte Welt*, Freiburg, Informationszentrum Dritte Welt, 1983.

Klinkert, G.P., 'De catastrofale overstroming op Java in het jaar 1861', *Tectona*, 30 (1937), pp. 731–6.

Koehler, K.G., 'Wood processing in East-Kalimantan', *BIES*, 7,3 (1972), pp. 93–129.

Köneke, A. (ed.), *Improving Environmental Soundness of Industrial Projects in Asian Countries*, Berlin, 1983.

Koentjaraningrat (ed.), *Villages in Indonesia*, Ithaca, NY, 1967.

Koorders, S.H., 'Beoordeeling der vruchtbaarheid van boschgronden naar den oorspronkelijken plantengroei', *Teysmannia*, 4 (1893), pp. 460–3.

——, 'Over de waarde van Albizzia moluccana, Miq., voor reboisatie op Java', *Teysmannia*, 5 (1894), pp. 276–322.

——, 'Iets over spontane reboisatie van verlaten koffietuinen', *Tijdschrift voor Nijverheid en Landbouw in Nederlandsch Indië*, 46 (1894a), pp. 355–66.

——, 'Waarnemingen over spontane reboisatie op Java', *Teysmannia*, 5 (1894b), pp. 467–78.

——, 'Wildernis in een 6 jaar geleden verlaten fort op Java', *Teysmannia*, 9

(1898), pp. 267–74.

Kositchotethana, B., 'Indonesia's pride: Arun gas field', *Indonesia Times*, 7 Nov. 1982.

Kostermans, A.J.G.H., 'The influence of man on the vegetation of the humid tropics', *Symposium on the impact*..., 1960, pp. 332–8.

Kötter, H., *et al.*, *Indonesien*, Tübingen and Basel, 1979.

Kovda, V., 'The management of soil fertility', *Nature and Resources*, 8,2 (1972), p. 3.

——, 'The world's soils and human activity', in Polunin (1972), p. 274.

Kramers, J.G., 'Vierde verslag omtrent de proeftuinen en andere mededeelingen over koffie', in *Meded.'s Lands Plantentuin*, 75 (1904).

Krausse, G., 'Economic adjustment of migrants in the city: The Jakarta experience', *International Migration Review*, 13,1 (1979), pp. 46–70.

Kroef, J.M. van der, 'Population pressure and economic development in Indonesia', in Spengler and Duncan, 1956, pp. 739–54.

Kronholz, J., 'Jakarta's push to move masses hits new snags', *The Asian Wall Street Journal*, 7 Sept. 1983.

Krug, H.J., 'Melanesien', in *Grosse Illustrierte Länderkunde*, vol.2, pp. 1423–7, Gütersloh, 1963.

Kündig-Steiner, W., 'Indonesien', in *Grosse Illustrierte Länderkunde*, vol.1, pp. 1279–1326, Gütersloh, 1963.

Lagerberg, K., *West Irian and Jakarta Imperialism*, London, 1979.

Lamb, A.F.A., 'Artificial regeneration within the humid lowland tropical forest', *Unasylva* 22,4, no.91 (1968), pp. 7–15.

Lamprecht, H., 'Zur ökologischen Bedeutung des Waldes im Tropenraum', *Forstarchiv*, 44 (1973), pp. 117–23.

Leake, J., 'The livestock industry', *BIES*, 16,1 (1980), pp. 65–74.

Leemann, A., 'Sozio-ökonomische Erhebungen zum Tourismus in Bali', *Zeitschr. f. Fremdenverkehr*, no.3 (1978), pp. 19–23.

Leeuwen, L. van, 'Aantekeningen over erosie', *Tectona*, 34/5 (1941), pp. 548–56.

Lehmann, H., 'Das Antlitz der Stadt in Niederländisch-Indien', *Länderkundliche Forschung* (Festschrift für Krebs), 1936, pp. 109–39.

Leiserson, M., *et al.*, *Indonesia: Employment and Income Distribution in Indonesia*, Washington DC, IBRD, 1980.

Leupold, W., and W. Rutz (eds), *Der Staat und sein Territorium*, Wiesbaden, 1976.

Lindauer, G., 'Indonesiens Engpässe: Infrastruktur und Übervölkerung', *Indo-Asia*, 11 (1969), pp. 335–40.

——, 'Urbanization and slums as socio-economic problems in the Far East and South-east Asia', *Internationales Asienforum*, 3 (1972), pp. 36–49.

Linden, P. van der, 'Contemporary soil erosion in the Sanggreman river basin related to the Quarternary landscape development', *Serayu Valley Project, Final Report*, vol.3, Amsterdam, 1978a.

350Land Use and Environment in Indonesia

——, 'A reconstruction of the Quarternary landscape development of the Serayu river basin, Central Java', *The Indonesian Journal of Geography*, 8,36 (1978*b*), pp. 1–16.

Lipscombe, L., *Overpopulation in Java: Problems and Reactions*, Canberra, 1972.

Litahalim, D., 'Flats, bright prospects, though rather costly', *Indonesia Times*, 30 May 1983.

——, *'Gelandangan* — who are they?' *Indonesia Times*, 25 July 1983.

Löffler, E., 'Übersichtsuntersuchungen zur Erfassung von Landressourcen in West-Kalimantan, Indonesien', *Erdkundliches Wissen*, 58 (1982), pp. 122–31.

Lowdermilk, W.C., 'Man-made deserts', *Pacific Affairs*, 8,4 (1935), pp. 409–19; see also 'Menschen maken woestijnen', *Tectona*, 29 (1936), pp. 613–17, and readers' correspondence in *Tectona*, 30 (1937), pp. 319–21.

Lutz, R., 'Klapperstorch und Spirale: Doktor Waluyos Problem', *Deutsches Allgemeines Sonntagsblatt*, 4 Oct. 1981.

MacAndrews, C., 'Transmigration in Indonesia: Problems and prospects', *Asian Survey*, 18,5 (1978).

——, 'Land settlement policies in Malaysia and Indonesia. A preliminary analysis', *Occasional Paper Series*, no.52, Singapore 1978.

——, and Chia Lin Sien (eds), *Developing Economies and the Environment*, Singapore, 1979.

MacComb, A.L., 'Land use and area development in the Upper Solo river basin' (Paper presented at workshop on regional development and transmigration, Solo), 1973.

——, and H. Zakaria, 'Soil Erosion in upper Solo river basin, Central Java', *Rimba Indonesia*, 16, 1–2 (1971), pp. 20–7.

——, and H. Soedarma, 'Some notes on social and economic background of regional land use in the upper Solo river basin', *Proceedings of the Workshop on Natural Resources*, Jakarta, 1972.

MacConnell, D.J., 'Land capability appraisal Indonesia: Farming systems and land use along a traverse of the Solo Valley', Bogor, FAO, 1974.

Machatschek, F., *Vergleichende Länderkunde der aussereuropäischen Erdteile*, Munich, 1941/2.

MacLeish, K., 'Java — Eden in transition', *National Geographic*, 139,1 (1971), pp. 1–43.

MacVey, R.T. (ed.), *Indonesia*, New Haven, Conn., 1963.

Mäder, U., 'Tourismus und Umwelt', in *Klar, schön war's, aber ...'*, Freiburg, 1983.

Manan, S., 'The effects of forest cutting on water and nutrient cycles', in Kartawinata and Atmawidjaja, 1974*b*, pp. 71–83.

Manning, C., 'The timber boom with special reference to East Kalimantan', *BIES*, 7,3 (1971), pp. 30–60.

Marschall, W., *Der Berg des Herrn der Erde. Alte Ordnung und Kulturkonflikt in einem indonesischen Dorf*, Munich, 1976.

Marsden, W., *The History of Sumatra*, London, 1783.

May, B., *The Indonesian Tragedy*, London 1978.

Meijer, W., 'Regeneration of tropical lowland forest in Sabah, Malaysia, forty years after logging', *Malay Forester*, 32 (1970), pp. 204–29.

——, *Indonesian Forests and Land Use Planning: Report of a fact-finding tour*, 1973, Lexington, 1975.

Meijerink, A.M.J., 'A hydrological reconnaissance survey of the Serayu river basin, Central Java', *Serayu Valley Project, Final Report*, vol.2, pp. 25–53., Enschede, 1978.

Messmer, M., 'Kampung-improvement in Indonesien', in *Marginalsiedlungen in Ländern der Dritten Welt*, Stuttgart, 1979, pp. 44–55.

Metzner, J.K., 'Mensch und Umwelt im östlichen Timor', *Geographische Rundschau*, 27,6 (1975), pp. 244–50.

——, 'Landschaftserhaltung und Möglichkeiten zur intensiveren Landnutzung durch Leucaena leucocephala in Kabupaten Sikka, Flores', *Erdkunde*, 30,3 (1976a), pp. 224–34.

——, 'Malaria, Bevölkerungsdruck und Landschaftszerstörung im östlichen Timor', *Geographische Zeitschrift*, Beihefte 43 (1976b), pp. 121–37.

——, 'Die Viehhaltung in der Agrarlandschaft der Insel Sumba und das Problem der saisonalen Hungersnot', *Geographische Zeitschrift*, 64,1 (1976), pp. 46–71.

——, 'Lamtoronisasi: An experiment in soil conservation', *BIES* 12,1 (1976), pp. 103–9.

——, 'Bodenrecht und Landschaftswandel in Sikka/Flores. Untersuchung zur Entwicklung der Grundbesitzverfassung in einem übervölkerten Agrarraum Ostindonesiens', *Geographische Zeitschrift*, 65,4 (1977a), pp. 264–82.

——, *Man and Environment in Eastern Timor*, Canberra, 1977.

——, 'Agrarräumliches Ungleichgewicht und Umsiedlungsversuche auf den östlichen Kleinen Sunda-Inseln. Konsequenzen für eine geoökologische Regionalplanung', *Giessener Beiträge zur Entwicklungsforschung*, I,4 (1978), pp. 29–47.

——, 'Eigenständige Weiterentwicklung traditioneller Formen bäuerlicher Zusammenarbeit in ihrer Wirkung auf die Agrarstruktur. Das Beispiel Adonara (Ost-Indonesien)', in Röll *et al.*, 1980, pp. 73–86.

——, *Agriculture and Population Pressure in Sikka, Isle of Flores: A contribution to the study of the stability of agricultural systems in the wet and dry tropics*, Canberra, 1982.

——, 'Innovations in agriculture incorporating traditional production methods: The case of Amarasi (Timor)', *BIES*, 19,3 (1983), pp. 94–105.

Meulen, G.F. van der, 'Diagrams of the biological agricultural method for the tropics', 1957, quoted by Thijsse, 1975b.

Michael, R., 'A comparative analysis of plans and problems: Jakarta', *Ekistics*, 45,266 (1978), pp. 4–12.

Michel, Th., *Interdependenz von Wirtschaft und Umwelt in der Eipo-Kultur von Moknerkon. Bedingungen für Produktion und Reproduktion bei einer Dorfschaft im Zentralen Bergland von Irian Jaya (West-Neuguinea)*, Indonesien, Berlin, 1983.

Missen, G.J., *Viewpoints on Indonesia: A Geographical Study*, Melbourne, 1972.

Mock, F.J., *Indonesia — Land capability appraisal, water availability appraisal*, Bogor, FAO/UNDP, 1973.

Möbius, K., 'Die Desa und ihre Bedeutung für die wirtschaftliche Entwicklung Indonesiens', in *Gegenwartsprobleme der Agrarökonomie* (Festschrift für Baade), pp. 284–300, Hamburg, 1958.

Mohr, E.C.J., 'Verslag eener excursie naar Banjarnegara in verband met het slibbezwaar, veroorzaakt door eenige rivieren in't Serajoe-dal', in *Jaarboek Dep. van Landbouw*, 1906.

——, 'Over het slibbezwaar van eenige rivieren in het Serajoedal', *Mededeelingen uitgaande van het Departement van Landbouw*, no.5, Batavia, 1908.

——, 'De bodem der tropen in het algemeen, en die van Nederlandsch-Indië in het bijzonder', *Koninklijke Vereeniging Koloniaal Instituut*, no.31, Amsterdam, 1933 (Part 1), 1935(Part 2).

——, 'The relation between soil and population density in the Netherlands East Indies', in *Comptes Rendus*, 1938, pp. 478–93.

Mounier, T., 'Ontginningen op Java in verband met den water-rijkdom en zorgen voor de bouwkruin', *Tijdschrift voor Nijverheid en Landbouw in Nederlandsch-Indië*, 26 (1882), pp. 42–58.

Müller, W., *Bibliographie deutschsprachiger Literatur über Indonesien*, Hamburg, 1979.

Müller-Darss, H., 'The significance of ergonomics to agro-forestry', *Agroforestry Systems*, 1 (1982), pp. 41–52 (1983), pp. 205–23.

Mulder, N., 'Individual and society in contemporary Thailand and Java — as seen by serious creative Thai and Javanese Indonesian authors', working paper, no.12, Bielefeld, 1981.

Muljadi, D., and F.J. Dent, 'Evaluation of Indonesian soil and land resources', *Indon. Agr. Res. and Dev. Journal*, 1, 1–2 (1979), pp. 21–3.

Munthe, G.N., 'Coal, the first alternative to oil', *Indonesia Times*, 9/11 Aug. 1983.

Myers, N., *The Sinking Arch*, Oxford: Pergamon Press, 1979.

National Academy of Sciences, *Firewood Crops: Shrub and Tree Species for Energy Production*, Washington DC, 1980.

NEDECO/SMEC, 'Cirebon-Cimanuk feasibility study', Part A., vol.1, Jakarta, 1973.

Neef, E., *Das Gesicht der Erde*, Frankfurt/Main, 1977.

Neubauèr, H.F., 'Die Pflanzenwelt Indonesiens und ihre Besonderheiten', in Imber and Uhlig, 1979, pp. 61–88.

Nibbering, W., *et al.*, *Firewood Trading and Consumption in the Kali Konto Project*

Area, East Java, Malang (Indonesia) etc., 1980.

Nirschl, J., *Die Forstwirtschaft in Niederländisch-Indien*, Leipzig, 1920.

Nitisastro, W., Population growth and economic development in Indonesia: A study of the economic consequences of alternative patterns of inter-island migration, unpubl. Ph.D. thesis, Berkeley, California, 1961.

——, *Population Trends in Indonesia*, Ithaca, NY, and London, 1970.

Nohlen, D., and F. Nuscheler (eds), *Handbuch der Dritten Welt*, Hamburg, various volumes and editions, 1978 ff.

Notohadiprawiro, T., 'Problems and perspectives of agriculture on alluvial soils of Indonesia', *Ilmu Pertanian*, 1,6 (1972), pp. 247–57.

Ooi, J.-B., 'Mining landscapes of Kinta', *The Malayan Journal of Tropical Geography*, 4 (Jan. 1955), pp. 1–57.

Opitz, P.J. (ed.), *Die Dritte Welt in der Krise*, Munich, 1984.

Oppenoorth, W.F.F., 'Homo (Javanthropus) soloensis, een Plisticeene mensch van Java', *Wet.Meded. Dienst Mijnbouw Ned. Indië*, no.20 (1932).

Ormeling, F.J., *The Timor Problem: A geographical interpretation of an under-developed island*, Groningen, 1956.

Palmer, I., *Rural poverty in Indonesia, with special reference to Java*, in ILO 1980, pp. 205–31.

Palte, J. (ed.), *see* Nibbering.

Papanek, G.F., 'The poor of Jakarta', *Economic Development and Cultural Change*, 24 (1975), pp. 1–27.

——, (ed.), *The Indonesian Economy*, New York, 1980.

Pauker, G.J., 'Political consequences of rural development programs in Indonesia', *Pacific Affairs*, 41,3 (1968), pp. 386–402.

Pelzer, K.J., 'Tanah Sabrang and Java's population problem', *Far Eastern Quarterly*, 5,2 (1946), pp. 132–42.

——, 'Physical and human resource patterns', in MacVey, 1963, pp. 1–23.

——, 'Man's role in changing the landscape in South-east Asia', *The Journal of Asian Studies*, 27,2 (1968), pp. 269–79.

——, 'The swidden cultivator as producer of agricultural exports: An Indonesian case study', *Wirtschafts-und Kulturräume der aussereuropäischen Welt'* (Festschrift für Kolb), *Hamburger Geographische Studien*, 24 (1971), pp. 261–75.

——, 'Planter and peasant: Colonial policy and the agrarian struggle in East Sumatera 1863–1947', *Verhandelingen van het Koninklijk Instituut voor Taal-, Land-en Volkenkunde*, no.84 (1978).

Penck, A., Das Hauptproblem der physischen Anthropogeographie, *Sitzungsberichte der Preussischen Akademie der Wissenschaften*, 1924 (repr. E. Wirth, 1969, pp. 157–80)

Pendleton, R.L., *Report to accompany the provisional map of the soils and surface rocks of the kingdom of Siam*, 1953 (Repr. in Sarot Montrakun, *Agriculture and Soils of Thailand*, Bangkok, 1964).

'Persuasion from peers', *Development Forum*, 8,1 (1980), p. 4.

Pescod, M.B., 'Impact of development on the rural environment', in Hill and Bray, 1978, pp. 257–83.

Pianka, H.G., 'Zur Problematik der Transmigration in Indonesien', *Internationales Asienforum*, 1 (1970), pp. 535–47.

Pickering, A.K., 'Soil conservation and rural institutions in Java', *IDS Bulletin*, 10,4 (1979), pp. 60–6.

Pitt, M.M., 'Smuggling and price disparity', *Journal of International Economics*, 11,4 (1981), pp. 447–58.

Pitty, A.F., *Geography and Soil Properties*, London, 1978.

Polunin, I., 'The effects of shifting agriculture on human health and disease', *Symposium on the impact . . .*, 1960, pp. 388–98.

Prabowo, D., 'Water development and farm production possibilities in the Solo river basin', Unpubl. diss. Washington State University, Seattle, 1977.

——, 'Allocation of farm resources in the Solo river basin', *BIES*, 14,1 (1978), pp. 45–62.

'Pre-Congress Conference in Indonesia'. *Symposium on Planned Utilization of the Lowland Tropical Forests*, Bogor, 1971.

Precht, F., *Stadtökologie*, Bonn, 1978.

Prescott-Allen, R. and Ch., *What's Wildlife worth?*, London, 1983.

——, 'The present status of fisheries in Indonesia', in *Proceedings of the International Seminar on Fisheries Resources and their Management in Southeast Asia*, Berlin, DSE/FAO, 1974, pp. 412–31.

——, 'Prolific, versatile sago palm attracts attention of science', *Ceres*, 17,4 no.100 (1984), pp. 3–4.

Pulunin, N., *The Environmental Future*, New York, 1972.

Raffles, Sir T.S., *History of Java*, 2 vols., London, 1817. — *Memoir of the life and public services of Sir Thomas Stamford Raffles, F.R.S.*, published by his widow, London, 1830.

Ramsay, D.M., and K.F. Wiersum, 'Problems of watershed management and development in the upper Solo river basin', paper presented at the Conference on ecologic guidelines for forest, land and water resources development in Indonesia, Bandung, 1974.

Reichling, J., 'Probleme der Entsorgung im urbanen und ruralen Bereich', in Egger *et al.*, 1972, pp. 242–50.

Reksodihardjo, W.S., 'Conditions of the forest reserves in Indonesia', in *Pre-Congress Conference . . .*, 1971, pp. 82-5.

Reksohadiprodjo, I., 'The accelerated growth of river deltas in Java', *The Indonesian Journal of Geography*, 4,7 (1964), pp. 1–15.

Rethwilm, D., *Zur Mobilisierung landwirtschaftlichen Potentials: der Fall West-Pasaman*, Göttinger Diss. 1980, Saarbrücken, 1981.

Revelle, R., 'The resources available for agriculture', *Scientific American*, 235,3 (1976), pp. 165–78.

Richter, G. (ed.), *Bodenerosion in Mitteleuropa*, Darmstadt, 1976.

Rieser, A., *Natural resources: Sumatra regional planning study, Bengkulu province*, draft final report, Jakarta, 1975*a*.

——, 'Natürliche Standortfaktoren tropischer Landwirtschaft (Sumatra/ Indonesien)', *Der Tropenlandwirt*, 76 (1975*b*), pp. 70–80.

——, 'Gezeitenbewässerungsprojekte in Indonesien', *Wasser und Boden*, 27 (1975), Beilage *ICID-Nachrichten*, 1/1975, pp. 2–3.

——, 'Beckenbewässerung in Indonesien', *Der Tropenlandwirt*, 77 (1976), pp. 9–18.

Rifai, M.A., 'Botanical exploration in Indonesia', in Kartawinata and Atmawidjaja, 1974, pp. 166–8.

Röll, W., 'Bevölkerungsdruck und Siedlungsaktivität in Zentral-Java. Ein Beitrag zur Landnahme unterbäuerlicher Sozialgruppen nach dem Zweiten Weltkrieg in Indonesien', *Geographische Rundschau*, 23,2 (1971), pp. 56–66.

——, 'Probleme der Infrastruktur und der ausseragraren Wirtschaft Indonesiens', in Imber and Uhlig, 1973, pp. 213–32.

——, 'Der Teilbau in Zentral-Java. Untersuchungen zur Grundbesitzverfassung eines übervölkerten Agrarraums', *Zeitschrift für ausländische Landwirtschaft*, 12,3-4 (1973), pp. 305–21.

——, 'Indonesien. Einführung in die Entwicklungsprobleme und den deutschen geographischen Forschungsstand', *Geographische Rundschau*, 27,4 (1975), pp. 137–8.

——, 'Probleme der Bevölkerungsdynamik, und regionalen Bevölkerungsverteilung in Indonesien', *Geographische Rundschau*, 24,7 (1975), pp. 139–50.

——, *Die agrare Grundbesitzverfassung im Raume Surakarta. Untersuchungen zur Agrar- und Sozialstruktur Zentral-Javas*, Wiesbaden, 1976.

——, *Indonesien. Entwicklungsprobleme einer tropischen Inselwelt*, Stuttgart, 1979.

——, 'Siedlung und Agrarwirtschaft von Pygmäen steinzeitlicher Kulturstufe im zentralen Bergland von Irian Jaya, Indonesien', in Röll *et al.*, 1980, pp. 111–20.

——, and A. Leemann, 'Lombok: staatlich gelenkte inner- und inter-insulare Umsiedlungsmassnahmen', *Erdkundliches Wissen*, 58 (1982), pp. 132–45.

——, and A. Leemann, 'Entwicklungsprobleme in Indonesien (Nusa Tenggara Barat.) Das Beispiel Lombok', *Asien*, no.12 (July 1984), pp. 72–8.

——, *et al.* (eds), 'Symposium Wandel bäuerlicher Lebensformen in Südostasien', *Giessener Geographische Schriften*, 48 (1980).

Roessel, B.W.P., 'Maximum afvoer en hydrologische empirie', *De Ingenieur in Nederlandsch-Indië*, 7 (1940), pp. 104–29.

Rosell, D.Z., 'Natural resources conservation — the geographer's view', in Hill and Bray, 1978, pp. 195–204.

Ross, M.S., 'The role of land clearing in Indonesia's transmigration program', *BIES*, 16,1 (1980), pp. 75–85.

356   Land Use and Environment in Indonesia

Rubinoff, I., 'Tropical forests: Can we afford not to give them a future?', *The Ecologist*, 16,6 (1982), pp. 253–8.

Rudin, W.F., 'Over *alang-alang* bestrijding', *Bergcultures*, 9 (1935), pp. 346–8.

Rutten, L.M.R., 'Over denudatiesnelheid op Java', *Verslag Kon. Ac. van Wetensch.*, Amsterdam, 1917, pp. 920–30.

Rutz, W., 'Indonesiens Gliederung nach Funktions-und Verwaltungs-räumen — Übereinstimmungen und Diskrepanzen', in Leupold and Rutz, 1976, p. 159–73.

——, 'Indonesien — Verkehrserschliessung seiner Ausseninseln', *Bochumer Geographische Arbeiten*, 27 (1976).

——, 'Javas Bevölkerungsdichte dargestellt durch eine Bevölkerungsdichte-karte auf der Grundlage von Verwaltungsbezirken', *Die Erde*, 108 (1977), pp. 115–23.

Ryanto, T., 'Jakarta's traffic jams . . .', *Indonesia Times*, 1 July 1982.

Sandy, I.M., *Atlas Indonesia. Buku Pertama Umum*, Jakarta, 1977.

Salim, E., 'The bursting tide', *Indonesia Times*, 1 June 1983.

Satoto, 'Glancing wood-processing industry in East Kalimantan', *Indonesia Times*, 29 Jan. 1982.

Schädle, W., 'Umweltprobleme in der Dritten Welt', in Optiz, 1984, pp. 208–26.

Schiller, B.L.M., 'The "green revolution" in Java: Ecological, socio-economic and historical perspectives', *Prisma*, no.18 (1980), pp. 71–93.

Schlereth, E., *Indonesien*, Berlin, 1983.

Schmidt, F.H., and J.H.A. Ferguson, 'Rainfall types based on wet and dry period ratios for Indonesia', *Verhandel. Djawatan Meteorol. dan Geofis*, 42 (1951).

Schmidt-Ahrendts, W., 'Verarbeitende Industrie' in Institut für Asienkunde, 1979, pp. 131–8.

Scholz, F., 'Der Herr des Bodens in Ost-Indonesien', unpubl. diss., University of Cologne, 1962.

Scholz, U., 'Permanenter Trockenfeldbau in den humiden Tropen. Beispiele kleinbäuerlicher Betriebe in Lampung, Süd-Sumatra', *Giessener Beiträge zur Entwicklungsforschung*, 1,3 (1977a), pp. 45–58.

——, 'Minangkabau. Die Agrarstruktur in West-Sumatra und Möglich-keiten ihrer Entwicklung', *Giessener Geographische Schriften*, 41 (1977).

——, 'Die Vegetation und Fauna Indonesiens', in Kötter *et al.*, 1979, pp. 54–61.

——, 'Land reserves in southern Sumatra/Indonesia and their potentialities for agricultural utilisation', *Geo Journal*, 4,1 (1980), pp. 19–30.

——, 'Die Ablösung und Wiederausbreitung des Brandrodungswander-feldbaus in den südostasiatischen Tropen. Beispiele aus Sumatera und Thailand', *Erdkundliches Wissen*, 58 (1982), pp. 103–21.

Schoorl, P., *De betekenis van grasland voor de veehouderij in Indonesië*, Bogor, 1953.

Schuitemakers, B., 'Maatregelen tot behaud van de bodem op Java', *Landbouw*, 21 (1949), pp. 153–76.

Schulz, U., 'Die Vegetation und Fauna Indonesiens', in Kötter *et al.*, 1979, pp. 54–61.

Scott van Veen, M., 'Some ecological considerations of nutritional problems on Java', *Ecology of Food and Nutrition*, 1,1 (1971), pp. 25–38.

Seibold, E., 'Die Entstehung der Ozeane und ihre Erforschung durch die "Glomar Challenger" ', in *Oceanographie*, Berlin, 1977, pp. 18–28.

Seidensticker, W., *See* 'Urbanisierung' and 'Slums', in Egger *et al.*, 1972, pp. 213–41.

Sethuraman, S.V., 'Urbanization and employment: A case-study of Djakarta', *International Labour Review*, 112,2–3 (1975), pp. 191–205.

Setten van der Meer, N.C. van, *Sawah cultivation in ancient Jawa: Aspects of development during the Indo-Javanese period, AD 1000–1400*, Canberra, 1979.

Shelton, N., 'The global rescue effort for the Java Rhino', *The Asian Wall Street Journal*, 6 June, 1983.

Sherman, G., 'What green desert? — The ecology of Batak grassland farming', in *Indonesia*, Ithaca, NY, 29 (1980), pp. 113–48.

SIDA/FAO, 'Effects of intensive fertilizer use on human environment', *Soil Bulletin*, 16, Rome, 1972.

Sievers, A., *Der Tourismus in Sri Lanka*, Wiesbaden, 1983.

Simanjuntak, P., 'Indonesia starts looking to its almost idle marine resources', *Indonesia Times*, 4–5 Nov. 1983.

——, 'Indonesia to develop sago as food and energy resource', *Indonesia Times*, 2 Feb. 1984.

——, 'Tulang Bawang ethanol plant to operate soon', *Indonesia Times*, 29 March 1984.

Snowy Mountain Engineering Company (SMEC), 'Serayu River Basin Study', 1974 (various reports).

Smits, M.B., 'Population density and soil utilisation in the Netherlands Indies', *Comptes Rendus . . .*, 1938, pp. 500–6.

Snodgrass, D., 'The family planning program as a model for administrative improvement in Indonesia', *Development Discussion Papers*, no.58, Harvard Institute for International Development, 1979.

Soedjarwo, 'The long-term effect of logging in Indonesia'. Keynote presented at the symposium on the long-term effects of logging in South-east Asia, Bogor, 1975.

Soemardjan, S., 'Zur Identität der indonesischen Gesellschaft', in Kötter *et al.*, 1977, pp. 181–201.

——, 'The deculturing effects of industrialisation on pre-industrial communities', in *Culture and Industrialisation*, Baden-Baden, 1978, pp. 33–50.

Soemarwoto, O., 'The soil erosion problem in Java', in *Proceedings of the International Ecological Congress*, pp. 361–4, The Hague, 1974.

——, 'Interrelations among population, resources, environment and development', *Economic Bulletin for Asia and the Pacific*, 32,1 (1981), pp. 1–33.

——, *et al.*, 'Ecology and health in Indonesian villages', *Ekistics*, 41,245 (1976), pp. 235–9.

Soepriyo, 'Better living conditions at resettlement areas', *Indonesia Times*, 21 April 1984.

Soeratman, and P. Guiness, 'Changing focus of the transmigration program', *Working Paper Series, no. 11*, Universitas Gadjah Mada, Yogyakarta, 1977: see also *BIES*, 12,2 (1977), pp. 78–101.

Soerjono, R., 'The flood of the Lamongan river', in *Rimba Indonesia*, 1968, pp. 30–48.

Soewardi, B., *et al.*, 'Improving the choice of resource systems for transmigration', *Prisma*, 18 (1980), pp. 56–70.

Sommer, V., 'Zurück auf die Bäume. Im Orang-Utan-Zentrum auf Sumatra werden zahme Affen an die Wildnis gewöhnt', *Die Zeit*, 26 Oct. 1984.

Soponkanaporn, T., *Onshore Tin Mining Waste water treatment*, Asian Institute of Technology thesis, Bangkok, 1979.

Speare, A., 'Alternative population distribution policies for Indonesia', *BIES*, 14,1 (1978), pp. 93–106.

Speed, F.W., *Indonesia Today*, Sydney, 1971.

Speelman, H., 'Geology, hydrogeology and engineering geological features of the Serayu river basin', in *Serayu Valley Project, Final Report*, vol.4, The Hague, 1979.

Spencer, J.E., *Shifting Cultivation in South-eastern Asia*, Berkeley, 1966.

Spengler, J.J., and O.D. Duncan (eds), *Demographic Analysis: Selected Readings*, Glencoe, Illinois, 1956.

Spooner, B., 'Ecology in perspective: The human context of ecological research', *International Social Science Journal* (UNESCO), 34 (1982), pp. 395–410.

Stadt, 'Iets over terrasseering', *Bergcultures*, 5,1 (1931), pp. 425–32.

Stanton, W.R., and M. Flach (eds), 'Sago: The equatorial swamp as natural resource', in *Proceedings of the Second International Sago Symposium*, Kuala Lumpur, 1979; The Hague, 1980.

Staudinger, M., 'Der Mensch zwischen Technosphäre und Biosphäre', in Precht, 1978, pp. 193–200.

Statistical Yearbook of Indonesia, Jakarta, 1982.

Steenis, C.G.G.J. van, 'On the origin of the Malaysian mountain flora', *Bull. Jard. Bot. Buitenzorg*, III, 13 (1935).

——, 'De invloed van den mensch op het bosch', *Tectona*, 30 (1937), pp. 634–52.

Stein, N., 'Die tropische Landwechselwirtschaft: Problematik und moderne Strukturveränderungen, dargestellt am Beispiel Nord-Sumatras', *Geographische Zeitschrift*, 60,4 (1972), pp. 322–40.

——, 'Der Dolok Sinabung: Vertikale Landschaftsgliederung eines Vulkans im nördlichen Batakhochland', *Die Erde*, 105,1 (1974), pp. 34–61.

——, 'Strukturelemente der Kulturlandschaft in Deli (Nordost-Sumatra) und deren räumliches Gefügemuster', *Geografisch Tijdschrift*, 8,2 (1974), pp. 117ff.

——, 'Geographische Analyse pazifischer Ökosysteme', *Erdkunde*, 30,2 (1976), pp. 152–6.

Stoddart, D.R., 'World erosion and sedimentation', in Chorley, 1969, pp. 43–65.

Strahler, A.N., *Physical Geography*, New York and London, 1951.

Strout, A.M., 'How productive are the soils of Java?', *BIES*, 19,1 (1983), pp. 32–52.

Struijk, H., 'Soil erosion and sediment yield of two small river basins in the Serayu river valley', in *Serayu Valley Project, Final Report*, vol.1, pp. 47–9.

Subardjo, A., 'Religiöse und spirituelle Aspekte', in Kötter *et al.*, 1979, pp. 116–23.

Sudiono, J., and L. Daryadi, 'Some impacts of the utilization of rain forests in Indonesia', *The Malaysian Forester*, 41,2 (1978), pp. 121–4.

Sukanto, M., 'Climate of Indonesia', *Arakawa*, vol.8, 1969, pp. 215–29.

Sumardja, E.A., *Report on UNESCO sub-regional seminar on ecology and environment sciences*, Bogor, 1974.

Sundrum, R.M., 'Inter-provincial migration', *BIES*, 11,1 (1976), pp. 70–92.

Suratman and Guinness, see Soeratman and P. Guinness.

Suryatna, M., and J.L. MacIntosh, 'Food crops production and control of *Imperata cylindrica* on small farms', paper presented at the workshop on *alang-alang*, BIOTROP seminar, Bogor, 1976.

Suwarnarat, K., 'Abwasserteichverfahren als Beispiel naturnaher Abwasserbehandlungsmassnahmen in tropischen Entwicklungsländern,' unpubl. diss., University of Darmstadt, 1979.

Swasono, S.E., 'The Land Beyond: Transmigration and Development in Indonesia', unpubl. Ph.D. diss., University of Pittsburgh, 1969.

Symposium on the impact of man on humid tropics vegetation, Goroka (Papua New Guinea), 1960.

TAD, 'Regional planning for East Kalimantan: Conclusions and recommendation', *Final Report*, Jakarta, 1980 (TAD Report no.17).

Temple, G., 'Migration to Jakarta', *BIES*, 11,1 (1975), pp. 76–81.

Terra, G.J.A., 'Some sociological aspects of agriculture in South-east Asia', *Indonesia*, 6 (1952/3), pp. 297–316, and pp. 439–55.

——, 'Mixed-garden horticulture in Java', *The Malayan Journal of Tropical Geography*, 3 (1954), pp. 33–43.

——, 'Farm systems in South-east Asia', *The Netherlands Journal of Agricultural Science*, 6 (1958), pp. 157–81.

——, 'Agriculture in economically underdeveloped countries, especially in equatorial and sub-tropical regions', *The Netherlands Journal of Agricultural Science*, 7,3 (1959), pp. 1–16.

360   *Land Use and Environment in Indonesia*

Teuscher, T., *et al.*, 'Livestock development in West Sumatra', *Schriften des Seminars für landw. Entwicklung*, no.24, Berlin, 1978.

Thijsse, J.P., 'Will Java become a desert?', working paper, 1974.

——, 'The magnitude of the population increase on Java and of transmigration as a possible remedy', working paper (1975*a*).

——, 'A plea for immediate activities to control the menacing deforestation and erosion in mountain areas on Java', paper prepared for the symposium in *Problems of Wasteland Control and Reclamation in Indonesia* (1975*b*); published in *Bio Indonesia*, 2 (1976).

——, 'Ontbossing en erosie op Java', *Landbouwkundige Tijdschrift*, 89,12 (1977), pp.443–7.

Thorenaar, A., 'Over hydrologische en orologische waarnemingen', *Tectona* 23 (1930), pp. 670–9.

——, 'Regeering, volk, bosch en woeste grond', *Tectona*, 30 (1937), pp. 817–56.

Tinal, U., and J.L. Palinewen, *A study of mechanical logging damage after selective cutting in the lowland dipterocarp forest at Beloro, East Kalimantan*, BIOTROP publication, Bogor, 1974.

Tinker, J., 'Is technology neo-colonialist?', *New Scientist*, 9 May 1974, pp. 302–5.

Tjondronegoro, S.M., *Land Reform or Land Settlement: Shifts in Indonesia's land policy*, Madison, Wisc., 1972.

Tobi, E., 'Reboiseering in het gebergte van Bagelen', *Tijdschrift voor Nijverheid en Landbouw in Nederlandsch-Indië*, 47 (1894), pp. 106–14.

Tobler, F., and H. Ulbricht, *Koloniale Nutzpflanzen*, Leipzig, 1945.

'Transactions of the Second International MAB/IUFRO Workshop on Tropical Rainforest Ecosystems Research', *see* Adisoemarto and Brünig, 1978.

Trumble, H.C., 'Bioclimatology of Indonesia in relation to vegetation', in *Proceedings of the Symposium on Humid Tropics Vegetation*, Tjiawi, 1958, pp. 218–39.

Uhlig, H. (ed.), *Fischer Länderkunde: Südostasien — Australien*, Frankfurt/Main, 1975.

Uhlig, H., 'Die Naturräume', in Imber and Uhlig, 1973, pp. 19–42.

——, 'Die Agrarlandschaft im Tropenkarst: Beispiele ihrer geo-ökologischen Differenzierung auf Java und Sulawesi', *Geografski Glasnik*, Zagreb, 38 (1976), pp. 113–35.

——, 'Bevölkerungsverteilung und ethnische Differenzierung', in Kötter *et al.*, 1977, pp. 259–67.

——, 'Erschliessung des ländlichen Lebensraums und Siedlungsformen', in Kötter *et al.*, 1977, pp. 267–74.

——, 'Oberflächenformen und Böden', in Kötter *et al.*, 1977, pp. 38–53.

——, 'Traditionelle Reisbausiedlungen in Südostasien', in *Recherches Géographiques* (Festschrift für Dussart), Lüttich, 1980, pp. 205–26.

——, (ed.), 'Spontaneous and planned settlement in South-east Asia', *Giessener Geographische Schriften*, 58, Hamburg, 1984.

Ulrich, J., 'Grossformen des Meeresbodens', in Dietrich, 1970, pp. 129–40.

United Nations, 'Review of water resources in Indonesia', *Session of the Regional Conference on Water Resources Development*, New York, 1975.

Utomo, K., 'Masjarakat Transmigran Spontan Didaerah W. Sekampung (Lampung)', doctoral thesis, Bogor, 1957.

——, 'Villages of unplanned resettlers in the subdistrict Kaliredjo, Central Lampung', in Koentjaraningrat, 1967, pp. 281–98.

Vacharapongpreecha, S., 'Don't go overboard for "fuel farms"', Asian nations warned', *Indonesian Times*, 31 March 1984.

Verhoef, L., 'Typen van stervend land in den Nederlandsch Indischen Archipel: Het Paloedal (Midden Celebes)', *Tectona*, 30 (1937), pp. 220–2.

Verstappen, H.Th., 'Geomorphologische Notizen aus Indonesien', *Erkunde*, 9 (1955), pp. 134–44. —

——, 'Quarternary climatic changes and natural environment in South-east Asia', *GeoJournal*, 4,1 (1980), pp. 45–54.

Versteeg, C., 'De lijdensgeschiedenis van een boschperceel uit de jaren 1866–75', *Tectona*, 26 (1933), pp. 597–609.

——, 'Iets over den cultuuraanleg uit de jaren 1866–75', *Tectona*, 29 (1936) pp. 768–76.

Vink, A.P.A., 'Hydrologie en bodembescherming op de bergcultuur-ondernemingen in Indonesië', *De Bergcultures*, 20,17 (1951), pp. 277–87.

Vogt, W., *Road to Survival*, New York, 1948.

Voogd, C.N.A. de, 'Ravijnbescherming op Bali', *Tectona*, 30 (1937), pp. 300–11.

Vorlaufer, K., *Ferntourismus und Dritte Welt*, Frankfurt Main, 1984.

Voss, F., 'Natural resources inventory, East-Kalimantan', *TAD Report* no.9, Hamburg, 1979.

Voss, J., *Die Bedeutung des Tourismus für die wirtschaftliche Entwicklung*, Pfaffenweiler, 1984.

Vries, C.A. de, 'New developments in production and utilization of cassava', *Abstracts on Tropical Agriculture*, 4,8–9 (1978), pp. 9–24.

Wahlert, G. von, 'Entwicklungshilfe: Alternativen zum Raubbau durch Dynamitfischerei', *Deutsches Allgemeines Sonntagsblatt*, 13 Jan. 1985.

Walandouw, P.H., 'Grassland in Indonesia', *Journal for Scientific Research*, 10–12 (1952), pp. 201–12.

Wanigasundara, M., 'Unplanned tourism wreaking havoc on Sri Lanka environment', *Indonesia Times*, 3 Oct. 1983.

Wardoyo, S.T.H. *et al.*, 'A case study on water pollution problems of the Surabaya river, East Java', in Furtado (ed.), 1980, pp. 629–33.

Weischet, W., *Die ökologische Benachteiligung der Tropen*, Stuttgart, 1977.

——, 'Das ökologische Handicap der Tropen in der Wirtschafts- und Kulturentwicklung', *Verhandlungen des 41. Deutschen Geographentags*, 1978, pp. 25–41.

Welte, E., 'Sind die Tropen wirklich im Nachteil?' *Umschau*, 78,20 (1978), pp. 634–8.

Wertheim, W.F., 'Sozialer Wandel in Java 1900–30', in *Moderne Kolonialgeschichte*, 1970, pp. 125–45.

White, B., 'Political aspects of poverty, income distribution and their measurement: Some examples from rural Java', *Development and Change*, 10 (1979), pp. 91–114.

White, P.T., 'Tropical rain forests: Nature's dwindling treasures', *National Geographic*, 163,1 (1983), pp. 2–47.

Whyte, A., 'The integration of natural and social sciences in the MAB Programme', *International Social Science Journal* (UNESCO), 34 (1982), pp. 411–26.

Wiersum, K.F., 'Protective forest and erosion control policy in Java, a comprehensive review of ideas and practices', FAO Upper Solo Watershed Management and Upland Development Project, Solo, 1976.

——, 'Bosbouw in Indonesië', *Nederlands Bosbouw Tijdschrift*, 50,10 (1978a), pp. 301–16.

——, *Erosie en bos op Java*, Wageningen, 1978b.

——, *Introduction to principles of forest hydrology and erosion, with special reference to Indonesia*, Bandung, 1979.

——, 'Possibilities for use and development of indigenous agro-forestry systems for sustained land use on Java', *Tropical Ecology and Development*, 1980a, pp. 515–21.

——, 'Erosie, plattelandsontwikkeling en bos op Java', *Landbouwkundig Tijdschrift*, 92 (1980b), pp. 338–45.

——, and S. Ambar, 'Tropical ecological forest research in Indonesia: Vegetation and erosion in the Jatiluhur area', *Final Report*, Bandung and Wageningen, 1981.

——, *et al.*, 'Influence of forest on erosion', report no.3 in *The erosion problem of the Jatiluhur area*, Bandung, 1979a.

Wigman, H.J., 'De bergwouden van Java', *Teysmannia*, 4 (1893), pp. 734–44.

Wijoyo, M., 'Jakarta's Chinatown', *Indonesia Times*, 9 July 1982.

Wind Hzn., R., 'Toenemend gebruik van houtgas', *Tectona*, 29 (1936), pp. 444–6.

Wirth, E., *Wirtschaftsgeographie*, Darmstadt, 1969.

Withington, W.A., 'Migration and economic development', *Tijdschrift voor Economische en Sociale Geografie*, 58,3 (1967), pp. 153–63.

Wittfogel, K., *Oriental Despotism*, New Haven, 1957.

Woelke, M., *Statistical Information on Indonesian Agriculture*, Singapore, 1978.

Wolterson, J.F., Soil erosion in the teak forest of Java, *Rapport Rijksinstituut Onderzoek Bos- en Landschapsbouw De Dorschkamp*, no.197, Wageningen, 1979.

World Tourist Organisation (WTO), 'Tourist carrying capacity', *UNEP*

Industry and Environment, 7,1 (1984), pp. 30–6.

Yeung, Y.M., 'The urban environment in South-east Asia: Challenge and opportunity', in Hill and Bray, 1978, pp. 17–33.

Yunus, H.D., 'Analysis of urban population density through air-photo. A case study of Yogyakarta city', *The Indonesian Journal of Geography*, 8,36 (1978), pp. 23–39.

Zahri, A., 'Regional development and migration policy', *Intereconomics*, 3,10 (1968), pp. 312–5.

Zimmermann, G.R., 'Transmigration in Indonesien: Eine Analyse der inter-insularen Umsiedlungsaktionen zwischen 1905 und 1975', *Geographische Zeitschrift*, 63,2 (1975*a*), pp. 104–22.

——, 'Die Viehhaltung in Indonesien', *Geographische Rundschau*, 27,4 (1975*b*), pp. 162–8.

——, 'Competitive use of land resources by different socio-ethnic groups. The example of Southern Sumatra', *New Zealand Geographical Society: Conference Series*, no.8 (1975*c*), pp. 97–108.

——, 'Ungeregelte Landerschliessung und planmässige Agrarkolonisation in den Waldgebieten Südostasiens. Das Beispiel Süd-Sumatra,' *Göttinger Geographische Abhandlungen*, 66 (1976), pp. 107–18.

——, ' "Landwirtschaftliche Involution" in staatlich geplanten indonesischen Transmigrationsprojekten', in Röll, *et al.*, 1980, pp. 121–30.

Zon, P. van, 'De beteekenis van den bosschen in de buitengewesten', *Tectona*, 22 (1929), pp. 436–51.

Zurek, M., 'Förderung der gewerblichen Fischerei und der Kleinfischerei auf genossenschaftlicher Basis in Java/Indonesien', *Integrierte Ländliche Entwicklung*, Munich and Mainz, 1980, pp. 279–92.

INDEX